MRE

Materials Research and Engineering
Edited by B. Ilschner and K.C. Russell

Horst Buhl (Ed.)

Advanced Aerospace Materials

Springer-Verlag
Berlin Heidelberg NewYork
London Paris Tokyo
HongKong Barcelona Budapest

Dr. HORST BUHL
DLR, German Aerospace Research Establishment
Institute of Materials Research
Linder Höhe
W-5000 Köln 90/Germany

Series Editors

Prof. BERNHARD ILSCHNER
Polytechnique Fédérale de Lausanne
Laboratoire de Métallurgie Mécanique
MX-D Ecublens Ecole
CH-1015 Lausanne/Switzerland

Prof. KENNETH C. RUSSELL
Department of Materials Science and Engineering and
Department of Nuclear Engineering
Room 8-411
Massachusetts Institute of Technology
Cambridge, MA 02139/USA

ISBN 3-540-55888-8 Springer-Verlag Berlin Heidelberg NewYork
ISBN 0-387-55888-8 Springer-Verlag NewYork Berlin Heidelberg

Library of Congress Cataloging-in-Publication Data
Buhl, Horst
Advanced aerospace materials / Horst Buhl.
(Materials researches and engineering)
Includes bibliographical references and index.
ISBN 3-540-55888-8 (alk. paper). -- ISBN 0-387-55888-8 (alk. paper: U.S.)
1. Airplanes -- Materials. 2. Composite materials.
I. Title. II. Series: Materials research and engineering (Unnumbered)
TL698.B84 1992 629.1--dc20 92-30704

Printed in Germany

Typesetting: Camera ready by author
Printing: Color-Druck Dorfi GmbH, Berlin; Bookbinding: Lüderitz & Bauer, Berlin
61/3020-543210 - Printed on acid-free paper.

Professor Dr. Wolfgang J. G. Bunk

Professor Dr. W.G.J. Bunk, Director of the DLR (German Aerospace Research Establishment) Institute of Materials Research since 1970, retires in 1992. With his retirement an epoch of the institute's history comes to an end, an epoch characterized by expansion in resources, competence, and size, and national as well as international appreciation.

As an expression of their gratitude for his leadership, the scientific staff of the institute devotes this book to Professor Bunk.

On behalf of the members of the staff

Horst Buhl

Foreword

Very light, very strong, extremely reliable - aircraft and aerospace engineers are, and have to be, very demanding partners in the materials community. The results of their research and development work is not only crucial for one special area of applications, but can also lead the way to new solutions in many other areas of advanced technology.

Springer-Verlag and the undersigned editor are pleased to present in this volume, an overview of the many facets of materials science and technology which have been the objective of intensive and systematic research work during past decades in the laboratories of the German Aerospace Research Establishment. Its contents shows clearly the interrelations between goals defined by the user, fundamentals provided by the scientists and viable solutions developed by the practical engineer.

The particular personal touch which has been given to this volume by its authors in dedicating it as a farewell present to Professor Wolfgang Bunk, inspiring scientist and director of the DLR Intitute of Materials Research for more than 20 years, has obviously given an added value to this important publication. Surely, this truly cooperative endeavour will render a valuable service to a large international community of interested readers, many of them having personal links to the Institute, its director and its staff.

Lausanne, June 1992 Bernhard Ilschner

Preface

During the period 1970 to 1992 the European aircraft industry experienced a big impetus and a worldwide breakthrough with the successful implementation of the family of Airbus aircraft. In the area of space launch vehicles, Ariane developed into an efficient and reliable carrier for satellites. Safe, reliable and durable materials were the precondition for this success in both cases. Columbus, the European contribution to the US space station Freedom, as well as the European efforts to create an autonomous space policy regarding such projects as Hermes and Sänger, are a big challenge for the materials research community, too.

The authors of this book, my collegues at the Institute of Materials Research of the German Aerospace Research Establishment, took part in this development from the beginning. The emphasis of this book, therefore, lies on the research activities of these authors. It is primarily conceived as a vista into the fascinating world of aerospace materials and may well serve as an introduction. On the other hand, its content is always embedded into the context of the international state of knowledge in materials science, and many topics are dealt with in detail, limited however by space considerations. A comprehensive bibliography yields access to the many specialized fields of aerospace materials.

The book is introduced by a visionary overview by W.J.G. Bunk on the present situation of aerospace materials and perspectives for the future. Thus, the ground is prepared to allow the reader to examine different aspects of the subject matter. The structure of the content is, as in many treatments of materials science and engineering, developed along two axes: First, groups of materials like metals, ceramics, polymers, and their composites are dealt with. Then properties as well as suitable test methods and the application related interpretation of test results form a second guideline. This holds for mechanical properties such as fracture and fatigue behaviour as well as for corrosion resistance and the field of microstructural and microanalytical techniques. After all, the goal is a cost effective and reliable aerospace structure, and these properties do not exclusively relate to one single material.

The book is designed as an introduction to materials research as well as a guide to anyone who comes in contact with advanced materials. It may be an introduction for graduate students of engineering science into the field of highly developed lightweight structural and high temperature turbine materials. The comprehensive survey may also supply materials scientists with information on the actual state of the art, new materials and test methods and quick answers to problems, as well as references to more profound information. Also, service engineers, not only from the aerospace industry, may find valuable details and references.

I am greatly indebted to my colleages at the Institute of Materials Research, who contributed to this book and who are listed below:

Braue, W.	Braun, R.	Bunk, W.G.J.
Döker, H.	Dudek, H.-J.	Eschweiler, J.
Fritscher, K.	Göring, J.	Kumpfert, J.
Lehnert, F.	Leucht, R.	Marci, G.
Nowack, H.[1])	Peters, M.	Peters, P.W.M.
Pleger, R.	Rätzer-Scheibe, H.-J.	Saruhan, B.
Schneider, H.	Schulte, K.[2])	Schulz, U.
Spiegelberg, M.	Staniek, G.	Trautmann, K.-H.
Ward, C.H.[3])	Welpmann, K.	

[1]) now Universität Gesamthochschule Duisburg
[2]) now Technische Universität Hamburg-Harburg
[3]) now Wright Laboratory, Wright-Patterson AFB, OH, USA

The support of Prof. B. Ilschner, the editor of the Springer Series, "Materials Research and Development", is gratefully acknowledged. He enabled the publication of this book and supported its preparation with much valuable advice. Special thanks owe given to Mrs. C. Kotauschek who typed most of the contributions and to Mrs. S. Giegerich who revised the text regarding the language. Thanks is due also to the following technical collaborators who were involved in the preparation of the numerous figures and diagrams: M. Alperth, K. Baumann, H. Hermanns, U. Krebber, H. Metternich, and H. Schurmann.

Cologne, June 1992 Horst Buhl

Contents

1. Aerospace Materials, Situation and Perspectives

W.J.G. Bunk

Since aluminium was first used about 1900 as a structural material for the Zeppelin, this light metal was further developed into a long list of alloys and properties tailored to fit the needs of aeroplane builders around the world. It seems to have come to a market saturation because of the advent of carbon fibre reinforced polymer matrix composites. Titanium and nickel base super alloys dominate compressor and turbine materials for jet engines. Intermetallics and ceramics may replace some alloys in the future. From nature engineers have learned to strengthen a component by particle or fibre reinforcement.

Quite a number of innovative processing techniques emerging from aerospace material targets have been designed and introduced. It is not surprising to recognize materials research and development efforts shifting to some extent from conventional materials processing companies to more or less integrated aerospace groups. This may even be more pronounced in the future for quality, safety and cost reasons. Transfer of advanced aerospace materials and processing technologies to other markets is possible and often wanted, but sometimes problematic from the economic point of view.

It takes 10 years or more until a new material produced with advanced processing technology reaches maturity and full commercialization. This fact is sometimes difficult to acknowledge by industrial management and governmental funding agencies. Strategic considerations lead to goals, requirements and limitations. Safety comes first, followed by performance, and finally cost of material and processing net shapes, including recycling: A system's approach.

One notices considerable competition between light metals and carbon fibre reinforced polymers (PMC). There are different extrapolation curves published in the literature concerning the balance of light metals and PMCs in airframe components.

One of the important requirements for aerospace materials is resistance to elevated temperatures. Figure 1.1 summarizes the experience of the aerospace industry.

For aerospace materials and components considerable development activity is required in order to correctly select and ratify their suitability. It is necessary to demonstrate performance achievement and acceptable manufacturing costs. Influencing factors include environmental conditions, ease of processing and ability to inspect and repair parts, to name just a few.

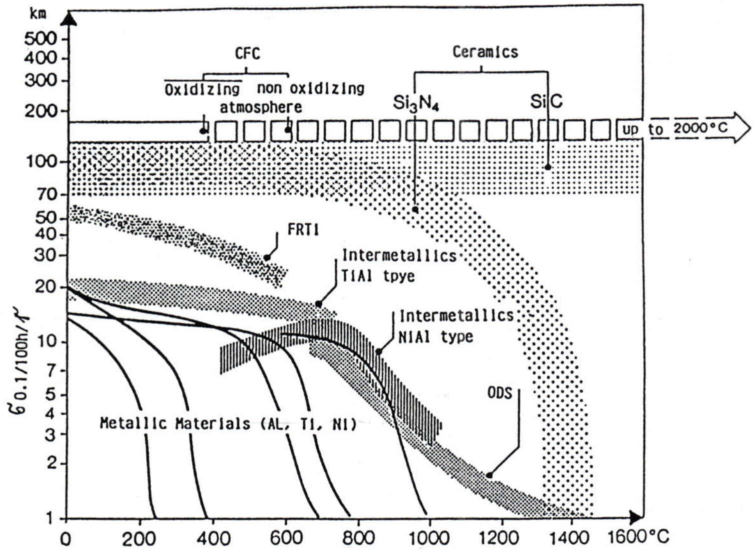

Figure 1.1 Temperature resistance of aerospace materials [1]

The evolution of new functional materials such as fibre optics and sensors resulted in the emergence of smart structures. Their response to heavy loads, etc., can be actively controlled and optimized.

In the following the situation and perspective of advanced aerospace materials will be outlined, covering airframes, engines and space vehicles.

Fuselage Materials

Figure 1.2 refers to a common impression in the transport aircraft field. Light metals are not as limited to elevated temperature exposure as are epoxy resin matrices which should not be exposed to more than 150 °C. Moisture sensitivity of polymer materials is another problem for PMCs. Ductility of today's resins, around 0.5%, is one order of magnitude lower than for light metals. Recycling of PMCs after service as aircraft components is an additional problem.

On the other hand no metal can compete with polymer composites in terms of specific strength and stiffness. Aircraft designers have the task to balance pros and cons from both performance and economic points of view. They are forced to increase the fly-to-buy ratio from one generation of aircraft to the other. Expensive lithium containing alloys should be differently processed by superplastically forming thin sheets and not by machining thick plates. It is inevitable to cut the cost of joining and assembling, the airbus fin being a convincing example for this strategy, which logically led to a PMC concept. For metallic structures the life cycle cost approach changed maintenance practices, e.g. also taking into account stress corrosion problems with respect to increased lifetimes of aeroplanes.

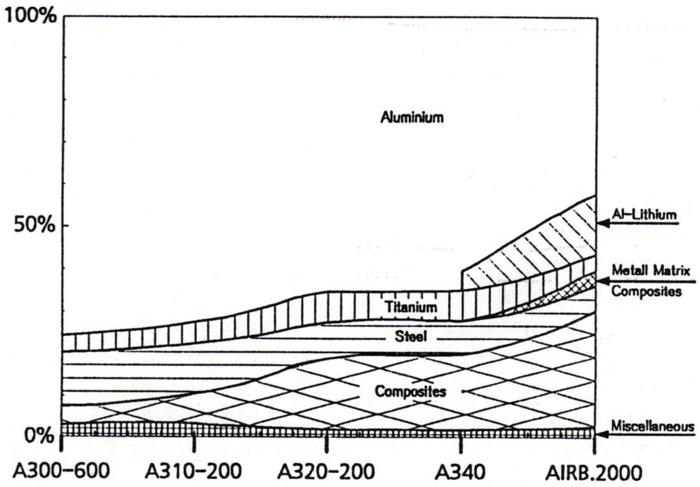

Figure 1.2 Airbus fleet material usage

Conventional isotropic and rather ductile metals can be contrasted with anisotropic and rather brittle polymer composites. Designers do well to acknowledge the impact of such material characteristics on the optimum design configuration.

Advanced materials generally are costly and the volume used in aerospace applications is lower compared to the consumer market industry. New alloys like those containing lithium, or metal matrix composites, are more expensive compared to conventional materials. Processing companies are forced to invest expensive equipment and to run large numbers of tests to establish material allowables. Knowing these barriers to the introduction of advanced concepts, one can easily understand industry's reluctance to take the high risks. Consequently, it takes a much longer time than sometimes expected until the desired breakthrough to serial production occurs.

Light Metals

Conventional 2XXX and 7XXX Al alloys have been used successfully in aeroplane fuselages, being optimized to fulfil the desired property profiles combining strength, fracture toughness, fatigue and corrosion resistance. But stiffness has always been a problem for designers because of the rather low Young's modulus of aluminium.

Conventional aluminium alloys have been taken to build short and middle range Airbus transport planes, e.g. for the Airbus A320. The need for further weight reduction and higher stiffness led to a massive effort to introduce lithium as major alloying element. Figure 1.3 explains the outstanding advantage of lithium as an alloying element of aluminium to reduce weight and to improve the Young's modulus in contrast to all other conventional alloying partners. It took some time for metallurgists to understand the microstructural and aging behaviour of these precipitation hardenable alloys, but success did not occur until casting engineers learned to directly chill cast under inert atmosphere rolling slabs of the dimensions needed by the industry. Al-Li alloys dominated the interests of aeroplane builders over the last 10 to 15 years. Taking into account the huge number of papers presented at the VI. International Al-Li Conference 1991 at Garmisch-Partenkirchen, Germany [2], most of the metallurgical uncertainties have been overcome with the exception of crack deviation problems and some corrosion phenomena. The big obstacle to a quicker introduction of this material with about 10% reduced specific weight and higher stiffness - designers calculate 15% advantage - is the 2 to 4 times higher price compared to conventional high strength Al alloys. It is hoped that this barrier to commercialization will be steadily lowered by rationalization of production in casting, rolling and extrusion mills and also by solving the scrap handling problem. At the beginning of the nineties the situation in Europe is characterized by intense qualification efforts of aeroplane builders and semifabricators to supply the Airbus company with the necessary data.

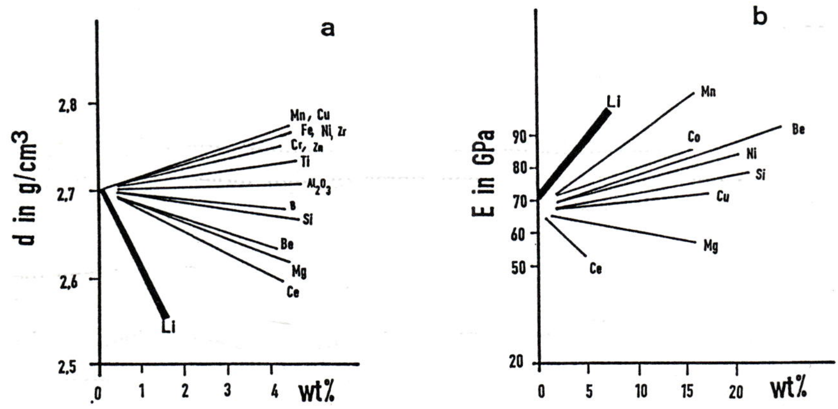

Figure 1.3 Significance of lithium as alloying element [3]

Sheet, plate, and extrusion and precision forgings of 8090 and 2091 are under investigation, including full scale fatigue and barrel tests representative for the Airbus A340. Depending on the outcome of these large scale qualification tests and the world fuel price level, the Airbus company plans to substitute for conventional high strength Al alloys ones containing lithium ones in a number of components, as indicated in Figure 1.4.

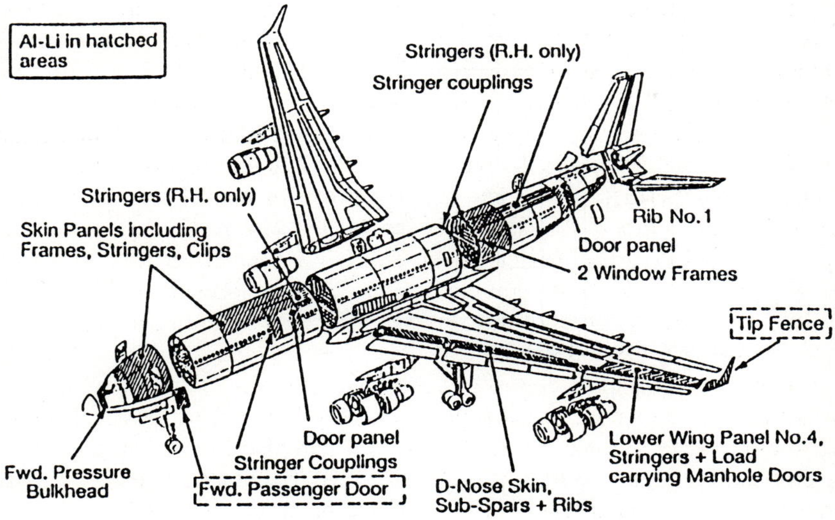

Figure 1.4 Airbus plans for substitution of conventional alloys [4]

Polymer Matrix Composites

About 30 years ago PMCs were introduced in aeroplane primary components. Since then prepregs with glass and carbon fibres in connection with epoxy resins have become standard materials for more and more structural parts, in competition to light metals. Covalently bonded carbon fibres of high molecular orientation show high strength and stiffness.

Figure 1.5 demonstrates the progress of pan and pitch carbon fibres during the last ten years. Not only strength and stiffness but also strain have been improved remarkably. Today, a range of different fibre and matrix types are available to the engineer, but problems continue with moisture and temperature degradation and with recycling.

Thermosets as a new class of PMCs reduce many of the problems still obvious with epoxy resin matrices. Higher strain and toughness of matrices result in improved impact and crack growth resistance, leading to higher damage tolerance.

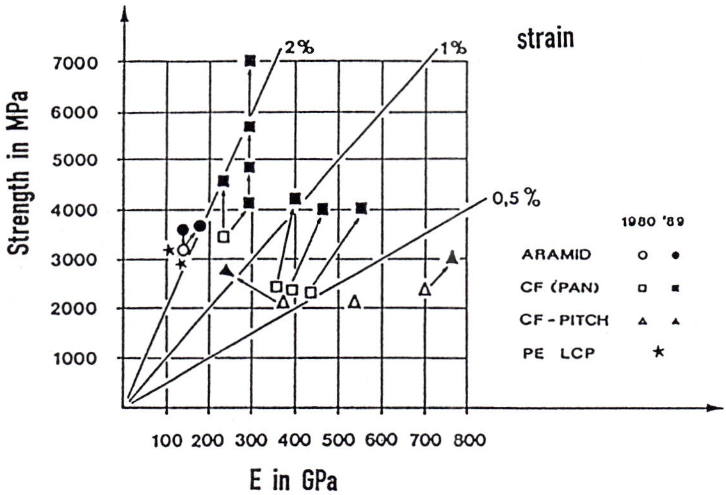

Figure 1.5 Increase of strain of carbon fibres [5]

Thermosets are attractive materials in comparison to thermoplastics because of their easier processing and handling. They are formable at elevated temperatures, weldable, and recycling is possible.

Today it seems to be too early to judge the potential of fibre reinforced thermosets in comparison to thermoplastics, especially with respect to cost considerations.

Polymer composites gained much interest in many aerospace areas because of their high stiffness and strength. The weak point is their rather low interlaminar strength, leading to unsatisfactory damage tolerance. Laminated 2D structures can be substituted by 3D woven structures which have improved interlaminar behaviour and acceptable damage tolerance performance. Aerospace development groups learned from textile techniques to design and manufacture 3D fibre preforms. In contrast to 2D ones, such 3D woven components are characterized by a local restriction of damage. Designers favour 3D configurations for certain design solutions because of the wide variety of fibre architectures which cannot be achieved with conventional 2D prepregs.

Some Promising Laboratory and Pilot Scale Developments

In the following, advanced material developments from the airframe and engine applications point of view are discussed.

Composites: Metal matrix composites were developed with the aim of using them in more demanding environments with requirements of higher specific properties and durability at elevated temperatures. Such composites could solve problems in the structural parts of supersonic flight planes or rotating parts of engines.

With the exception of some missile fins, most Al and Mg fibre reinforced composites did not leave the pilot scale project stage because processing techniques had not been optimized enough. Designers encourage materials engineers to intensify the development of endless fibre MMC for fuselage components of high stiffness values. Therefore, a breakthrough of MMC applications in special areas is probable in the near future: fibres of the spun SiC type are available and will come down from the present price range with increasing production rates. For some years short fibre Al-MMCs of superior tribological quality and increased thermoshock resistance have been standard practice for automotive pistons [6]. This breakthrough in a large market for light metal matrix composites will encourage aerospace designers to think more in terms of metal matrix composites.

Recently jet engine producers expressed interest in developing SiC fibre (140 μm thickness) reinforced titanium compressor parts because of their high strength levels for more efficient engines. Figure 1.6 indicates the progress achieved controlling the interface stability of SiC-fibres and TiAl6V4 matrix. These pilot scale results should inspire compressor designers to use new design concepts taking advantage of the much improved strength in fibre orientation. One realizes the importance of a special fibre coating to control the fibre matrix composite. Particle reinforced light metals as isotropic metal composites are ready to be introduced, primarily in aerospace parts.

Rapidly solidified transition element alloyed Al-X materials showed an astonishing high temperature resistance up to 400 °C, a temperature level unknown for light metals. One example for large scale production of rapidly solidified Al alloys is planar flow cast Al-12Fe-1V-2Si for elevated temperature applications [7].

Spray casting as a unique PM process was invented in Europe [8]. In order to optimize the rather complicated traditional PM processing technology, full scale slabs of Al alloys, including particle reinforced composites, have been successfully produced. Powder metallurgical processing seems to have gained more acceptance via spray casting of preforms which can be extruded or forged to final shapes without any intermediate processing steps.

Mechanical alloying has been successfully used to generate new compositions of artificial alloys for special purposes. In USA highly corrosion resistant 9052 components for naval applications are on the market. In Europe carbon has been introduced to the long range of alloying elements of Al: mixing of Al powders with carbon powder forms Al_4C_3 particles in extremely fine dispersion, leading to up to 3% C/Al alloys and temperature resistance higher than 400 °C.

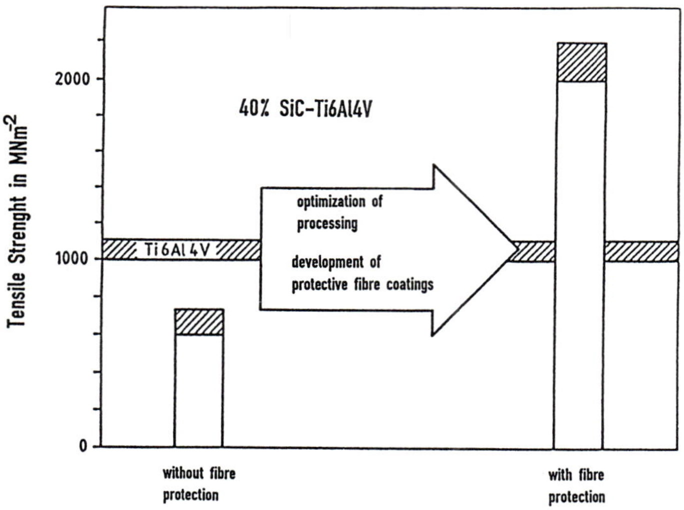

Figure 1.6 Optimization of processing and fibre/matrix interface of SiC-Ti6Al4V by special fibre protective coatings (DLR)

Metal matrix composites may well become a promising alternative to conventional alloys. Figure 1.7 plots specific strength to specific modulus comparing PMCs and MMCs with conventional materials. The advantage of composites over monolithic metals is evident. Figure 1.8 plots specific strength to temperature, demonstrating the dominance of polymer composites up to about 250 °C and leaving the field to fibre reinforced metals for higher temperatures. It should be remembered, however, that fibre reinforced metals lack satisfying fracture toughness because of low ductility. Furthermore, fibre orientation limitations must be taken into account.

Various kinds of composites with high-performance reinforcements and temperature resistant matrices are currently under consideration. Figure 1.9 classifies advanced fibre reinforced composites with reference to their upper temperature limits. Apart from partially overlapping the temperature scale, this figure only indicates present experience. Many R&D projects are funded in this area with optimistic expectations not only to improve reliability but also to eliminate expensive processing routes. Ceramic fibre reinforced glass and glass ceramic matrix composites have been investigated with considerable success since the beginning of the eighties. It seems to be possible to introduce them in advanced propulsion systems with good dimensional stability up to 1000 °C. Ceramic composites for very high temperature applications are discussed in section "Ceramic and Ceramic Matrix Composites". Figure 1.10 plots realistic service temperatures against the foreseeable year of introduction in advanced engines.

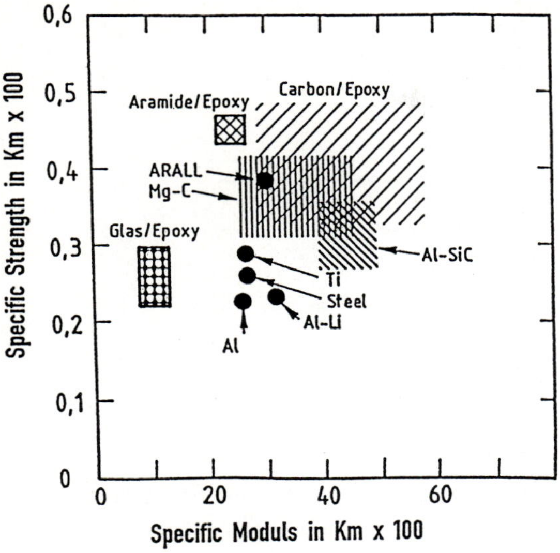

Figure 1.7 Composites surpass metals in specific strength and modulus [9]

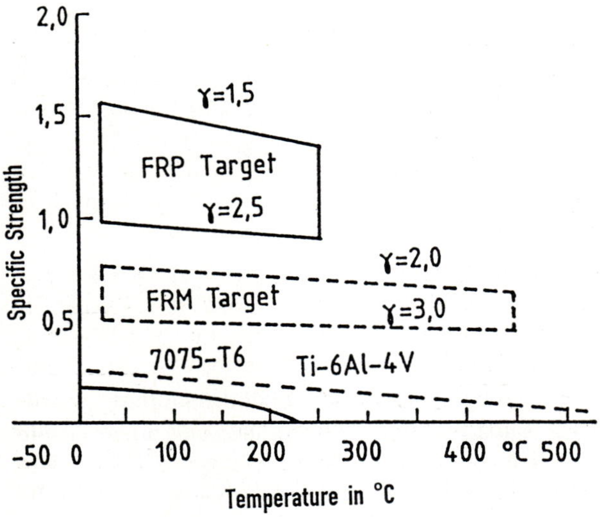

Figure 1.8 Targets for fibre reinforced polymers and metals [10]

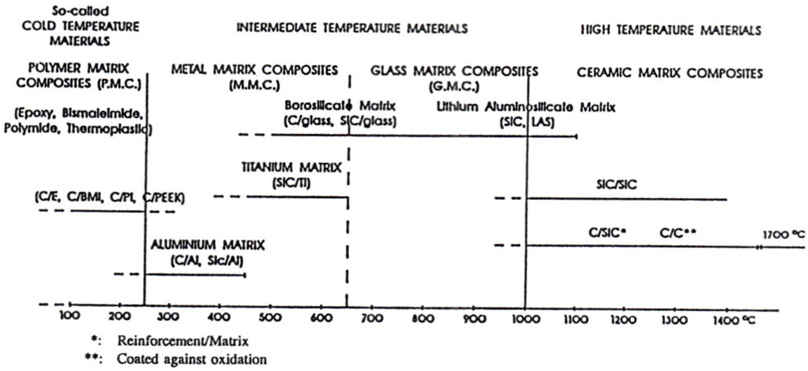

Figure 1.9 Classification of composites [11]

Intermetallics: Due to their excellent strength-to-weight ratio for temperatures well above 700 °C, titanium aluminides could substitute for conventional jet engine compressor materials. Many attempts are underway to improve their low ductility and rather poor toughness at lower temperatures. Already commercially available are Ti_3Al base intermetallics. TiAl also has some potential for future research and development. Figure 1.11 summarizes recent results of TMT super alpha 2, proving the advantage of these processing techniques for improving microstructure and temperature resistance.

Aluminium Laminates: Unidirectional or cross-ply layers of preimpregnated aramid, glass or carbon fibres can be laminated with thin metal alloy sheets to form a new laminated sheet material with outstanding fatigue and fracture toughness properties (ARALL, Glare). Figure 1.12 demonstrates how fibre prepregs and metal sheets are built up. The excellent fatigue properties of ARALL are obtained by crack bridging. The laminate types differ with respect to aluminium sheet thickness, type of alloy and adhesive and residual stress state.

Propulsion Materials

Compressor Materials: Advanced jet engine concepts demand materials capable of sustaining high static and dynamic loads at high temperatures and under hostile environments. The need for a light metal concept in the first stage compressor encouraged the development of titanium alloys because of their better specific strength. More and more compressor stages turned from heavier metals to Ti alloys with improved temperature and oxidation resistance.

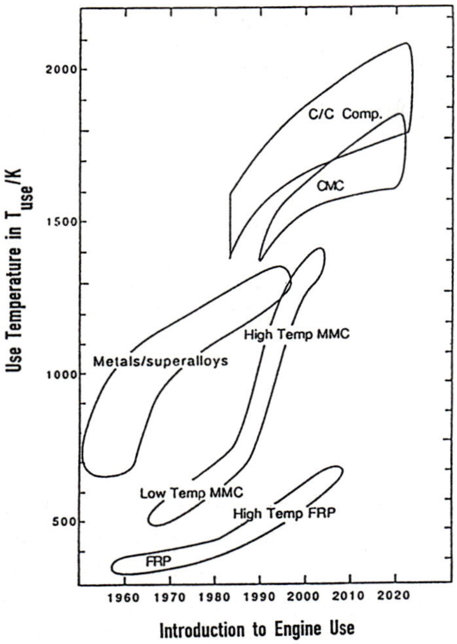

Figure 1.10 Introduction in engine use [12]

The dramatic substitution of conventional metals by advanced Ti and Ni alloys is demonstrated in Figure 1.13. For many years both groups dominated the field until intermetallics and MMCs recently became candidates for future disks and blades.

Figure 1.14 assesses today's choices of metallic and ceramic materials and tomorrow's options from another source. For compressor parts Ti alloys dominate the scene. Lower compressor stages have been successfully equipped for a long time with well documented alloys. High compressor stages are targets for conversion to advanced Ti alloys with improved temperature capability up to 600 °C, and in the future even higher with appropriate oxidation coatings.

Compressor and turbine designers hope to be able to introduce advanced material concepts into new generation jet engines. A lot of test efforts will have to be undertaken to approach this ambitious goal.

Turbine Materials: Materials for turbines and combustion chambers have to meet requirements for high temperatures and hot gas corrosion environments. Because of these requirements material scientists and engineers rather soon acknowledged

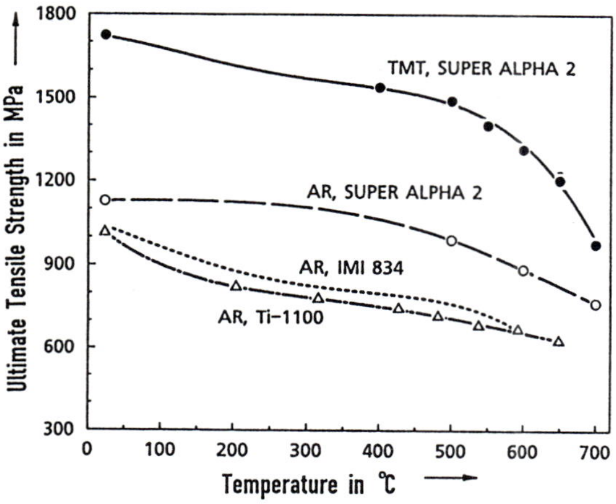

Figure 1.11 Temperature resistance of Ti alloys and intermetallics [13]

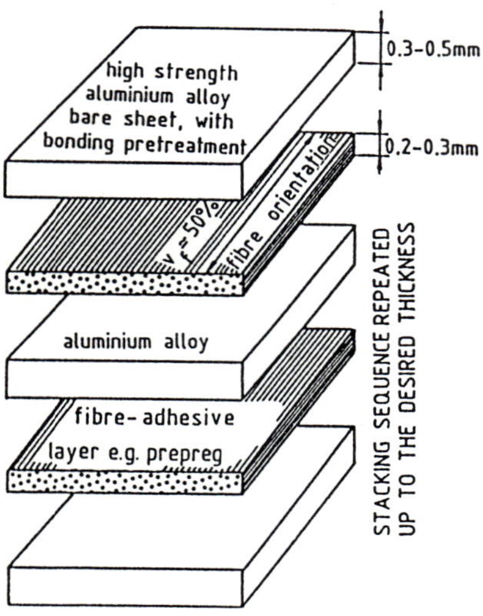

Figure 1.12 Build-up of ARALL [14]

that there is not any Ni based super alloy to fulfil the needs and hopes reflected in advanced turbine concepts. Designers of blades helped a lot in developing complicated cooling devices. Since then a gap of a few hundred °C exists between

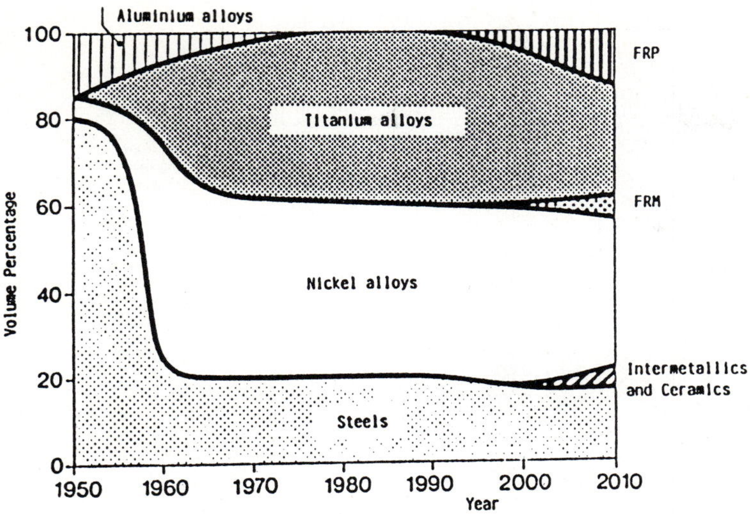

Figure 1.13 Situation of jet engine materials [15]

gas temperatures and the metallic part. Consequently, advanced turbine designers demand materials with higher and higher temperature capability. One definite step in the proposed direction has been the processing of blades of special Ni super alloy single crystals. The absence of grain boundaries significantly improves high temperature deformation behaviour. In this high technology field cross fertilization from materials scientists to component designers and vice versa became obvious.

It is not easy to predict which of the current R&D projects will lead to the next step of higher turbine efficiency. May be, intermetallic Ni aluminides will substitute partly for Ni or Co alloys. May be, a hybrid stator or combustor combining metal with ceramic will leave the desk of a turbine designer.

Figure 1.15 demonstrates a hybrid turbine blade test with promising results. Test rig data confirm the hybrid metal-ceramic concept to be of some relevance for turbine blades, stator segments and compressors.

For both important engine parts - compressors as well as turbines - intermetallic components are candidates for substitution of rather heavy super alloys because of their lower specific weights, remarkably high strength and improved oxidation resistance. Although not all questions concerning their long service time exposure qualitites have been answered, it is appropriate to continue extended test runs to prepare intermetallics for next generation engines.

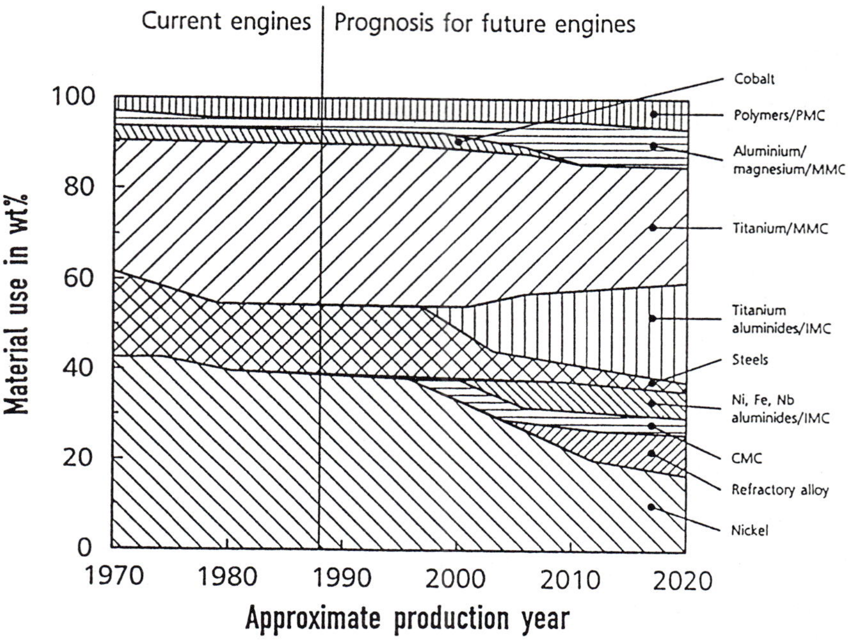

Figure 1.14 Today's choices and tommorrow's options of jet engine materials (NASA)

Figure 1.16 summarizes the visions of advanced materials developers for high performance engine generations. Metal or intermetallic composites as well as ceramic or even polymer composites could well be used in certain parts of the engine. From a more conservative point of view, it appears to be a great challenge to substitute advanced composites for monolithic materials.

Space Vehicle Materials

Requirements and Design Considerations: Space programs are characterized by ambitious goals. Minimum weight requirements for mechanically and thermally heavily loaded structures are common. Consequently, new design and processing concepts as well as an optimum choice of temperature resistant materials are required. It is not surprising to recognize a continuous improvement of metals, composites and ceramics motivated by scientific, technical and economic considerations for future space vehicle projects.

The motivations for innovative design concepts for space transportation systems are the challenging requirements for hypersonic flight. The design goal is the development of highly efficient and safe structures. During extended cruises at

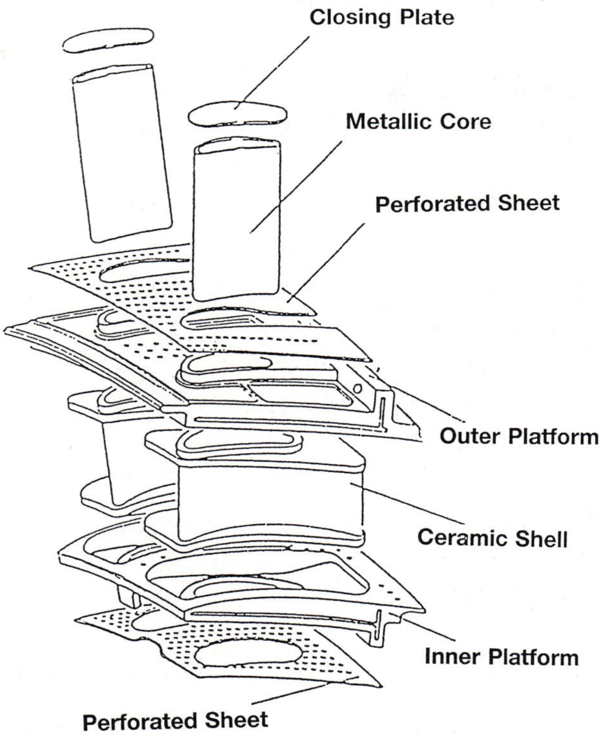

Figure 1.15 Experimental hybrid metal-ceramic turbine blade [16]

hypersonic speeds temperatures up to 700 °C can be expected over major areas of the airframe with local excursions of short duration to 1100 °C. During the reentry and descent phase, aerothermodynamic interactions will produce 1600 °C and more.

The material selection depends to a large extent on the designer's decision whether it is intended to operate in a hot environment without thermal protection up to the material's limit or whether shielding devices are necessary. These two design concepts are commonly referred to as hot and cold structure concepts, respectively.

As for metals, their mechanical properties reflect less than 30% of their theoretically possible values due to defects, etc. Carbon fibres are very attractive because 70% of theoretical strength is achievable. However, more than strength, temperature capability is the dominating requirement.

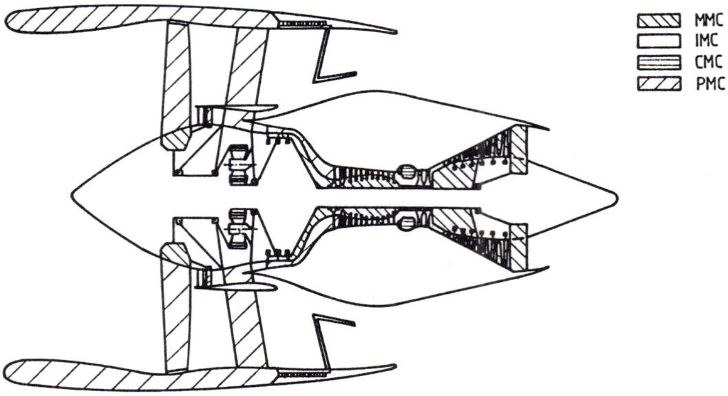

Figure 1.16 21st century ultra-high bypass engine (NASA proposal)

Around stagnation areas, ceramic matrix composites such as C/SiC or SiC/SiC or carbon/carbon offer optimal response concerning strength and modulus, however not taking into account oxidation resistance. Therefore, critical surfaces have to be protected by coatings. The airframe substructure is expected to be constructed of light metal sections possibly specially insulated.

Space vehicle application oriented materials and structures are one of the great challenges for materials engineers. Although modified high temperature metallic structures may meet some requirement profiles, fibre reinforced composites with metallic or ceramic matrices seem to become candidate materials for next generation systems. Figure 1.17 lists just a few of many other proposals for airframe and engine components.

Proposed Structural Components: At extremes of performance, i.e. very high temperatures, ceramic composites are candidate materials for jet engine parts as well as reusable space vehicle components. CMC with high strength and high toughness responds quite well to the increasing requirement of the aerospace industry for materials capable of operating at high temperatures up to 1600 °C and above, in aggressive environments with acceptable mechanical properties. Good impact strength, damage tolerance, resistance to creep and thermoshock are requirements which can be fulfilled by some C- or SiC-fibre reinforced SiC matrices processed via CVI or silicon organic precursors. But progress is needed in the temperature resistance of fibres in oxidizing atmospheres and in the reduction of matrix porosity. In the case of carbon-carbon oxidation protection via special multilayer or functionally graded coatings must be further developed to withstand thermal cycling.

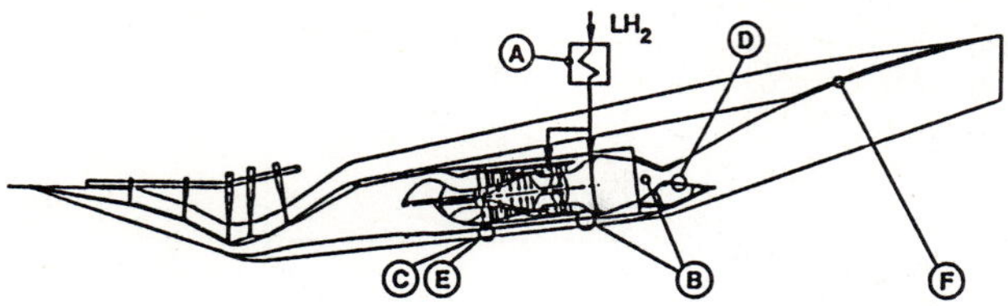

Structures	Material Candidates	Materials Surface Temperatures
Ⓐ H_2-Heat Exchanger	C/Cu – alloy	600°c
Ⓑ Nozzle Casing	SiC/Ti-alloy	700°C
Ram Duct Casing	SiC/Ti_3Al	800°c
Ⓒ Hot Strut Shell	TiAl⁺	900°C
Ⓓ Nozzle Flaps Nozzle Cooling Structure	Superalloy	700°C – 1200°C
Ⓔ Hot Strut Shell	SiC/SiC, SSiC	1250°C
Ram Duct Hot Structure	C/SiC⁺	1700°C
Ⓕ Expansion Ramp	C/C⁺	2000°C

⁺ Oxidation Resistant Coating

Figure 1.17 Candidate space vehicle materials [17]
Source: MTU Focus, 1 (1992)

The US space shuttle introduced reusable high temperature resistant SiO_2 ceramic composite tiles to withstand the reentry phase of space missions more than 10 years ago. European developments concentrated on carbon and silicon carbide fibre reinforced silicon carbide matrices by chemical vapor infiltration. Fibre reinforced ceramic composites of 2D or 3D structures have been successfully tested in oxidizing atmospheres under thermal and mechanical cycling conditions in order to simulate practical space mission operations (Figure 1.18).

This rigid external insulation by a ceramic fibre reinforced ceramic matrix composite has to be complimented by some internal insulation design. One infiltrates preforms of SiC-fibre fabrics with SiC-matrix materials using a modified chemical vapor infiltration process characterized by thermal and pressure gradients and resulting in comparatively short processing cycles of up to 50 hours. Specially coated fibres assure a definite low bond strength between fibre and matrix. The fibres take responsibility for load transfer by crack bridging, deflection and fibre pull out, leading to the desired damage tolerance behaviour.

Functionally graded metal-ceramic structures could become the answer to certain parts of space components where the ceramic surface serves as a heat shield and the metal substrate gives enough structural strength.

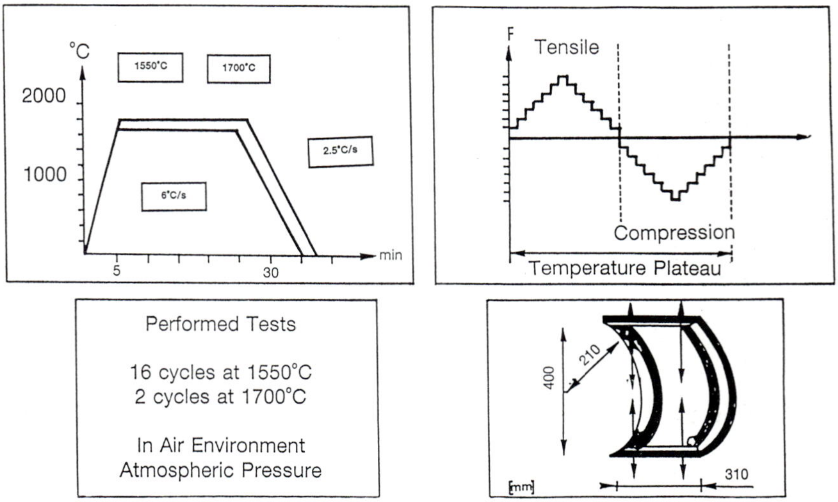

Figure 1.18 Thermomechanical qualification of ceramic composite components [18]

An internal multiscreen insulation for the HERMES space shuttle is under development for protecting the aluminium frame structure between -165 °C and 1450 °C in oxidizing environments. A large number of gold coated alumina foils isolate all internal structural elements.

The main differences between the US SiO_2-based tile concept and current European plans on the other hand are the rigid external structure of the CMC and the internal multilayer insulation of the cold metal structure. One still unsolved problem is the resistance of coated carbon-carbon to oxidation at very high temperatures. Figure 1.19 schematically explains the direction of R&D concerning this frontier of materials development.

It seems that ceramic parts could fill quite a number of requirements for advanced jet and rocket engines as well as space vehicle components. Hypersonic planes require thermostructural materials like CMC.

Up to 1000 °C, glass-ceramic matrices with carbon or SiC fibres have a market; up to 1400 °C, endless fibre reinforced ceramic matrices. Above 1400 °C coated carbon-carbon meets aerospace requirements.

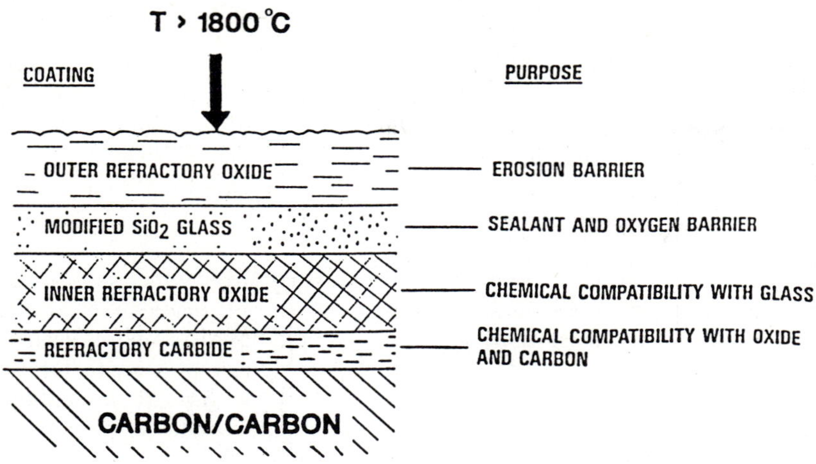

Figure 1.19 Concept of coating system for C-C [19]

Quality and Safety Aspects

For a long time the safety and durability of fatigue critical aircraft structures has been one of the main topics of fundamental and applied research and development at many universities, national research institutes, and industrial laboratories. Two basic philosophies of fatigue oriented design have been discussed: the safe-life and the damage tolerance approach, the main difference being the inspectability for significant cracks during service. Repeated inspection is necessary for damage tolerant components, which dominate today's structures.

Fatigue of aerospace materials attracted great interest from both the scientific and engineering points of view. This fact resulted in many international meetings where important achievements in understanding crack initiation, crack propagation, environmental effects, the significance of thresholds, etc. have been reported. This concerns fuselage and engine materials. The ultimate aim of all efforts is to assure the quality of alloys and composites. From extended case studies and detailed research programmes, including variable-amplitude and spectrum loading for military and civil airplanes, a number of specifications and standards have been agreed upon by all relevant parties. Today it seems possible to predict the lifetime of metallic materials and components to a large extent. But still more research projects are in progress to fill existing gaps in the science of fatigue.

Fracture mechanics is an indispensable tool to analyse the damage tolerance of aerospace materials. This discipline has been introduced to the design stage of materials which will be used in structural components. During recent years many discussions between materials scientists thinking in terms of lattice defects, dislocations and plasticity on one hand and engineers thinking mainly in terms of continuous mechanics on the other hand resulted in mutual understanding and joint efforts to overcome not yet fully understood phenomena. Although a lot of knowledge has been generated during the years to improve the reliability of materials and component tests, much remains to be done to avoid catastrophic failures and to optimize design and testing methodology.

Fracture properties are presently used in many industrial areas, either as material selection criteria or as material quality requirement. In aerospace, it is also a design parameter to avoid catastrophic failures. The common assumption used is that the presence of defects in materials is due to the rigours of manufacturing, testing and service. All engine and structural component designs are susceptible to degradation. Future progress in understanding degradation and its prevention depends largely on the successful integration of continuum mechanics with the scientific disciplines of materials science, chemistry, and physics.

Conclusions

The most important requirements for aerospace materials are lower density and higher modulus. Of equal relevance are temperature resistance, reliability and safety. Fuel cost considerations also influence advanced design concepts.

The fuselage of future transport aircraft will be built from light metals with an increasing amount of PMC structures. Jet engines will see intermetallics, MMCs and CMCs substituting partly for today's titanium and nickel alloys.

How to select appropriate of structural materials for space vehicles is still an open issue. Reinforced metals for inner structures and ceramic composites for outer components seem to attract most designers.

2. Metallic Materials and Metal Matrix Composites

2.1 Metallic Materials

2.1.1 Aluminium Alloys

2.1.1.1 High-Strength Aluminium Alloys

K. Welpmann and M. Peters

From the beginning of aviation people have searched for lighter and stronger aircraft materials. The first generation of aircraft was mainly built of wood and fabric. But soon aluminium became an important structural material, especially after A. Wilm had discovered the age hardening capability of Al-Cu alloys in 1906. Nowadays high strength aluminium alloys are still the most important structural materials for aircraft and aerospace vehicles.

In recent years several excellent books have been written which cover the entire scope of aluminium alloys [20-22]. This comprehensive survey will confine itself to high strength aerospace aluminium alloys.

Pure aluminium is a very soft metal which cannot be used for structural applications. When alloyed with certain other elements, it gains strength by solid solution hardening, but the most effective strengthening mechanism is age or precipitation hardening. The essential principles of this phenomenon shall be briefly described.

Principles of Age Hardening

The basic requirement for an alloy to be age hardenable is a decreasing solid solubility of an alloying element at decreasing temperature. Age hardening then involves the following steps: 1. Solution heat treatment: annealing above the solid/solid phase boundary within the single phase region to dissolve the alloying elements. 2. Quenching: rapid cooling below the phase boundary to obtain a supersaturated solid solution (SSSS) of the alloying elements in aluminium. 3. Aging: storing the alloy at room temperature (naturally aged T4 condition) or at elevated temperatures (artificially aged T6 condition) to decompose the SSSS. The most frequently used alloying elements which fulfil the above requirement are copper, zinc, magnesium and lithium.

The decomposition process of the SSSS during aging is very complex and the resulting precipitation sequence depends upon the respective alloy system. Usually it includes the precipitation of different metastable intermediate phases before - at longer aging times or higher aging temperatures - the respective equilibrium phases are formed. The precipitation sequence often begins with the homogeneous nucleation of very small coherent particles like GP zones in the Al-Cu system or δ' in Al-Li. In Al-Cu the zones are followed by the formation of semicoherent Θ' particles which either nucleate homogeneously within the aluminium matrix or heterogeneously at lattice defects such as dislocations. Maximum age hardening, i.e. maximum hardness or yield strength, occurs when the coherent particles obtain a critical diameter (Al-Li system) or when a critical combination of coherent and semicoherent particles is present (Al-Cu and other systems).

Commercial high strength aluminium alloys mostly contain two or more primary alloying elements (copper, zinc, magnesium, lithium), at least one minor alloying element (maganese, chromium, zirconium), plus various impurities (iron, silicon). All contribute to the very complex particle microstructure of the alloy.

Microstructure of Commercial Al Alloys

Age Hardening Phases: The nature and amount of age hardening phases depend on the properties and percentage of the alloying elements and on the age hardening condition of the alloy. By changing the age hardening conditions, a variety of different precipitation populations can be produced. The Al-Cu-Mg alloy 2024, e.g., is mostly used in the damage tolerant T3 condition. In this naturally aged alloy state the coherent GPB particles are the predominant age hardening phases. When this alloy is used in the peak aged T6 condition, the microstructure is dominated by the semicoherent S' platelets.

Dispersoid Particles: A second category of particles is always present in commercial high strength age hardenable Al alloys, namely intermetallic compounds of the transition metals manganese, chromium, or zirconium which are added as minor alloying elements. These intermetallics effectively control the grain structure by preventing recrystallization and/or grain growth. Their distribution in the matrix which is often inhomogeneous cannot be influenced by any heat treatment. They contribute only slightly to the strength of the alloys.

Chromium and manganese bearing dispersoids, which are incoherent and mostly of the dimensions 0.1-0.5 μm, have a detrimental effect on the quench sensitivity of Al alloys. Therefore, in certain alloys chromium has been replaced by zirconium, e.g. 7075 and 7050. Lithium bearing alloys always hold zirconium which forms very small (0.02 μm) coherent Al_3Zr particles.

Constituent Particles: Coarse constituent particles are always found in commercial high strength aluminium alloys. They represent intermetallic compounds of the impurity elements iron and silicon with aluminium and/or the alloying elements. They form by eutectic decomposition during ingot solidification. During subsequent working of the ingots the larger particles are fractured, which reduces their sizes to the range of 1-10 μm and causes them to become aligned as stringers in the working direction. In semifinished products their size and distribution cannot be further modified.

The coarse constituents have a detrimental effect on fracture toughness. Therefore, minimizing the size and number of these particles by reducing the iron and silicon contents and thus increasing fracture toughness was one important effort in the development of 2XXX and 7XXX alloys. It led in several steps from relatively impure 2024 and 7075 alloys to the low iron and silicon versions 2324 and 7475, respectively.

Precipitate Free Zones (PFZ) at Grain Boundaries: Grain boundaries represent a sink for vacancies as well as for solute atoms. Thus, during age hardening the areas adjacent to the grain boundaries are depleted of vacancies and solute, and nucleation and growth of precipitates cannot occur. Therefore, all age hardenable aluminium alloys exhibit precipitate free zones (PFZ) along grain boundaries. The formation of PFZs is accompanied by the nucleation and growth of relatively large particles of the equilibrium phases at the grain boundaries.

PFZs mostly have a harmful effect on the properties of aluminium alloys. They are softer than the neighbouring matrix since they are free of precipitates, and thus - assisted by the equilibrium phases at the grain boundaries - can lead to premature failure under an applied stress. Besides this, PFZs along with grain boundary precipitates are believed to increase susceptibility to stress corrosion cracking. Consequently, heat treatment and processing procedures are often employed to minimize PFZ width and grain boundary precipitation.

Grain Morphology and Crystallographic Texture: During hot deformation (rolling, forging, extrusion) of the annealed ingots the grains are always stretched in the deformation direction. Since recrystallization does not occur due to the presence of manganese, chromium, or zirconium bearing dispersoids, the elongated grain structure is mostly retained even after cold deformation procedures. Thus, for the solution heat treated condition of rolled products highly elongated and very flat "pancake"-shaped grains are quite typical. This grain structure leads to a directionality of properties. Especially fracture toughness and resistance to stress corrosion cracking (SCC) vary with stress direction, the minimum K_{IC} values and maximum SCC susceptibility occurring in the short transverse direction, i.e. with the stress direction normal to the flat grains.

Besides this grain anisotropy, various crystallographic textures are always found in wrought products of high strength aluminium alloys, which also may lead to an anisotropic behaviour of properties. Generally, they do not offer any problems in applications such as aerospace sheet or plate material. Recently, however, lithium containing aluminium alloys were reported to exhibit pronounced crystallographic textures [23]. They are probably the cause of a marked crack deviation during fatigue crack propagation experiments [24], which is unacceptable to the aircraft industry. Modified thermomechanical treatments were proposed to overcome this phenomenon [25].

Hardening Mechanisms

The strength of an age-hardened alloy is controlled by the interaction of moving dislocations with precipitates. The strength increase is essentially a function of the particle radius. This is schematically shown in Figure 2.1, where the volume fraction f of the particles is considered as being constant. Generally this condition is fulfilled during age hardening. In the early stage of age hardening the precipitates are coherent with the aluminium matrix and can be sheared by dislocations during plastic flow. The critical resolved shear stress due to the interaction of moving dislocations with the particles, is then proportional to the particle radius, i.e. the strength increases with the particle size.

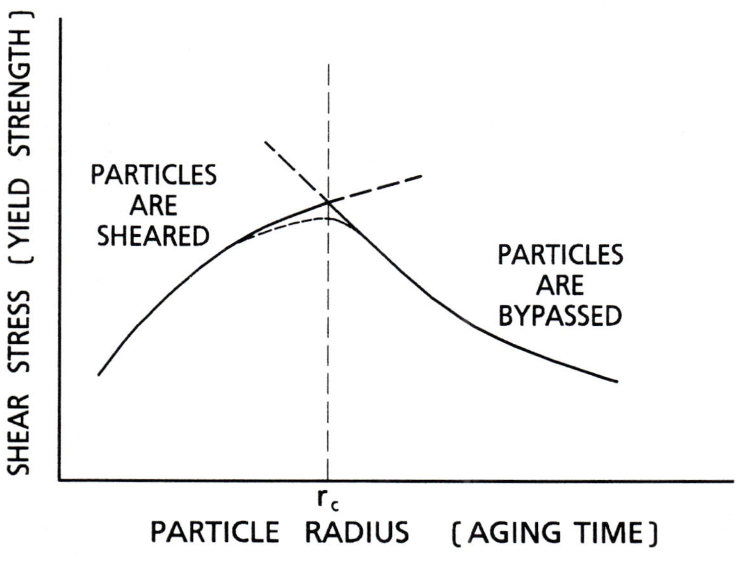

Figure 2.1 Schematic diagram of critical resolved shear stress (yield strength) as a function of particle radius (aging time) when the particles are sheared or by-passed by dislocations, at constant particle volume fraction.

As the particles grow, they lose their coherency with the matrix and can no longer be cut by dislocations; or coherent particles like δ' in Al-Li alloys attain a critical size r_c at which shearing is energetically no longer favourable. In both cases the particle-dislocation interaction changes from shearing to looping or by-passing the particles by a mechanism first proposed by Orowan. The strength is then inversely proportional to particle distance. Thus, the shear stress now decreases at increasing particle size and increasing particle distance for f = constant. This is the situation with overaged alloys. It follows that the curve given in Figure 2.1 can as well describe the typical age hardening behaviour of an alloy with the maximum hardness at the transition from shearing to by-passing the particles.

Corrosion Properties

Aluminium alloys rapidly form a protective oxide layer of alumina on their surface, making them quite corrosion resistant. However, the oxide barrier can be destroyed under certain environmental conditions such as the presence of halide ions, which cause localized corrosion. The occurrence of various types of localized attack like pitting, intergranular corrosion, exfoliation corrosion or stress corrosion cracking (see also section "Stress Corrosion Testing") depends upon the microstructure of the alloys. The mechanism of the corrosion process is usually associated with a local cell action between the solid solution of the matrix and the intermetallic second phases with a different electrode potential. The susceptibility to localized corrosion like exfoliation corrosion or stress corrosion cracking can be reduced and virtually eliminated by appropriate heat treatments such as T7X tempers for Al-Zn-Mg-Cu alloys or T8X tempers for 2XXX alloys, however with some sacrifice in strength and ductility [26]. The use of Alclad products is another concept to improve the corrosion behaviour. The core alloy is cathodically protected by the cladding alloy due to their different corrosion potentials. Further ways to enhance the corrosion resistance of high strength aluminium alloys are the powder metallurgical production route or special heat treatments such as retrogression and reaging.

Alloys and Temper Designations

The Aluminum Association (AA) of the United States has introduced a four-digit numerical system for designating wrought aluminium and aluminium alloys. This system has been accepted by most countries and is now called the International Alloy Designation System (IADS). The first of the four digits indicates the alloy group as given in Table 2.1. The last two digits are arbitrary numbers identifying the specific alloy in the alloy group, except group for 1XXX where they indicate the aluminium purity. The second digit indicates modifications of the original alloy, e.g. 2124 and 7475 are higher purity versions of 2024 and 7075, respectively. Alloys which are still experimental have the letter X preceding their designation number.

The high strength age-hardenable aluminium alloys used for aerospace applications are found in the following alloy groups: The Al-Cu and Al-Cu-Mg alloys in the 2XXX group; a few Al-Mg-Si alloys in the 6XXX group; the Al-Zn-Mg-Cu alloys in the 7XXX group; some lithium containing alloys with lithium as the major alloying element in the 8XXX group.

Designation	Al content or major alloying elements
1XXX	99.00% aluminimum[2])
2XXX	copper
3XXX	manganese
4XXX	silicon
5XXX	magnesium
6XXX	magnesium and silicon
7XXX	zinc
8XXX	lithium or others
9XXX	unused

Table 2.1 Designations for wrought aluminium alloys

In order to specify the mechanical properties of an alloy, an Aluminum Association temper designation system is used which is explained in the American National Standard ANSI H 35.1. It takes the form of letters and digits that are added to the alloy number. This system has also been internationally adopted.

The temper designation system for heat-treatable aluminium products employs the letter T followed by a number from 1 to 10 for the basic alloy conditions. Often additional digits are added to the basic designation in order to give a more detailed description of the respective condition. They describe, e.g., whether a product is stress-relieved, and in which way; whether a product requires heat treatment by the user; or they designate a specific heat treatment of the producer or user which may be the subject of a patent.

For high strength aerospace aluminium alloys only a few of the basic alloy designations are used. They can be briefly described as follows:

T3 indicates that a product is solution heat treated, cold worked and naturally aged. It applies to alloys which achieve a stable condition during natural aging and whose strength is improved by cold deformation.

T4 designates a product that is solution heat treated and naturally aged. It applies to products which achieve a stable condition during natural aging and which are not (or cannot) be cold worked after solution heat treatment.

T6 designates a product that is solution heat treated and artificially aged.

[2]) If not otherwise stated, all concentrations are given in mass percentages

T7 means that a product is solution heat treated and stabilized. It applies to products that are overaged during artificially aging to provide specific characteristics such as improved resistance to stress corrosion.

T8 indicates that a product is solution heat treated, cold worked and artificially aged.

Aerospace Structural Alloys

For the specific requirements of the aircraft and aerospace industries a variety of aluminium alloys has been engineered during the past decades of aerospace developments. Alloys are in use with combinations of properties such as high strength and fracture toughness, excellent fatigue and corrosion resistance, low density, high temperature capability and cryogenic strength. Some of these alloys like 2024 have become general purpose alloys, but due to costly purification procedures and/or highly sophisticated fabrication routes and thus high prices most of them are restricted to aerospace applications.

A selection of presently used wrought high strength structural aluminium alloys is listed in Table 2.2 along with their nominal compositions. Table 2.3 shows some of the typical mechanical properties of the same alloys. Table 2.4 gives some examples for the use of aluminium alloys and tempers in major components of commercial and military transport aircraft. The following description of aerospace aluminium alloys will consider also the specific alloy tempers which are highly important for the required properties and their use in major aircraft components.

Alloy	Al	Cu	Mg	Zn	Mn	Cr	Zr	Fe	Si	others
2014	Bal	3.9-5.0	0.2-0.8	0.25	0.4-1.2	0.1	-	0.7	0.5-1.2	
2024	Bal	3.8-4.0	1.2-1.8	0.25	0.3-0.9	0.1	-	0.5	0.5	
2324	Bal	3.8-4.4	1.2-1.8	0.25	0.3-0.9	0.1	-	0.12	0.1	
2219	Bal	5.8-6.8	0.02	0.1	0.2-0.4	-	0.1-0.25	0.3	0.2	
2618	Bal	1.9-2.7	1.3-1.8	0.1	-	-	-	0.9-1.3	0.1-0.25	0.9-1.2 Ni
X2095	Bal	3.9-4.6	0.25-0.6	0.25	0.1	-	0.04-0.18	0.15	0.12	0.25-0.6 Ag 1-1.6 Li
6013	Bal	0.6-1.1	0.8-1.2	0.25	0.2-0.8	0.1	-	0.3	0.6-1.0	
7075	Bal	1.2-2.0	2.1-2.9	5.1-6.1	0.3	0.18-0.28	-	0.5	0.4	
7475	Bal	1.2-1.9	1.9-2.6	5.2-6.2	0.06	0.18-0.25	-	0.12	0.1	
7050	Bal	2.0-2.6	1.9-2.6	5.7-6.7	0.1	0.04	0.08-0.15	0.15	0.12	

Table 2.2 Composition of some high strength aerospace aluminium alloys. For Al-Li alloys see Table 2.5

Alloy Temper	Product	YS MPa	UTS MPa	El %	Ref.
2014-T6	sheet	400	455	7	[27]
2024-T361[1]	sheet	330	440	15	[27]
2324-T39	plate	386	455	10	[28]
2219-T81	sheet	345	455	12	[27]
		421[2]	572[2]	15[2]	[27]
2618-T61		372	441	10	[27]
X2095-T8	extrusion	696	715	5.5	[29]
6013-T6	sheet	324	359	8	[30]
7075-T6	sheet	469	538	8	[27]
7075-T6[1]	sheet	427	496	8	[27]
7475-T61	sheet	496	552	12	[27]
7475-T7351	plate	393	475	16	[28]
7050-T73651	plate	455	510	11	[27]

Table 2.3 Typical mechanical properties of some high strength aerospace aluminium alloys. For Al-Li alloys see Table 2.6.
[1]) Alclad, [2]) at -196 °C

Aircraft	Upper Wing Skin	Lower Wing Skin	Fuselage Skin
BAe 146	7075-T6	2024-T351	2024-T3[1]
A300/310	7075-T651	2024-T351	2024-T3[1]
A320/330/340	7150-T651	2024-T351	2024-T3[1] -T351[1] -T42[1]
B747	7075-T651	2024-T351	2024-T3[1]
B757/767	7150-T651	2324-T39	2024-T3[1]
B777	7055-T7751	2324-T39	2XXX[2]-T3[1]
C-5B	7175-T73	7175-T73	7475-T6[1]

Table 2.4 Aluminium alloys for major components of present and future civil and military transport aircraft. [1]) cladded, [2]) C-188, proprietary to Alcoa

Al-Cu(-Mg) Alloys (2XXX Alloy Group)

The first age-hardenable aluminium alloy was an Al-Cu alloy. And the first commercially produced age-hardenable alloy was the German Duralumin alloy which was used for the frames of Count von Zeppelin's lighter-than-air vehicles. It had 4% copper, 0.5% magnesium, and 0.5% manganese. This alloy can be regarded as the predecessor of most of the 2XXX alloys. Copper and magnesium bearing alloys are still widely used for aircraft and space vehicles.

At room temperature Al-Cu(-Mg) alloys decompose by the formation of disc-shaped Guinier-Preston zones on {100} matrix planes which produce a considerable strength increase. An additional contribution to strength comes from the 1.5 to 3% stretch applied to rolled or extruded products for straightening the material and minimizing the effects of residual quenching stresses.

The naturally aged and cold worked 2024-T3 is the most popular alloy of this group. It has high ductility, excellent fatigue crack propagation resistance, and high fracture toughness. The good combination of these properties is mainly due to the lack of coarse grain boundary phases and the absence of precipitate free zones. The alloy was introduced in the 1930s, and as a medium strength, damage tolerant alloy it is still an outstanding alloy for components in which fatigue and fracture toughness is critical. The fuselage skin and the lower wing skins of nearly every commercial passenger airplane are made of 2024-T3, as can be seen from Table 2.4, and it is used almost exclusively for the smaller general aviation aircraft. Because of corrosion problems, 2024-T3 sheet and plate is cladded with pure aluminium.

Two higher purity versions of 2024 were developed, especially for the Boeing airplanes 757 and 767 namely 2224 and 2324, which are both also used in the naturally aged T3 condition. They are stronger compared to 2024-T3 and they reveal a higher toughness by limiting the iron and silicon contents and thus reducing the volume fraction of coarse constituent particles.

The aging sequence for Al-Cu alloys during artificial aging can be described as

$$\text{SSSS} \rightarrow \text{GP zones} \rightarrow \Theta' \rightarrow \Theta\ (Al_2Cu)$$

and for alloys which additionally contain magnesium as

$$\text{SSSS} \rightarrow \text{GPB zones} \rightarrow S' \rightarrow S\ (Al_2CuMg).$$

The transition phases Θ' and S' are semicoherent with the aluminium matrix but have already the same chemical composition as the equilibrium Θ and S phases. Θ' and S' often nucleate on dislocations since the dislocations can help to provide the extra energy needed to form a semicoherent phase boundary between the precipitate and the matrix. Thus, a homogeneous distribution of dislocations produces a very fine and uniform distribution of the transition phases. 2XXX alloys, therefore, respond very well to deformation prior to aging.

During artificial aging the yield strength of 2024 is considerably increased, but not the ultimate tensile strength. Because of the small difference between the YS and the UTS and the disproportionate decrease in ductility and fracture toughness, aircraft designers hesitate to use 2024 in T6 or T8 conditions.

For the construction of the space shuttle a higher purity version of the 2024 base alloy was developed, namely 2124. Less iron and silicon to minimize the constituent particles and closely controlled thermomechanical treatments make this alloy suitable for use in the artificially aged T8 temper.

Another 2XXX alloy is even older than 2024 but still in use. It is alloy 2014, which has less magnesium compared to 2024. In the United States airplane wheels are still made of 2014 forgings; in the United Kingdom it is still extensively used in naturally and artificially aged conditions as extrusions or forgings for aircraft.

For improving properties at elevated temperature, in the United Kingdom a nickel containing Al-Cu-Mg alloy was developed. The latest version, 2618, known as RR58 in the UK, was the major structural material for the first supersonic transport "Concorde". In this alloy iron is an alloying element, because it forms the $Al_9(Fe-Ni)$ phase which provides creep resistance at elevated temperatures.

Two alloys belonging to the 2XXX alloy group both serve the purpose of cryogenic tankage, i.e. of storing liquid fuels such as hydrogen and oxygen for space applications. Both alloys are easily weldable. The first is designated 2219 and is an Al-Cu alloy with small additions of manganese, vanadium and zirconium. In the T8 condition it is presently the alloy for expendable tanks for space vehicles such as the space shuttle. The second is an Al-Cu-Li alloy with small amounts of magnesium and silver. Its preliminary registration number is X2095, but it is better known as 049 "Weldalite" made by Reynolds Metals, USA [29, 30]. It is a lightweight, ultrahigh-strength alloy which has been developed to replace 2219.

Al-Mg-Si Alloys (6XXX Alloy Group)

The age hardening phase of this alloy group is Mg_2Si, i.e. both elements must always be present for successfully age hardening this alloy type. The aging sequence can be given as

$$\text{SSSS} \rightarrow \text{GP zones} \rightarrow \beta' \rightarrow \beta\ (Mg_2Si).$$

The 6XXX alloys are medium strength alloys which are mostly corrosion resistant and readily weldable. One of them, 6061, is a general purpose alloy which is used for aircraft parts where no particular design characteristics are required. The alloy 6013, which has a higher copper content, is considerably stronger than 6061. Its properties in the T6 condition are comparable to 2024-T3; it is about 3% lighter and has excellent stretch forming characteristics. Thus, it has recently been regarded as a potential replacement alloy for 2024-T3 sheet [31].

Al-Zn-Mg(-Cu) Alloys (7XXX Alloy Group)

Among the heat treatable aluminium alloys, zinc and magnesium containing alloys have the greatest potential for age hardening, and their natural aging response is very pronounced. The hardness, however, does not approach a saturation value within reasonably short aging times. It rather increases indefinitely, i.e. naturally aged tempers are unstable. Therefore, 7XXX alloys are not used in the naturally aged T3 or T4 conditions.

Other than for 2XXX alloys, deformation prior to aging has little or no effect on the precipitation behaviour of 7XXX alloys. Thus, alloys of this alloy group are always used in either the T6 (peak aged) or T7 (overaged) tempers.

During artificial aging Al-Zn-Mg alloys with a high Zn:Mg ratio (all 7XXX aerospace alloys) decompose according to the following aging sequence:

$$\text{SSSS} \rightarrow \text{GP zones} \rightarrow \eta' \rightarrow \eta\ (MgZn_2).$$

The coherent GP zones which are approximately spherical form between room temperature and about 160 °C. Extended aging transforms them into the semicoherent transition phase η' which is the precursor of the equilibrium η phase.

Copper additions up to 1% increase the strength by adding a component of solid-solution hardening, but do not alter the basic precipitation mechanism. At higher contents copper participates in the precipitation process by substituting copper and aluminium atoms for zinc atoms in the semicoherent and incoherent $MgZn_2$ phases. Improvement in resistance to stress corrosion cracking by copper additions is partly related to this process.

The chromium bearing dispersoids, present in many 7XXX alloys, retain the dynamically recovered structure and prevent excessive growth of recrystallized grains. In thin sheet the elongated grain structure prevents the development of high stresses along grain boundaries and thus improves the stress corrosion properties. Alloy 7075 in the peak aged T6 temper corresponds to this condition. It is still often used for aircraft in the form of thin sheet or thin extrusions.

Unacceptably low stress corrosion resistance in the short transverse direction of thick plate or forgings led to the application of a duplex aging treatment designated T73. The first aging step at about 120 °C is followed by an overaging step at a higher temperature of 160-170 °C. Thick sections of 7075-T73 have good stress corrosion resistance, however, at the expense of strength. Very large die forgings are produced according to this procedure.

To combine the high strength of T6 with the stress corrosion resistance of T73, another heat treatment has been developed which is known as retrogression and re-aging (RRA) and which for 7075 involves the following stages: (1) Conventional T6 treatment (24 h at 120 °C), (2) short heat treatment (a few seconds to a few minutes) at an intermediate temperature of 200-260 °C, and water quench, (3) final re-aging treatment 24 h at 120 °C. By this treatment the size and volume fraction of the grain boundary precipitates is increased. The RRA treatment cannot readily be applied to thick sections because of the short time interval in stage (2).

The need for high strength and higher toughness alloys stimulated extensive development programmes which led to alloy 7475. The main cause for the high fracture toughness of this alloy is a considerable reduction of the levels of iron and silicon impurities and thus a decrease in the volume fraction of coarse constituent particles, which are always detrimental for high fracture toughness. Increasing the ratio of zinc to magnesium by lowering the magnesium content, and specific thermomechanical treatments provide balanced properties in 7475. This alloy is now widely used in aircraft in the form of thick plate and forgings, but also as the fuselage skin on the C-5B military cargo airplane.

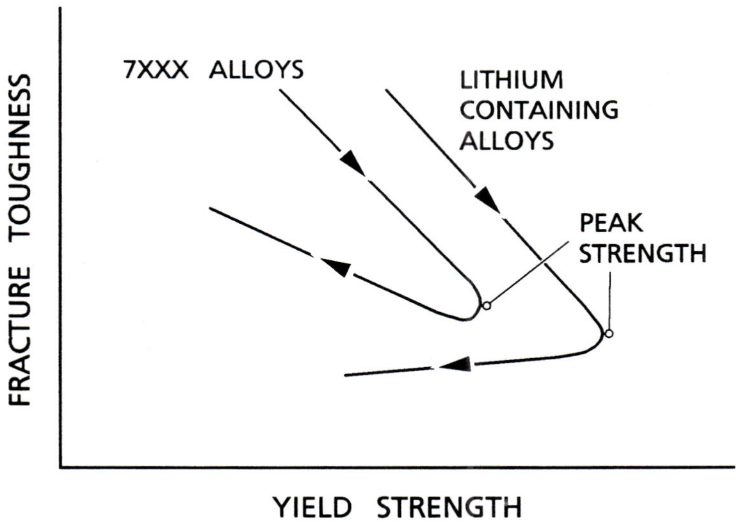

Figure 2.2 Schematic diagram of the relationship between fracture toughness and tensile yield strength for 7XXX [32] and Li containing [33] alloys. The arrows indicate the direction of increasing aging time.

The chromium bearing dispersoids are thought to produce a phenomenon called quench sensitivity, which occurs in thick sections like forgings or plate. It produces lower strength compared to what can be achieved in rapidly quenched thin sections. The slower quench rate in thick sections results in a depletion of copper and magnesium near the chromium dispersoids, forming a PFZ around them and nucleating the η phase on them. Thus, during artificial aging less solute is available for the precipitation of GP zones or η' particles and lower strength results. As a consequence, for thicker sections the alloy 7050 was developed in which chromium is substituted by zirconium. Additions of 0.1% zirconium are as effective as 0.25% chromium in preventing the recrystallization of hot-worked products during solution heat treatment. 7050-T7 is not quench-sensitive and thus is used for thick sections in many aircraft, mainly military. A European version of this zirconium bearing alloy called 7010 contains somewhat less copper.

The slightly modified version 7150 was developed by Alcoa as an improved performance upper wing skin alloy for the Boeing 757/767 aircraft where it is used in the T651 temper. The same alloy in a T6151 temper has slightly better corrosion properties at the same strength levels and is used on the Airbus A310 and the McDonnell Douglas MD11.

Generally, the fracture toughness of age-hardenable aluminium alloys decreases when the strength is increased. This is schematically shown in Figure 2.2. Aging 7XXX series alloys beyond the peak strength makes the toughness increase again [32]. The overaged strength-toughness curves always lie lower. Nevertheless, this behaviour of increased overaged toughness is advantageous for 7XXX alloys, which are used in the overaged T7 tempers because of their better corrosion properties. 2XXX alloys behave similarly. For lithium containing alloys, however, the fracture toughness decreases further during overaging, as can be seen from Figure 2.2 [33].

With the development of damage tolerant, high strength, high toughness, and corrosion resistant 2XXX and 7XXX aerospace aluminium alloys a standard has been achieved at which considerable improvements cannot be expected for the same alloy systems. One example for a new promising alloy system is the use of lithium as a major alloying element to provide substantial weight savings.

Outlook and New Developments

New developments in the area of aluminium alloys are focussed on either improving the performance of the material or reducing the cost of the final component. In most cases the progress made is being judged by comparing the advancement to the properties of the two most dominant aerospace variants, the damage-tolerant alloy 2024 and the high strength alloys 7075/7475.

Improved Properties: The tolerance of how alloys handle cracks under static or cyclic loading conditions, i.e., the damage tolerance, is indicated by fracture toughness and fatigue crack growth resistance data, respectively. Compared to the damage tolerant standard 2024-T3, a new 2XXX-T3 alloy - C188, recently proposed by Alcoa - is said to be 20% superior in toughness and 15% in crack growth resistance. Macrocomposites based on bonded arrangements of alternating layers of thin aluminium sheet and fibres reveal the lowest overall fatigue crack propagation rates. ARALL (aramid reinforced aluminium laminates) - developed in the early 1980s at Delft University - and Glare (glass fibre reinforced aluminium laminates) are extensively investigated.

At the upper end of conventional high strength aluminium alloys a new 7XXX alloy was recently selected for the Boeing 777. Designated 7055, the alloy provides a 10% increase in specific strength over 7150-T61 and -T77, without a corresponding loss in corrosion resistance.

Reducing weight and increasing stiffness via conventional ingot metallurgy can be achieved with Al-Li alloys (see section "Aluminium-Lithium Alloys"). Further improvements in stiffness lead to composite materials with non metallic reinforcements (see section "Metal-Matrix Composites").

Rapid solidification techniques have the potential to further improve mechanical properties since new alloys can be developed beyond the limitations of ingot metallurgy (see "Powder Metallurgy"). Examples are 7XXX alloys with yield strength values beyond 800 MPa; Al-Fe base alloys with superior elevated temperature stability; corrosion resistant, Cu-free alloys produced via mechanical alloying [34].

Cost Effective Processing Techniques: Near-net shape technologies have been developed to reduce the cost and weight of aerospace components. One of the most challenging is superplastic forming/diffusion bonding (SPF/DB). By choosing the optimum combination of high temperature and slow strain rate, sheet material can be superplastically deformed 1000% and more. Combined with diffusion bonding, SPF/DB allows the manufacture of complex multilayer structures.

Since the high deformation degrees are largely controlled by creep processes, one prerequisite for superplasticity is an ultra-fine grain size, which has to be stable during the high temperature exposure. Several approaches were followed to achieve this goal. The first was to develop new alloys particularly designed for superplasticity like the Supral series [35]. These Al-Cu-base alloys contain zirconium, forming fine stable Al_3Zr precipitates which stabilize the grain structure. The second approach was to refine the pan-cake shaped grain structure of conventional high strength aluminium alloys by recrystallization. This was first demonstrated for 7075 by proposing a four-step thermomechanical processing. After solution heat treatment, overaging at a relatively high temperature leads to precipitation of a high density of large equilibrium phase particles. The deformation zones that form around these particles during subsequent rolling serve as nucleation sites for recrystallization when the alloy is finally heated above the recrystallization temperature [36]. Total elongations as high as 1200% were achieved with about 15 μm grain size [37]. In other efforts, powder metallurgy proved to be a viable tool to achieve ultra-fine grain structures with the potential for substantial increases in strain rate during superplastic forming. Finally, Al-Li alloys with their inherent finer grain size are particularly attractive for SPF.

Compared to titanium alloys, wide exploitation of SPF/DB for aluminium alloys has been prevented so far mainly for two reasons. First, these alloys tend to develop cavitation with increasing strain mainly due to the presence of large iron or silicon constituent particles as impurities in these alloys. Providing back pressure during SPF has, however, been found to effectively suppress cavitation. The second drawback for aluminium alloys is linked to their tendency to form tenacious oxide layers on the surface which inhibits solid state diffusion bonding. New bonding methods are being developed to overcome this limitation. In transient liquid-phase bonding the oxide film is destroyed by roll cladding a layer of zinc onto the surfaces to be joined. The zinc interlayer also protects the aluminium surface from reoxidizing and forms a low-melting-point phase that is reasonably soluble in the aluminium alloy matrix. In other efforts an attempt is made to remove the oxide layer by ion etching before DB, both performed in vacuum [38].

Further activities in net-shape processing include advanced casting methods and new forging techniques. For example, the inner flap tracks of the Airbus A330/340 are high performance investment castings of the alloy A357, reducing the production costs substantially [39]. Incremental forging is primarily aimed at reducing tool cost and machining steps as well as the amount of material, which particularly pays off for more expensive material like Al-Li alloys [38]. Powder metallurgy has not only the potential for new alloy compositions with improved properties; new processing routes are developed like spray forming, which turns out to be a viable process for particulate-reinforced MMC production. Combinations of new alloys and processing techniques are discussed, like SPF of particulate reinforced MMC sheet material or hybrid sheet structures made out of aramid fibres and MMC sheet material.

2.1.1.2 Aluminium-Lithium Alloys

R. Braun

Weight reduction is a great impetus to employing materials with improved structural performance in aircraft and space vehicle applications. The threat of composites has induced efforts to develop metallic materials which are more attractive and competitive than the conventional materials used. By adding lithium a new class of aluminium alloys has been developed which offers 8 - 10% less density and an approximately 10% higher stiffness. Some of these aluminium-lithium based alloys have now attained commercial status. The usual semi-fabricated product forms, e.g. sheet, plate, extrusions and forgings will be offered [40].

The most advanced alloys are the quaternary Al-Li-Cu-Mg alloys 8090, 8091, both developed by British Alcan Aluminium, and 2091, registered by Pechiney, as well as Alcoa's Al-Li-Cu alloy 2090. In the Soviet Union a family of lithium containing alloys has been developed since the early 1970s which includes Al-Mg-Li alloys designated 1420 and 1421 [41]. Compositional ranges of several Al-Li alloys are listed in Table 2.5 [41-44]. Recently, activity on Al-Li alloys has started in Japan [45]. Further developments are the weldable Al-Cu-Li-Ag-Mg alloys coded "Weldalite", with high strength at cryogenic temperatures [46] and the dispersion strengthened Al-Mg-Li alloy 905XL manufactured by mechanical alloying techniques [47]. For detailed information the reader is referred to the proceedings of six International Al-Li Conferences which have taken place at Stone Mountain (1980) [48], Monterey (1983) [49], Oxford (1985) [50], Paris (1987) [51], Williamsburg (1989) [52], and Garmisch-Partenkirchen [2].

Alloy	Al	Li	Cu	Mg	Zr
8090	Bal	2.2-2.7	1.0-1.6	0.6-1.3	0.04-0.16
8091	Bal	2.4-2.8	1.6-2.2	0.5-1.2	0.08-0.16
2091	Bal	1.7-2.3	1.8-2.5	1.1-1.9	0.04-0.16
2090	Bal	1.9-2.6	2.4-3.0	0.0-0.25	0.08-0.15
1420	Bal	1.8-2.2		4.9-5.5	0.05-0.3

Table 2.5 Composition of selected lithium-bearing aluminium alloys [41-44].

Precipitates in Al-Li Alloys

The strength of aluminium-lithium based alloys is associated with age hardening due to precipitation of finely dispersed phases. The microstructure of Al-Li alloys was reviewed recently [53, 54]. The main aspects are summarized below.

Associated with the decomposition of a supersaturated solution, lithium addition to aluminium results in the $\delta'(Al_3Li)$ phase which precipitates homogeneously throughout the aluminium matrix and the $\delta(AlLi)$ phase which is nucleated heterogeneously primarily at grain boundaries. The δ' precipitates are spherical, ordered and coherent with the matrix and impede the motion of dislocations, thus enhancing strength. Due to the coherency, the δ' particles are sheared by moving dislocations giving rise to planar slip. The effect of slip inhomogeneity is enhanced by the ordered structure of δ' causing the dislocations to move in pairs. The resulting slip bands lead to stress concentrations at grain boundaries. Therefore, further alloying elements are required for the formation of additional precipitates to homogenize deformation and prevent strain localization.

The addition of copper leads to formation of $\Theta'(Al_2Cu)$ and $T_1(Al_2CuLi)$ phases, the latter being the predominant copper bearing strengthening precipitates in the commercial Al-Li-Cu alloy 2090. T_1 nucleates heterogeneously on dislocations and low angle grain boundaries and mitigates strain localization occurring in binary Al-Li alloys. The addition of copper also results in the icosahedral $T_2(Al_6CuLi_3)$ phase which nucleates predominantly on high angle grain boundaries. Due to the associated precipitate free zones, this grain boundary phase reduces ductility and fracture toughness.

Besides δ' precipitates, in ternary Al-Li-Mg alloys the incoherent Al_2LiMg phase is observed which nucleates at grain boundaries and dislocations. This phase does not contribute to precipitation hardening and does not reduce coplanar slip behaviour. It reduces ductility, fracture toughness and corrosion behaviour due to coarse grain boundary precipitates and associated PFZs. However, in quaternary Al-Li-Cu-Mg alloys magnesium additions introduce the precipitation of the S′(Al_2CuMg) phase which is partially coherent and which nucleates heterogeneously on matrix dislocations, low angle grain boundaries and other structural inhomogeneities. Thus, in quaternary Al-Li-Cu-Mg alloys three hardening phases can precipitate: δ', T_1, and S'. Which of the copper bearing phases is formed preferentially depends upon the relative concentrations of the three alloying elements. In alloys with high copper and lithium contents relative to magnesium the T_1 phase dominates, whereas high copper and magnesium concentrations relative to lithium result in the preferential precipitation of S'. If the S' phase has grown to sufficient size and distribution by aging, it is effective in slip dispersion.

The mechanical properties of commercial aluminium alloys depend considerably on the grain structure which is controlled by additions of the transition elements chromium, manganese or zirconium. Zirconium has proved to be very effective in retarding recrystallization attributed to the formation of submicron coherent and ordered Al_3Zr dispersoids. Furthermore, quench sensitivity is lowered utilising zirconium compared to chromium and manganese [55].

Tensile Properties

To replace today's alloys used in aircraft structures in a variety of product forms including sheet, plate, extrusions, forgings, the new Al-Li alloys have to meet various specifications required for damage tolerant, medium or high strength structural materials. Because the required levels of mechanical properties can not be achieved by one alloy alone, several categories of lithium bearing alloys were developed differing in grain structure, temper and composition. For sheet, damage tolerance is provided by a fine recrystallized structure, whereas an unrecrystallized grain structure is necessary for medium strength applications [56]. In high strength Al-Li alloys the copper content is increased compared to the lower strength variants.

Tensile properties for sheet of lithium-bearing and conventional aluminium alloys are given in Table 2.6 [24, 40, 57-59]. The Al-Li alloys achieve strength equivalent to the incumbent alloys. However, elongation is inferior approaching the minimum values of the conventional alloys (2091-T8X sheet showing high ductility values is quite susceptible to stress corrosion cracking, which is not observed for 2024-T3). Strong crystallographic texture and pronounced anisotropy of mechanical properties are associated with the unrecrystallized grain structure in Al-Li alloys revealing strength minima at angles in the range of 40 - 70 °C to the rolling direction [60]. Recrystallization diminishes the texture and reduces the anisotropy. Tensile properties for plate materials are given in Table 2.7 [40, 56].

Fracture Toughness

The fracture behaviour of Al-Li alloys is characterized by the weakness of the grain boundaries [61]. Failure is almost completely intergranular with concomitant

Alloy	Test Direction	YS MPa	UTS MPa	El. %
		Damage tolerant alloys		
2091-T8X	L	348	441	18
	T	335	459	17
2091 CPHK-T8X	L	335	419	14
	T	326	435	13
8090-T81	L	336	437	11
	T	309	445	13
2024-T3	L	369	498	20
	T	331	486	22
		Medium and high strength alloys		
8090-T8	L	436	503	5.0
	T	428	511	7.5
2090-T83	L	496	552	6.3
	T	496	538	7.8
7475-T761	L	450	510	13.1
	T	435	504	13.3
7075-T6	L	486	548	12.5
	T	480	552	11.3

Table 2.6 Tensile properties of 1.6 mm thick sheet of lithium-bearing and conventional aluminium alloys [24, 40, 57-59]

Alloy	Test direction	YS MPa	UTS MPa	El. %
8090-T8771	L	483	518	4.3
	T	451	509	4.1
	S	367	453	3.2
8090-T8151	L	397	449	5.6
	T	354	447	7.5
	S	288	421	5.6
2024-T351	L	365	470	17
	T	335	472	15
	S	298	414	5.8

Table 2.7 Tensile properties for medium strength 8090-T8771 and damage tolerant 8090-T8151 as well as 2024-T351 plate material, 38 - 65 mm thick [40, 56]

low ductility and fracture toughness, if loading is applied in ST direction, i.e. perpendicular to the pancake shaped grains of wrought products. With in-plane test directions intergranular delamination and ductile shear fracture are observed, too.

Several mechanisms were taken into considerations to promote grain boundary failure including: (a) planar slip giving rise to stress concentrations at grain boundaries, (b) strain localization in precipitate-free zones, (c) embrittlement caused by precipitation of constituent particles and equilibrium phases such as T_2 and δ at grain boundaries, (d) embrittlement due to segregation of impurity elements like Na, K, Ca, and the alloying element Li to grain boundaries [61].

To improve the fracture toughness the propensity of Al-Li alloys to grain boundary fracture has to be reduced. The extent of strain localization is decreased by coprecipitation of additional hardening phases such as S' or T_1. Because these phases nucleate heterogeneously on substructural inhomogeneities such as dislocations or subgrain boundaries, the deformation behaviour can be modified by mechanical working prior to ageing [62]. Minimizing the content of impurity elements (iron, silicon, alkali metals), effects of constituent particles and impurity segregation are reduced. In commercial alloys, however, the total content of these elements can not be decreased below 0.1 wt% for economical reasons. Aging at low temperature reduces precipitation of equilibrium phases at grain boundaries and the associated precipitate free zone width, and refines the matrix precipitation. To shorten the longer ageing time periods necessary to achieve equivalent strength at lower temperature, the level of stretch is increased to enhance S' precipitation within the matrix. For 8090 plate material a stretch level of 6-8% is used [62].

Furthermore, the fracture toughness depends upon the grain structure of the semi-fabricated products. Under plane strain conditions a coarse unrecrystallized grain structure is required to achieve high fracture toughness and strength with plate and thick section extrusions, whereas for sheet a fine recrystallized grain structure exhibits optimum plane stress fracture toughness [56].

Alloy/Test direction	LT	TL	SL
Al-Li alloys			
2091-T8X51	44	41	
8090-T8151	43	35	23
8090-T8771	42	38	19
2090-T8E41	35	24	17
Conventional Al alloys			
2024-T351	36	33	26
2014-T651	24	22	19
7075-T651	29	25	20
7475-T651	43	37	32

Table 2.8 Fracture toughness K_Q in $MPa\sqrt{m}$ for plate material of lithium bearing and conventional aluminium alloys, depending upon test direction [40, 57, 63, 64]

In Table 2.8 fracture toughness values for plates of Al-Li and conventional Al alloys are listed [40, 57, 63, 64]. With regard to in-plane fracture toughness Al-Li alloys compete well with baseline alloys. However, the short transverse fracture toughness is generally inferior compared with that of conventional alloys at the same strength level, though acceptable levels of S-L fracture toughness have been obtained (exceeding 18 $MPa\sqrt{m}$) [65]. Toughness as well as ductility shortcomings in short transverse direction were also found with 8090 hand forgings [66]. Figure 2.3 shows the fracture toughness of damage tolerant Al-Li sheets which approach the levels of 2024-T3 alloy [24, 57].

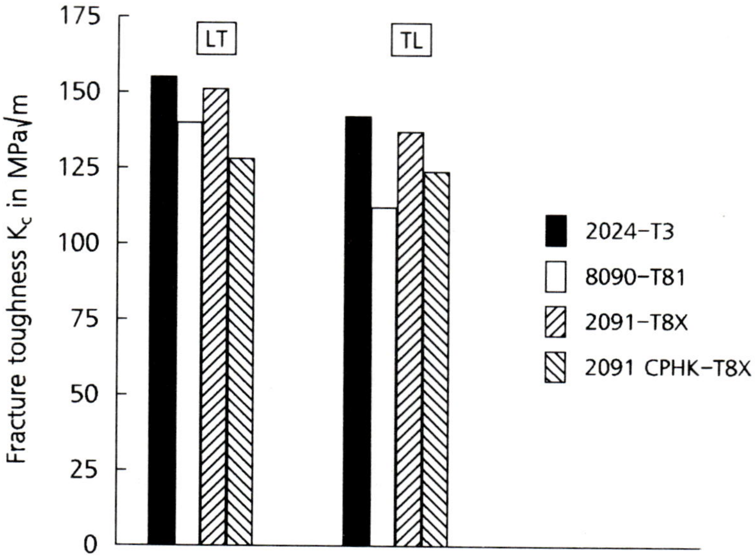

Figure 2.3 Fracture toughness K_c of damage tolerant sheet of Al-Li and 2024-T3 alloys [24, 57]

Fatigue Behaviour

To characterize the fatigue behaviour of a material the resistance to both initiation and growth of fatigue cracks has to be evaluated. High cycle fatigue tests have shown that Al-Li alloys behave as conventional alloys [24, 67] (Figure 2.4). Crack initiation occurs at constituent particles and is controlled by surface quality [65, 67]. Compared with conventional alloys the fatigue life of Al-Li alloys is less sensitive to surface treatments, e.g. pickling, anodising or shot peening [68]. Lithium depletion on the surface seems not to reduce the fatigue strength significantly [69].

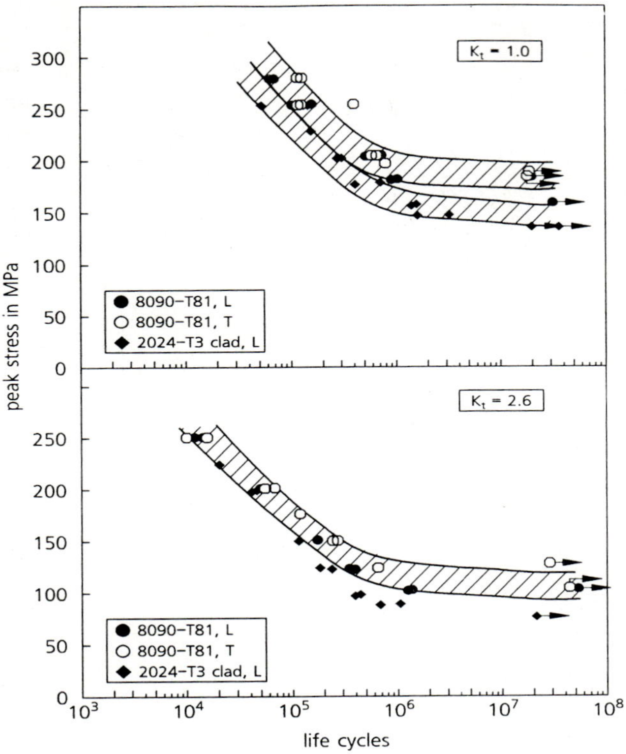

Figure 2.4 S-N data for damage tolerant 8090-T81 and 2024-T3 clad sheet using smooth ($K_t = 1$) and notched ($K_t = 2.6$) specimens (R=0.1, f=80 Hz) [24]

The fatigue crack propagation behaviour of Al-Li alloys under constant amplitude loading is superior compared to that of conventional aluminium alloys [65] (Figure 2.5). Besides the contribution of the higher Young's modulus, the improved resistance to fatigue crack propagation is primarily caused by a strong texture and propensity to planar slip resulting in a crack path along crystallographic planes [67]. Crack deflection and crack branching reduce stress intensity at the crack tips. The tortuous crack path induces crack closure effects associated with rough fracture surfaces and fretting debris. The superiority of Al-Li alloys with regard to crack-growth resistance is not retained compared to baseline alloys if small cracks (1 to 1000 μm) are considered [70]. Crack growth rates are 2-3 orders of magnitude higher than those of long cracks at the same stress intensity levels. This is attributed to reduced crack closure effects due to the limited wake of small cracks.

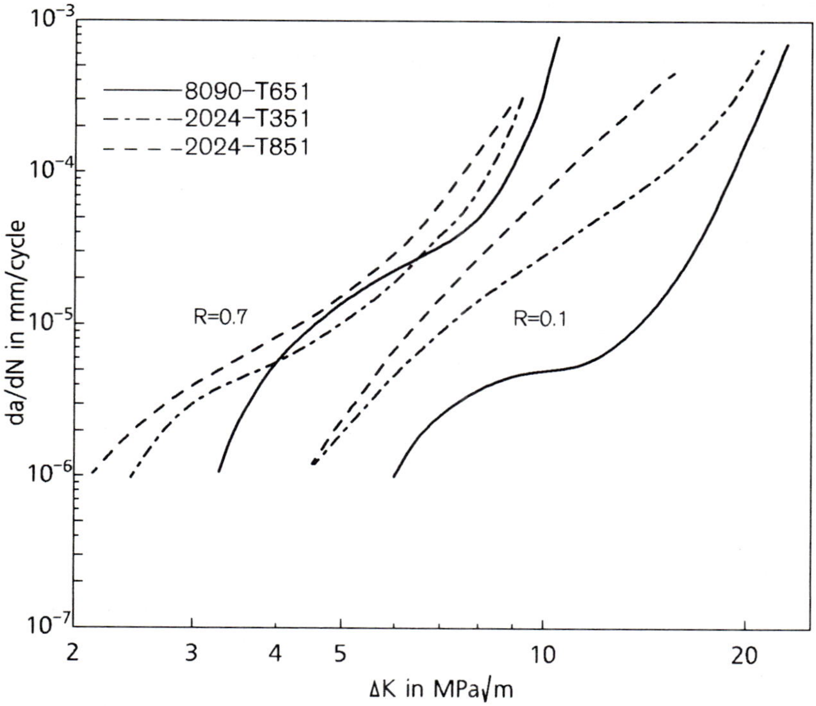

Figure 2.5 Fatigue crack growth curves of 8090-T651, 2024-T351 and 2024-T851 tested in laboratory air (LT orientation) [74]

Corrosion Behaviour

Similarly to other aluminium alloys used in aircraft applications the corrosion behaviour of Al-Li alloys depends upon grain structure, heat treatment and alloy composition. As found in the limited number of outdoor exposure tests, the corrosion performance of Al-Li alloys is equal or better than that of conventional alloys to be replaced. With the Al-Cu-Li-Cd alloy 2020, which was commercially produced for 10 years (1959 - 1969) and used in the military aircraft RA-5C, no problems regarding corrosion or stress corrosion cracking were reported [71, 72]. In the Soviet Union, no corrosion damage has occurred with the 1420 alloy used in aircraft for more than 20 years [41]. Thus, Al-Li alloys do not have poor corrosion behaviour, which might be anticipated due to the high reactivity of lithium.

Al-Li alloys are galvanically compatible with conventional aluminium alloys [73]. The lithium bearing alloys were found to corrode preferentially when galvanically coupled with conventional alloys; however, the current generated was not significantly enhanced compared with that measured in galvanic couples of conventional alloys. The pitting corrosion resistance evaluated from salt spray tests and immersion tests in chloride solutions is better than that of 2XXX and 7XXX alloys. Exposed to marine atmosphere or urban environments 8090 was attacked by intergranular corrosion. Furthermore, depending upon the aging treatment, a propensity to exfoliation corrosion was observed, resulting in blisters on the surface [73, 75, 76].

To examine whether a modified temper, production route or alloy chemistry has influenced the corrosion behaviour of the material investigated, quick reliable tests are needed. To evaluate the exfoliation corrosion resistance of 2XXX and 7XXX alloys the EXCO test (ASTM G34) and the MASTMAASIS test (ASTM G85 ANNEX 2) are used. In the EXCO test specimens are continuously immersed in highly acidified medium, whereas the MASTMAASIS test involves a cyclic intermittent salt spray technique. Both tests are also applied to exfoliation corrosion testing of Al-Li alloys. Results are given in Table 2.9 for sheet and plate products [77-81].

Alloy	EXCO		MASTMAASIS	
	48 h	96 h	2 weeks	4 weeks
		Sheets		
8090-T6	P	EC/ED	P	P/EA
8090-T81	B	B/EA	B	EA
2091-T8X	P/EA	EA/EB		
2091 CPHK-T8X	N/B	B	N/P	B
2024-T3	P	P/EA	P	P
		Plates		
8090-T651	ED			
2090-T81		ED		P
2091-T8X51	B	EA	EA	EA
2024-T351	EA	EB/EC		

Table 2.9 Results of EXCO and MASTMAASIS tests from visual inspection (N = no attack, P = pitting, B = blistering, EA - ED = exfoliation corrosion with increasing severity according to ASTM G34) [77-81]

The reliability of accelerated corrosion testing techniques is judged from the correlation between their predictions and the long-term performance of materials in natural environments. Contrary to results in Table 2.9 the 2090 alloy in peak aged tempers exhibits an excellent exfoliation corrosion resistance in seacoast atmosphere [81], and 8090-T651 performs better than 2024-T351 plate in rural environments [82]. Due to this poor correlation the EXCO test is generally considered to be unreliable for predicting the exfoliation corrosion behaviour of Al-Li alloys, although good correlation was found between results of the 48 h EXCO test and two years exposure to marine atmosphere for 8090 extrusions [83].

Better predictions are obtained performing MASTMAASIS tests. To be severe enough, the exposure period has to be extended to four weeks [82]. However, this test failed to discriminate between different underaged tempers which are less resistant to exfoliation corrosion [81]. The evaluation of corrosion damage by visual assessment is a deficiency of both accelerated tests. Contrary to visual rating 2091-T8X51 specimens were severely damaged by exfoliation corrosion in the MASTMAASIS test as found by metallographic examination [79]. Therefore, further work is needed to provide a reliable accelerated exfoliation corrosion test for lithium bearing aluminium alloys.

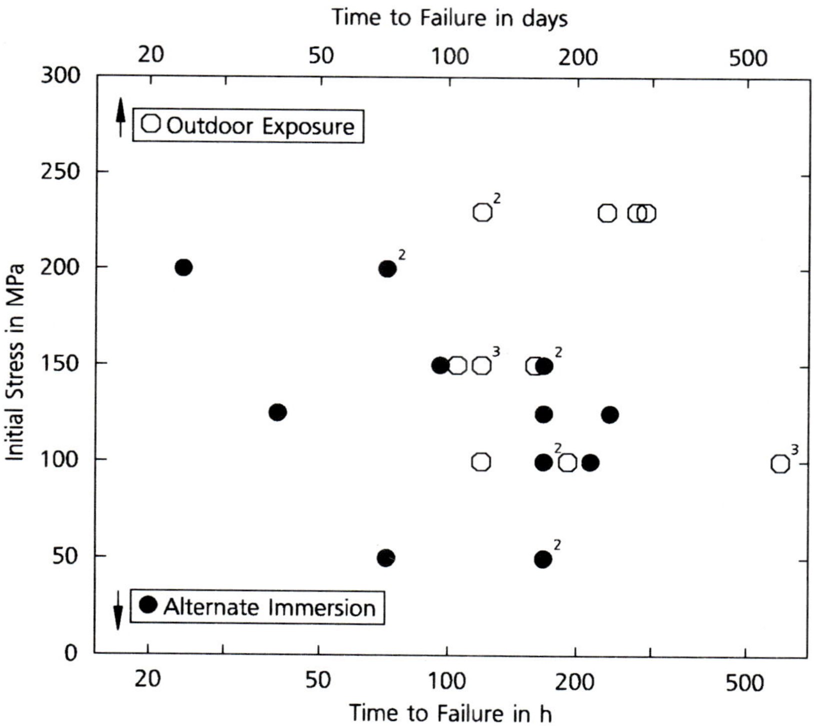

Figure 2.6 Time-to-failure data of 2091-T8X51 plate material obtained from outdoor exposure and alternate immersion tests in 3.5% NaCl solution. Tensile specimens were strained in ST direction. The arrows indicate the upper and lower scale, respectively [79].

With regard to stress corrosion cracking (SCC) alternate immersion tests in 3.5% NaCl solution give reliable predictions for Al-Li alloys in natural environments. As shown in Figure 2.6, time-to-failure data of 2091-T8X51 tensile specimens which were exposed to urban industrial environments correlate well with those evaluated from alternate immersion test in 3.5% NaCl solution. Whereas in longitudinal and long transverse direction unrecrystallized Al-Li alloys are virtually immune to SCC (threshold stresses exceeding 75% of the 0.2% proof stress), the short transverse SCC resistance is very low. With 8090-T651 and 2091-T8X51 plate threshold stresses below 50 MPa are found in alternate immersion tests in ST direction [78, 79]. In these accelerated tests plates of Al-Li and conventional Al alloys behave similarly (Figure 2.7). In agreement with Figure 2.7, the SCC initiation resistance of 8090-T651 plate, evaluated from exposure tests in rural environments, is similar or slightly better than that of 2024-T351 and 2324-T39 plates [82].

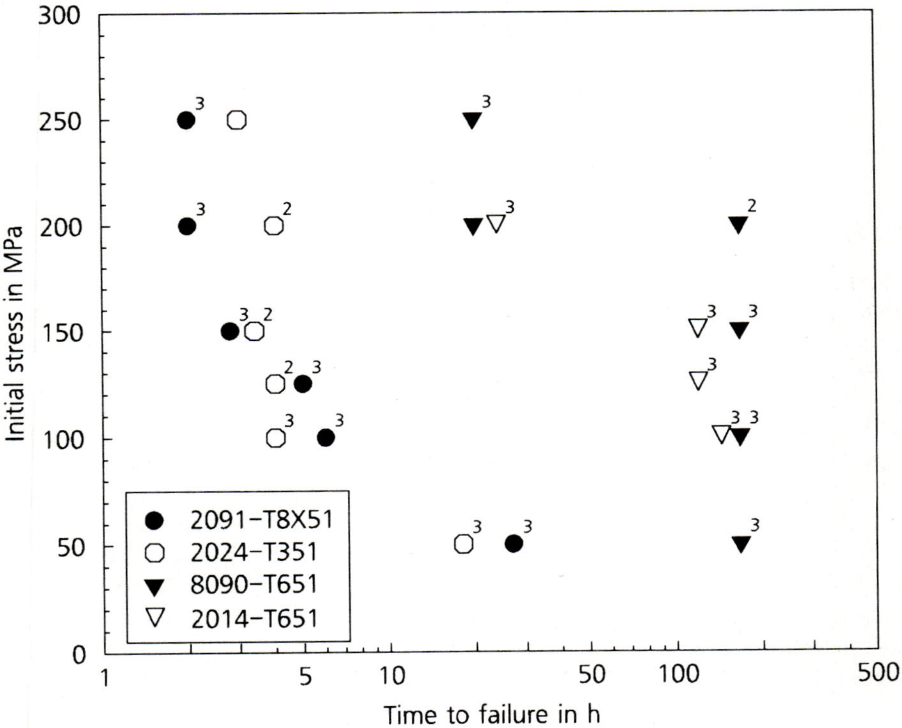

Figure 2.7 Time-to-failure data of plate material of Al-Li and conventional Al alloys. C-ring specimens were strained in ST direction and alternately immersed in 3.5% NaCl solution

The SCC behaviour of Al-Li alloys depends upon the aging treatment. Whereas underaged 2090 plate material is quite susceptible to stress corrosion cracking in short transverse direction, SCC resistance improves significantly upon aging to near peak strength conditions [84]. Similar behaviour is observed for 8090, which is most resistant in the overaged temper [85]. The SCC resistance can be further enhanced with more complex heat treatments, which influence precipitation within the matrix, thereby reducing the potential differences between matrix and grain boundary phases [86]. Zinc addition to Al-Li-Cu-Mg-Zr alloys also increases the SCC resistance to a threshold stress of 250 MPa in the peak aged temper [86].

With Al-Li sheet the SCC behaviour depends upon grain structure and temper. Unrecrystallized sheet of the medium strength 8090-T6 alloy is not susceptible to SCC in the rolling plane [77]. However, damage tolerant recrystallized sheet of 2091-T8X and 8090-T81 are prone to stress corrosion cracking [59, 65]. The low SCC resistance of 2091-T8X (threshold stress below 100 MPa) is improved by metallurgical adjustments which are proprietary to Pechiney (CPHK version). For 2091 CPHK-T8X as well as 8090-T81 sheet a threshold stress of 225 MPa in long transverse direction is found in alternate immersion tests in substitute ocean water [87].

The corrosion fatigue behaviour of Al-Li alloys is similar to that of 2XXX alloys and their resistance is improved compared to 7XXX alloys [88].

Outlook

Problems with aluminium-lithium alloys which are preventing their extensive application are shortcomings of ductility, fracture toughness and stress corrosion resistance particularly in short transverse direction. The phenomenon of "crack deviation" has to be cleared for 8090 sheet showing cracks which propagate away from the normal of applied stress during fatigue crack growth tests [65]. Investigations performed recently have shown that the properties of Al-Li semi-fabricated products can be improved. Using a thermal mechanical processing route prior to hot rolling results in significant improvement in short transverse fracture toughness and SCC resistance for 8090 plate [89]. Thin extrusions of 8090 fulfil requirements for use in the Airbus A330/340 by aging at increased temperature and by narrowing the magnesium and copper contents close to the upper specification limits [90]. Applications of Al-Li alloys are proceeding most extensively in the construction of the Westland-Agusta EH101 helicopter [91]. Recrystallized 8090 sheet with thicknesses ranging from 0.6 to 4.0 mm will replace sheet of conventional damage tolerant and medium strength alloys. Different profiles will be manufactured in unrecrystallized 8090. Evaluation of thin walled extrusions showed that 8090 extrusions are apparently ready for serial production [66]. Furthermore, die forgings amenable to effective cold compression will be made from this alloy.

Since zinc additions enhance the SCC resistance, it is possible that a new generation of Al-Li alloys containing zinc will be developed [58, 92]. First evaluation of the Soviet alloys 1420 and 1421 in the United States revealed their potential for applications in western airframe structures [93]. Finally, the costs of these new alloys, being two to four times higher than those of conventional alloys, are crucial for widespread introduction as structural materials for future aircraft generations.

2.1.1.3 Powder Metallurgy of Aluminium Alloys

G. Staniek and F. Lehnert

Powder-metallurgical processing of aluminium alloys has certain advantages when compared to the ingot metallurgical route, such as finer and more homogeneous microstructure with better mechanical properties and the possibility of achieving near net shape production. Liquid phase sintering of Al-Cu based alloy powder has been in use for many years to produce net shape parts by powder metallurgy (PM) [94]. But there are only a few aluminium alloys with moderate properties that form a liquid phase and that permit a sintering in spite of the oxidation of the powder surface. Because of this limitation, in the last three decades many consolidation methods other than sintering such as extrusion, forging etc. have been developed to process the material. This has allowed the use of powder metallurgy for more advanced aluminium alloys and production routes such as rapid solidification (RS) and mechanical alloying (MA) [95, 96].

2XXX and 7XXX alloys were among the first to be manufactured by the PM route [97]. However, in the case of these alloys, where the IM route has been optimised, the gain in mechanical properties by resorting to PM processing without alloy modification is only marginal. The real advantage of Al/PM is in the production of new alloys that cannot be fabricated by IM. Depending on the purpose of the alloy, several PM routes have evolved.

Rapid solidification has been used to extend the solubility of alloying elements, particularly transition and rare earth elements, and refine the structure of intermetallic phases responsible for mechanical properties. A number of techniques have been developed to achieve the desired high solidification rates. Mechanical alloying is a dry, high energy milling process that made it possible to produce dispersions of insoluble oxides and carbides which stabilise the microstructure. As in the case of RS, supersaturated solutions, fine distributions of intermetallic and intermediate phases are generated, but without utilising a liquid phase. By blending the alloy powder with a strengthening phase, discontinuously reinforced alloys containing insoluble oxides or carbides or fibre reinforcement phases are also produced by PM.

Particulate Solidification and Processing

There are many different routes of powder metallurgical processing of aluminium alloys, like performing every step in vacuum or reducing the process to the essential steps, saving time and money. The necessity of degassing the powder in vacuum has often been debated and there are numerous attempts to avoid this cumbersome treatment. However, PM processing can only be simplified to a certain extent. A procedure frequently used for aluminium alloys [98] is shown in Figure 2.8.

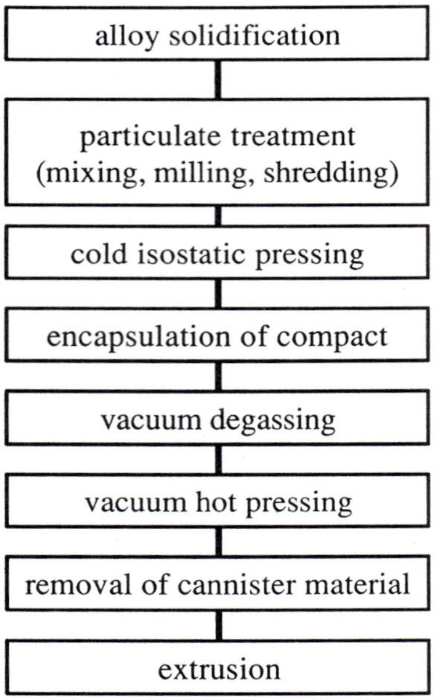

Figure 2.8 Processing sequence for aluminium alloy particulate

Alloy Solidification by gas atomisation is one of the two methods preferably applied to aluminium alloys. Solidification rates are intermediate to high, and the powder shape offers advantages in handling and compaction if compared to other particulate shapes like ribbons and flakes. Optimisation of the process aims at smaller particles with finer solidification structure and distribution of second phase particles and higher supersaturation of alloying elements. Melt spinning and planar flow casting as the other methods achieve generally higher solidification rates. However, for optimum properties the ribbon shaped particulate has to be shredded before further processing [99]. Melt extraction produces fibre shaped particulates that are more rapidly solidified than powders because of contact with the serrated disc that performs the extraction from the melt. The fine particulate can be processed without any further comminutation [100].

Processing of Solidification Products may include powder blending, shredding of ribbons or mechanical alloying. Mixing of alloy powder and reinforcement phases and consolidation of the mixture can be used to produce composites by powder metallurgy. In addition to rapid solidification, mechanical alloying gives a possibility for new compositions by mixing and milling aluminium alloy powder with element or compound powder of metallic or ceramic phases in high energy mills.

Oxides and hydroxides that grow on Al alloy surfaces during and after solidification or that are generated during mechanical alloying generally impede the compaction of the particulate into a bulk material. For this reason, degassing at elevated temperatures is applied before powder compaction in most instances to minimise the detrimental effect of surface layers on properties of the compacted material [101].

Consolidation of Aluminium Alloy Particulate according to Figure 2.8 [98] starts with cold isostatic pressing to a density that still allows subsequent degassing. After hot pressing, a certain amount of shear deformation is necessary to weld the powder particle surfaces. This deformation is given to the particulate during extrusion or forging. In hot isostatic pressing, the shear deformation normally is not large enough to distribute the oxides and hydroxides in the desired way.

Aluminium Powder Alloys and Properties

Alloying aluminium with oxygen, carbon, transition metals, rare earth metals and Li offers new potentials for improved room temperature and elevated temperature strength or low density/high stiffness and has led to the development of a number of new aluminium based PM alloys. Table 2.10 shows the designations and compositions of the alloys and Table 2.11 gives examples of their mechanical properties [98, 102-104].

High Strength Aluminium Powder Alloys for advanced aerospace materials have been developed and processed since the 1950s starting with powder processing of 2XXX and 7XXX aviation alloys. Different solidification methods have been used to obtain finer structures in the particulate than with ingot metallurgy. The benefit of structure refinement by rapid solidification and the addition of alloying elements such as Co, Zr or Ni has led to improvement in a number of mechanical properties. The effect of increasing solidification rate on fatigue properties is shown in Figure 2.9 for an Al-6Fe alloy. On the other hand, properties such as creep rupture strength at elevated temperatures may be favoured by a coarser microstructure and lower solidification rates [105].

As in ingot metallurgy, precipitation hardening is still used as a strengthening mechanism in high strength Al PM alloys. In addition, solid solution hardening by supersaturation of the alloy matrix can be extended to high solute concentrations, and dispersions of intermetallic phases in very fine distributions can significantly contribute to the strength of the alloy (7090, 7091, X7093 in Table 2.11). Fracture toughness is a property that has been improved by specially designed PM alloys, e.g. X7093, but is still a problem with many of the MA alloys for elevated temperature applications.

Mechanical alloying of high strength alloys has been applied to Al-Mg based alloy powders, as well as to RS powders with a larger number of alloying ele-

Alloy	Composition	Developer
	High strength alloys	
IM7075	Al-5.6Zn-2.5Mg-1.6Cu-0.23Cr	[102]
7090	Al-8.0Zn-2.5Mg-1.0Cu-1.5Co	Alcoa
7091	Al-6.5Zn-2.5Mg-1.5Cu-0.4Co	Alcoa
X7093	Al-9.0Zn-2.2Mg-1.5Cu0.14Zr0.10Ni	Alcoa
9052	Al-4.0Mg-0.8O-1.1C	Inco
	Alloys for elevated temperatures	
Dispal 2	Al-1.0O-2.0C	PEAK
X8019	Al-8.3Fe-4.0Ce	Alcoa
Al-Fe-Mo	Al-8.0Fe-2.0Mo	Pratt & Whitney
8009	Al-8.5Fe-2.4Si-1.3V	Allied Signal
Al-Cr-X	Al-5.0Cr-2.0Zr-1.0Mn	Alcan
Al-Ti-X	Al-3.0Ti-3.0Ce	[109]
	High strength/low density alloys	
905XL	Al-4.0Mg-1.3Li-0.4O-1.1C	Inco
Al-Be-X	Al-22.6Be-10.8Li	[108]
	High stiffness alloys/SiC reinforced	
2014/20p	Al-4.4Cu-0.5Mg-0.2Zn-0.1Cr	SC, Alcan
2080/15p	Al-3.8Cu-1.8Mg-0.2Zr	PM, Alcoa
6061/15p	Al-1.0Mg-0.6Si-0.3Cu-0.2Cr	SC, Alcan
7049/15p	Al-7.7Zn-2.5Mg-1.5Cu-0.2Mn	SC, Alcan
7075/15p	Al-5.6Zn-2.5Mg-1.6Cu-0.2Cr	SC, Alcan
8090/15p	Al-2.3Li-1.2Cu-0.7Mg-0.1Zr	SC, Alcan

Table 2.10 Chemical composition of powder metallurgical aluminium alloys [98, 102-104]. SC=spray cast, PM=powder metallurgy

ments. According to Benjamin [95], the major applications for mechanically alloyed aluminium alloys are at ambient rather than elevated temperature. IN-9052 (Table 2.11), a solid solution alloy, has been shown to give combinations of tensile strength, fatigue strength, and toughness normally only seen in precipitation-hardened 7XXX aluminium alloys. Moreover, this alloy exhibits excellent corrosion behaviour.

High Temperature Aluminium Powder Alloys usually obtain their thermal stability from intermetallic phases in RS alloys or from dispersoids in MA materials. Sintered aluminium powders (SAP) were among the first aluminium alloys to take advantage of an oxide dispersion [106] and to offer good strength at elevated temperatures, but insufficient strength at low and intermediate temperatures.

With the exception of Al-Cr based alloys, most elevated temperature alloys do not make use of age hardening. Microstructure stability is an important factor for

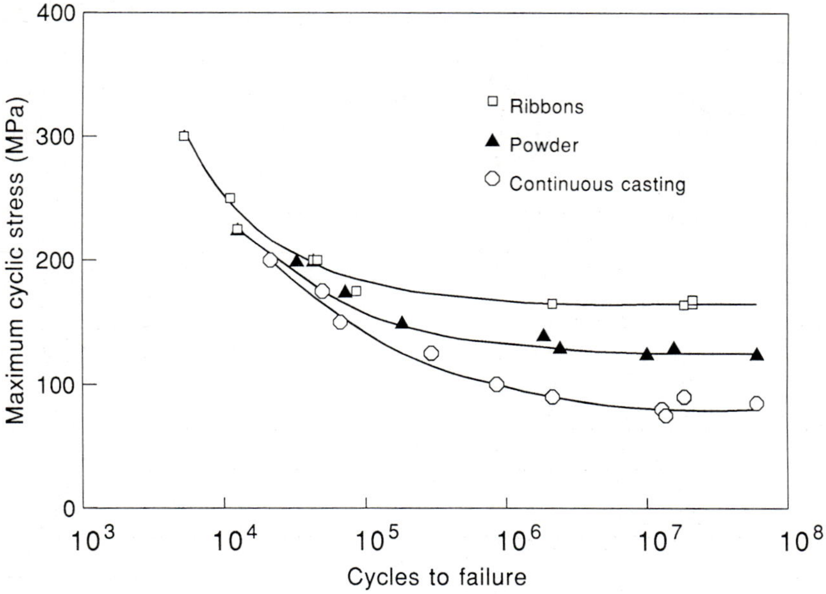

Figure 2.9 Notched fatigue properties of Al-6Fe extrusions from continuous cast, air atomized and melt spun alloy [105]

their application. Because of thermal degradation of the microstructure during consolidation and particulate surface effects, not all the benefits gained from rapid solidification can be retained. Alloy modification is necessary to improve the distribution of intermetallic phases and increase room temperature strength and thermal stability of the microstructure. Thermal stability of RS alloys is achieved by alloying with elements such as Fe, Cr or Zr with low diffusivity and equilibrium solubility; many of the high temperature alloys are based on the Al-Fe system. The interfacial energy of second phases and their distribution can be influenced by further additions. Well-known examples of these types of alloys are Al-Fe-Ce, Al-Fe-V-Si and Al-Cr-Zr. As a measure of elevated temperature strength, tensile yield strength values after 1000 hour exposure at temperature are plotted in Figure 2.10 for a number of IM and PM Al alloys [98, 103].

Mechanical alloying has been applied to improve the oxide distribution obtained in SAP and, additionally, to introduce carbides into the microstructure, e.g. in Al-O-C and Al-Mg-O-C, with the result of extended thermal stability [107]. More recently, RS powder has been mechanically alloyed and the resulting material has been shown to have improved tensile properties. Figure 2.11 shows the effect of mechanical alloying on the tensile yield strength of RS Al-8Fe-4Ce [108]. However, MA generally increases the hydrogen content in the powder and this can lead

Alloy	YS MPa	UTS MPa	Elongation %	K_{IC} $MPa\sqrt{m}$	El. Mod. GPa
		High strength alloys			
IM7075	456	530	13	34	72
7090	579	621	9	31	74
7091	545	593	11	46	72
X7093	579	614	12	47	75
9052	380	450	13	44	76
		Alloys for elevated temperatures			
Dispal 2	340	380	12	-	80
X8019	338	427	5	-	79
Al-Fe-Mo	393	512	10	-	82
8009	605	636	9	11	-
Al-Cr-X	426	460	14	-	86
Al-Ti-X	255	294	22	-	-
		High strength/low density alloys			
905XL	450	520	9	30	-
Al-Be-X	483	510	20	-	96
		High stiffness alloys/SiC reinforced			
2014/20p	466	493	2	-	100
2080/15p	365	483	7	33	100
2080/20p	393	517	6	22	110
7049/15p	598	643	3	-	90
7075/15p	556	601	4	-	95
8090/15p	499	547	3	14	101

Table 2.11 Mechanical properties of PM aluminium alloys [98, 102-104]

to lower fracture toughness and ductility as compared with the as-solidified alloy (Figure 2.11 and Table 2.11).

For certain alloy compositions and particle size distributions in systems such as Al-Ti and Al-Zr, RS is not sufficient to produce the desired high content of alloying elements, and MA offers a production method to extend the solubility limit. Ternary RS alloys based on Al-Ti with intermediate Ti content possess good properties up to at least 300 °C [104].

Low Density/High Stiffness Alloys have been developed by adding Be or Li to aluminium, which increases both strength and elastic modulus by precipitation of finely distributed phases and simultaneously reduces density. Today, Al-Li alloys are candidate materials for applications where weight considerations play a major role because of their high stiffness to weight ratio. A maximum of only 4.2% Li can be retained in solid solution from conventional ingot metallurgy. The powder

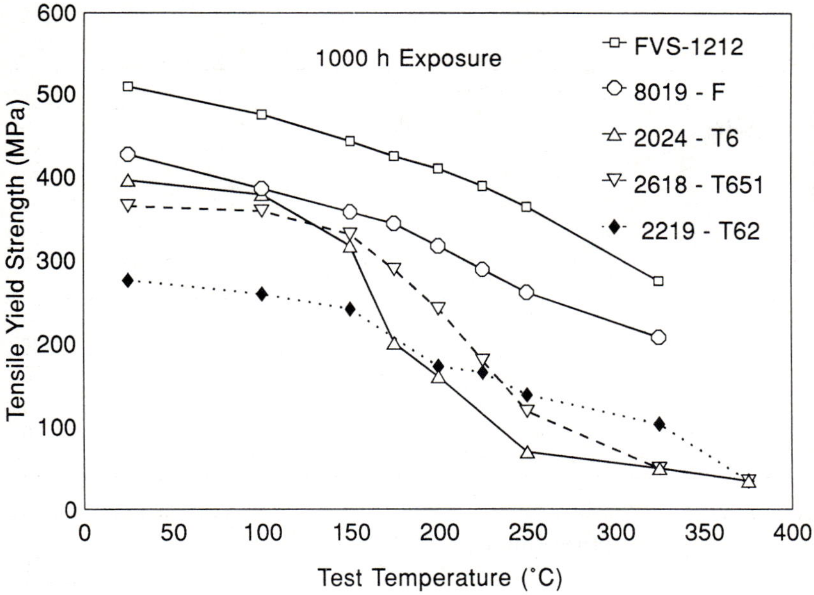

Figure 2.10 Elevated temperature tensile yield strength of Al IM and PM alloys after 1000 hr exposure [98, 103]

metallurgical approach (RS and MA) can extend the solubility range in order to obtain large increases in stiffness. Unfortunately, these highly alloyed materials exhibit poor fracture toughness levels. Mechanical alloying of Al-Li alloys is at an advanced stage and the INCO alloy 905XL in Table 2.10 and Table 2.11 shows good mechanical properties at a moderate Li level [109].

PM Al-Be alloys also offer a substantial reduction of density, since high concentrations of Be can be obtained by rapid solidification. Precipitation hardening of Be is not as effective as that of Li, and Al-Be-Li alloys take advantage of the high solubility of Be and the high strengthening effect of Li (Table 2.11). However, the toxicity of beryllium seems to restrict the application of these alloys [103].

High Stiffness Composites based on discontinuously reinforced aluminium powder alloys combine the properties of PM aluminium alloys with the discrete reinforcement phase that can consist of hard particles, whiskers or chopped fibres. Although fabrication by ingot metallurgy is possible, finer distributions of reinforcement phases with higher volume content can be achieved by powder metallurgy or spray deposition [110].

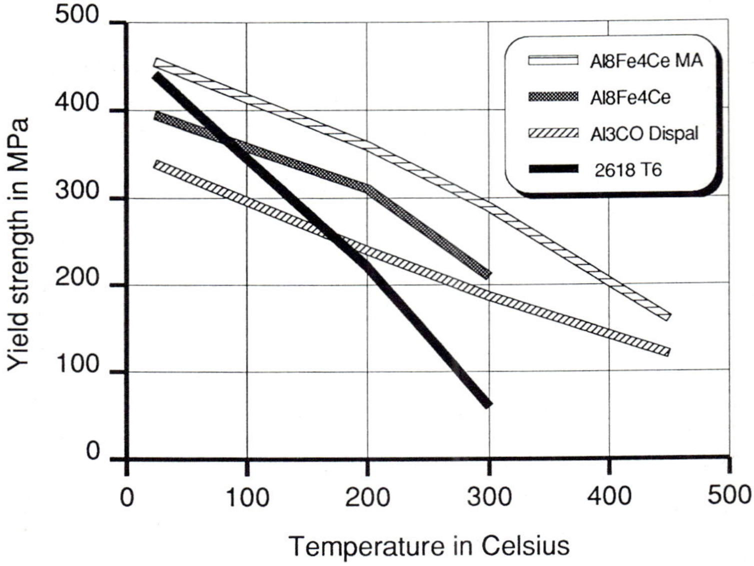

Figure 2.11 Attritor milling of Al-8Fe-4Ce with O and C [108]

Stiffness is an important property for aircraft structures where high buckling resistance is demanded [111]. The addition of SiC particles to aluminium alloys raises the elastic modulus with increasing volume content and provides the basis for high stiffness structures. Up to 20 vol% SiC particles have been added to 2014, 6061, 7075 and 7049 alloys by PM or spray deposition (Table 2.10), increasing the elastic modulus well beyond 100 GPa (Table 2.11).

Alloy 8090 is a typical low density particle reinforced material based on an aluminium-lithium alloy for application in products where low weight and high stiffness are required. A specific application would be the replacement of titanium alloy tubing by a 8090 alloy reinforced with 15% SiC which has equivalent stiffness at half the weight. Spray deposition as a powder metallurgical variant has a high potential as a method for the production of MMCs, but at present this process still has to be optimised in respect to production costs and capacity. The majority of the powder metallurgy-processed MMCs are reinforced with SiC. These particulates vary in size between 3-12 mm and their content is determined by the composite properties desired [110]. On the other hand, MMC properties are also considerably determined by the matrix characteristics. The common 2XXX matrix is well-known for a combination of strength and damage tolerance, while 7XXX matrix systems provide a higher strength potential. The 6XXX alloys provide improved corrosion resistance and some of the 8XXX can be used for service at elevated temperatures. Primarily, the matrix alloy composition and the heat treatment determine the strength level, while reinforcement content controls the elastic modulus. For a 2080 alloy given in Table 2.11 [112], an increase of the SiC content from 15 to 20% raises the modulus from 100 to 110 GPa with negligible effect on density. Furthermore, wear resistance also increases with the volume content of SiC particles. Since this composite alloy is heat treatable, a range of strengths can be produced (Table 2.11).

Particulate Surface Effects

Consolidated aluminium PM alloys may show properties which are inferior to those of the original solidification product, i.e. those of individual melt spun ribbons. Part of the loss in certain properties can be attributed to coarsening that occurs due to extended exposures to elevated temperatures during powder processing. The main cause of inferior properties, however, is the surface film that covers the particulate particles. This film is composed of aluminium oxides and hydroxides [113]. These form during and after atomisation. Being very thin, on the order of a few nanometers, they can be deformed plastically or elastically and are easily separated from the particle surface during consolidation. The film fragments tend to concentrate in aggregates which are distributed throughout the structure and negatively influence properties.

A second cause of inferior properties is hydrogen [114]. Concentrations of hydrogen are particularly high in PM and rapidly solidified products due to the much higher solubility of hydrogen in liquid aluminium and slow diffusion out of the solid metal. The source of hydrogen is the reduction of water by molten aluminium. Water also reacts with solid powder particles to form aluminium oxide hydroxide or trihydroxide films. At elevated temperatures, e.g. during hot pressing or extruding, the hydroxides decompose, forming molecular water which reacts with the metal to form hydrogen and aluminium oxide. This hydrogen diffuses into the metal. Furthermore, the high hydrogen activity at the metal surface resulting from the ongoing surface reaction

$$2Al + 3H_2O = Al_2O_3 + 6H \tag{2.1}$$

retards the diffusion out of the solid metal of previously dissolved hydrogen due to a lowered or reversed concentration gradient.

To lower the hydrogen concentration, the powder is degassed by heating in vacuum at a temperature close or equal to the highest temperature used in subsequent processing steps as described before. The beneficial effect of the degassing treatment is most likely caused by the thermal decomposition of voluminous aluminium oxide hydroxide to aluminium oxide and gaseous water. The aluminium oxide not only has a significantly smaller volume, but also breaks up into small particles during working. Being finely dispersed in the metal structure, these do not affect mechanical properties as much as do the voluminous, soft aggregates of oxide hydroxide. Removal of water prevents hydrogen embrittlement and additional oxide formation. Figure 2.12 shows the reactions occurring during thermal outgassing.

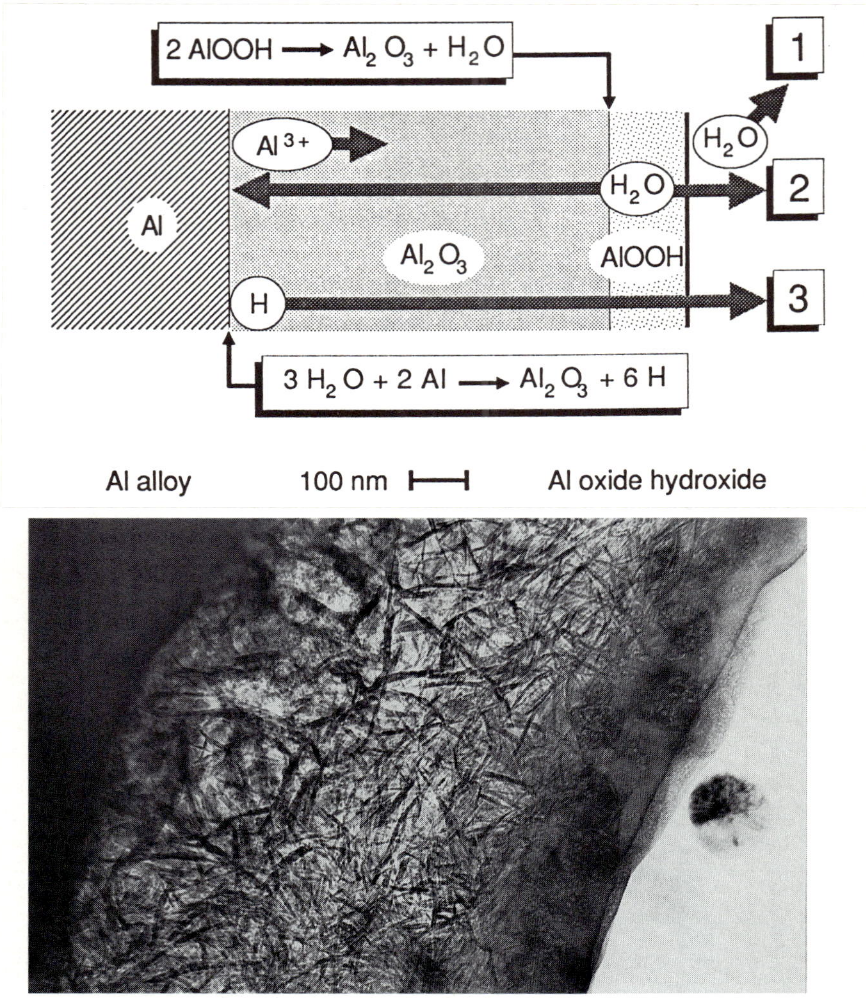

Figure 2.12 Schematic of chemical reactions at elevated temperatures in aluminium alloy surfaces and TEM micrograph of water atomized Al-Cr-Zr-Mn powder. Numbers relate to peaks in figref refid=stan6 page=no..

Although degassing generally improves mechanical properties, the improvement is still far from being satisfactory. The properties of PM extrusions also decrease with the length of storage of the alloy powder. This process can not be reversed completely by degassing, most likely not even after considerably longer degassing times.

The potential for unwanted structural changes, precipitation or coarsening of intermetallic particles severely limits the time and temperature of the vacuum treatment. Mechanically alloyed powder also does not respond well to thermal outgassing because most of the oxide and hydroxide has been worked into the interior of the particles.

Chemical powder treatments were evaluated as a substitute for thermal vacuum degassing [115]. The goals were to find an effective treatment for temperature sensitive alloys and to improve the results generally obtained with thermal degassing. Wet chemical treatment of powders is performed before extrusion for the purpose of dissolving the existing oxide or hydroxide layers and replacing them with modified surface films.

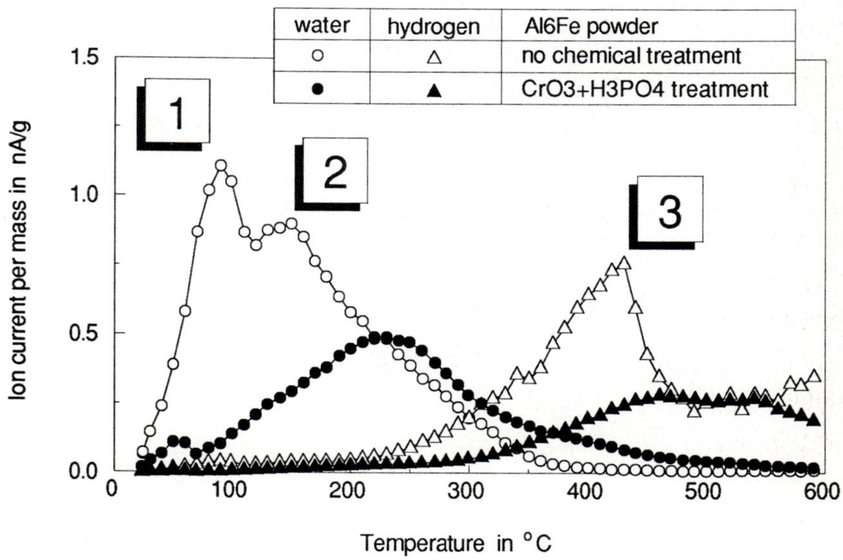

Figure 2.13 Mass spectrum of water and hydrogen for atomized Al-6Fe after four months of storage and after chemical treatment with chrome phosphoric acid [115]

With chemical powder treatment, the same mechanical properties could be obtained as with conventional degassing of Al-6Fe and 7075 powder. In some cases, depending on shapes and extrusion ratio, fracture toughness was superior to that of degassed specimens. These results are attributed mainly to the uniform distribution of smaller surface oxide fragments in powder extrusions generated from chemically treated powders, indicating a more effective bonding of powder particles. In addition, diminished or modified gas development during powder processing is also assumed to positively affect consolidation. The effect of chemical treatment on gas content is illustrated in Figure 2.13. The thermal dehydration of a powder sample stored for four months in ambient atmosphere indicates that most of the water is either physisorbed (peak no. 1) or weakly chemisorbed (peak no. 2). The maximum of the hydrogen evolution (peak no. 3) is delayed relative to the second peak by about 300°C, suggesting that the rate controlling step in reaction (2.1) is the diffusion of water through the oxide layer (Figure 2.12). After the powder surface has been treated with a solution of chromic and phosphoric acid, the water content is considerably lower.

The investigations have shown that chemical surface treatment of aluminium alloy powders can replace thermal vacuum degassing. Properties obtained by either pretreatment are compatible. Further optimisation of the chemical methods is assumed to have the potential of achieving properties that exceed those obtainable by degassing.

Powder metallurgy has been sucessful in the development of new aluminium alloys that exhibit superior properties in certain desired combinations. Alloy classes have been evaluated that meet the requirements for high temperature service conditions, stiffness and strength. However, high manufacturing costs limit the application of Al PM techniques to areas, where an increased price is justified by the demand for improved properties such as aviation and space applications. Candidate alloys are predominantly those that can only be produced by PM techniques such as RS or MA.

2.1.2 Titanium Alloys and Aluminides

2.1.2.1 Titanium Alloys

M. Peters and H. Buhl

In 1795, Martin Klaproth, a German chemist, examined rutile and discovered an oxide of an unknown element. Greek mythology helped him to find a name: he called it titanium (Ti) after the first children of the gods of earth and sky, the Titans.

A lot of efforts were made to isolate the metal, but to turn it into a ductile material remained unsuccessful for more than a hundred years. This changed in 1910 when the American M.A. Hunter used sodium for the reduction of $TiCl_4$ to titanium sponge. He found the metal to be ductile at elevated temperatures. The major breakthrough, however, for commercial production was pioneered by the work of W.J. Kroll from Luxembourg in the 1930s. Kroll, who used magnesium to reduce $TiCl_4$, is generally considered to be the father of the titanium industry. Still today the primary methods of producing titanium metal utilize either the Kroll or Hunter processes, although next to these two thermochemical processes, electrochemical processes may gain more attention in the future, especially since they provide the capability of producing a higher purity titanium sponge [116, 117].

There is probably no other material which is closer related to aerospace than titanium. With a density of 4.5 $g \cdot cm^{-3}$, Ti alloys are only about half as heavy as steel and Ni-base superalloys and thus yield an excellent strength-to-weight ratio. Furthermore, they reveal exceptional corrosion resistance. Titanium is blessed with abundant resources. It is the tenth most abundant element and - following aluminium, iron and magnesium - the fourth most plentiful structural metal in the earth's crust. However, due to its high reactivity and the energy- and cost-intensive reduction technology, Ti alloys are relatively expensive [21, 118].

In the following, Ti alloys will be briefly reviewed. For a more detailed study the reader is particularly referred to the proceedings of the series of International Titanium Conferences taking place every four years, starting 1968 in London [119] followed by meetings in Boston [120], Moscow [121], Kyoto [122], Munich [123], Cannes [124], and San Diego [125].

Physical Metallurgy

Titanium undergoes an allotropic phase transformation at 882 °C (β transus), where the body centered cubic high temperature β phase transforms into the low temperature hexagonal close packed α phase. The alloying elements of Ti alloys are roughly classified into whether they increase or decrease this transformation temperature, as schematically shown in Figure 2.14. α stabilizers raise the β transus temperature whereas β stabilizers reduce it. The latter are further subdivided into β-isomorphous elements with a high solubility in titanium and β-eutectoid elements, which have limited solubility and form intermetallic compounds. Finally, there are elements with essentially no influence on the transus temperature, which are classified as neutral. Since they primarily strengthen the α phase they are also referred to as α strengtheners. Among the typical alloying elements of titanium, Al and interstitial O are α stabilizers, V, Mo, Nb and Ta are β-isomorphous elements, Fe and Cr are β-eutectoid formers and Sn and Zr are neutral elements.

Classification and Commercial Alloys

Usually Ti alloys are classified as α, near-α, (α+β), metastable β and β alloys according to their predominant alloying elements [126]. This is shown in Figure 2.15, which schematically illustrates a three-dimensional phase diagram composed of Ti-Al (α stabilizer) and Ti-V (β stabilizer) binary diagrams.

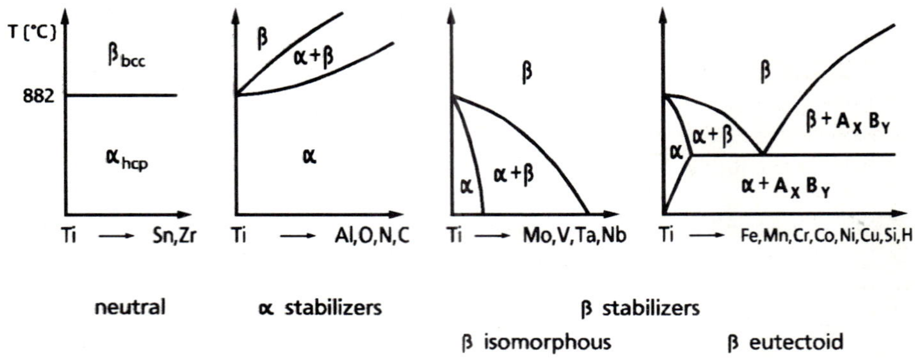

Figure 2.14 Schematic phase diagrams of binary Ti alloys

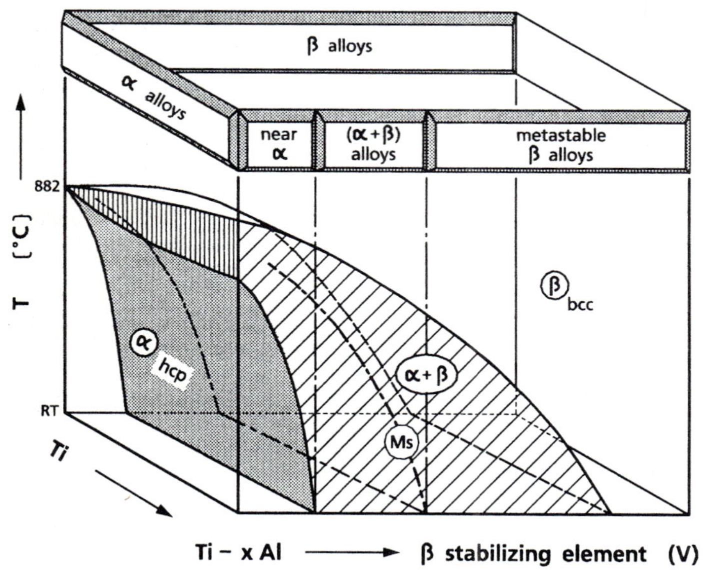

Figure 2.15 Schematic three-dimensional phase diagram for classification of Ti alloys

Al is by far the most important alloying element of Ti. Not only has Al a high solubility in Ti and thus a large strengthening capability, with only about half the density of Ti it also substantially reduces the specific weight of Ti alloys. However, in the design of Ti alloys care has to be taken not to exceed 9% of the so-called Al equivalent,

$$Al \quad eq. = \%Al + 1/3\%Sn + 1/6\%Zr + 10\%O < 9\% \tag{2.2}$$

in order to avoid embrittling the alloy due to the formation of the ordered Ti_3Al intermetallic compound [127]. For this reason the Al content of nearly all conventional Ti alloys does not exceed 6%. For some time, however, condiderable activities have been devoted to develop new alloys primarily consisting of titanium aluminides based on Ti_3Al and TiAl intermetallic compounds - alloys with Al concentrations substantially beyond this limit characterized by equation (2.2) (see section "Titanium Aluminides"). In the following, the various classes of conventional Ti alloys will be briefly described, examples of which are given in Table 2.12.

Alloy	Chemical Composition	T_β in °C	Year
	α alloys		
	Ti-5Al-2.5Sn	1040	1953
	Near-α alloys		
IMI 685	Ti-6Al-5Zr-0.5Mo-0.25Si	1020	1969
Ti-6242S	Ti-6Al-2Sn-4Zr-2Mo-0.1Si	1000	1970
IMI 829	Ti-5.5Al-3.5Sn-3Zr-1Nb-0.25Mo-0.3Si	1015	1976
IMI 834	Ti-5.8Al-4Sn-3.5Zr-0.7Nb-0.5Mo-0.35Si-0.06C	1045	1984
Ti-1100	Ti-6Al-2.75Sn-4Zr-0.4Zr-0.45Si	1020	1987
	(α+β) alloys		
Ti-6-4	Ti-6Al-4V	995	1954
IMI 550	Ti-4Al-4Mo-2Sn	975	1956
Half 6-4	Ti-3Al-2.5V	935	1958
Ti-6-6-2	Ti-6Al-6V-2Sn	945	1962
Ti-6-2-4-6	Ti-6Al-2Sn-4Zr-6Mo	930	1966
	Metastable β alloys		
Beta III	Ti-11.5Mo-6Zr-4.5Sn	760	1969
Beta C	Ti-3Al-8V-6Cr-4Zr-4Mo	795	1969
Ti-17	Ti-5Al-2Sn-4Mo-2Zr-4Cr	890	1972
Ti-10-2-3	Ti-10V-2Fe-3Al	795	1976
Ti-15-3	Ti-15V-3Cr-3Al-3Sn	760	1981
β-CEZ	Ti-5Al-2Sn-4Mo-4Zr-2Cr-1Fe	890	1988
Beta-21S	Ti-15Mo-2.7Nb-3Al-0.2Si	815	1990

Table 2.12 Commercial Ti alloys

α *Alloys* only contain α-stabilising and strengthening elements. They are mostly used where strength is not a prime design criterion but corrosion is a problem. Therefore, applications in aerospace are limited. Ti-5Al-2.5Sn is primarily used for welding applications.

Near-α Alloys usually contain about 5 to 6% Al as their main alloying element with additions of neutral elements (Sn, Zr) but also small portions of β stabilizers. They are the classical elevated temperature materials, due to the good creep and oxidation resistance of the α phase while small portions of β add to strength as well as ductility. Si additions lead to precipitation of fine silicides which effectively hinder dislocations from climbing, thus further improving creep. Alloying with Si particularly allowed the upper use temperature of near-α alloys to be raised close to 600 °C over the years, as is reflected by the two latest near-α alloys IMI 834 and Ti-1100 [128-130].

(α+β) Alloys represent the class of Ti alloys with the most balanced mechanical properties. This explains why by far the most widely used Ti alloy - Ti-6Al-4V - belongs to this class. Developed in the 1950s this alloy still today represents more than 50% of all Ti alloys shipped and is thus considered to be the workhorse of the Ti industry. Other (α+β) alloys have only found limited application.

Metastable β Alloys - also called near-β, super-β or simply β alloys - are so rich in β stabilizers that the β phase can be completely retained at room temperature if appropriate cooling is applied. Due to the cubic crystal structure metastable β alloys can be processed at much lower temperatures than near-α or (α+β) alloys; in fact, some of them are developed for forming operations at room temperature, e.g. Ti-15-3. Metastable β alloys can be hardened to the overall highest strength levels of Ti alloys and still retain high toughness. Their major drawbacks are the increased density due to the high volume of their alloying elements like Mo, Ta, V or Nb, their reduced ductility at peak-strength levels and their limited weldability due to segregation and hardening in the welded condition. Recently trends are observed to develop metastable β alloys for elevated temperatures (β-CEZ) or with improved oxidation resistance (Beta 21S) for use as matrix for fibre reinforcement [131, 132].

β Alloys only contain β-stabilising elements. Since such alloys are difficult to melt, have high density, low stiffness and strength as well as poor corrosion resistance, they are not found in aerospace applications (if at all).

Microstructures and Mechanical Properties

The mechanical behaviour of Ti alloys is mainly determined by two factors, the chemical composition and the microstructure. To some extent both influence each other [126, 133].

The chemical composition primarily determines the volume fractions of the α and β phases. Due to the limitation of slip systems in the hexagonal close-packed crystal structure, the α phase is less ductile and more difficult to deform than the non-close packed β phase. Furthermore, the diffusion coefficient of α Ti is about two orders of magnitude lower than that of β Ti. Thus, increasing Al contents effectively improves the creep and oxidation resistance of Ti alloys, while simultaneously leading to reduced ductility and poor deformation behaviour (see section "Titanium Aluminides"). Si, Sn, Zr and interstitial oxygen effectively strengthen the α phase. Si atoms tend to segregate to dislocations and inhibit dislocation climb, thus further improving the creep resistance. Zr tends to homogenize silicide precipitation; upon aging $(Ti(Zr))_5Si_3$ may form.

Solid solution strengthening of the β phase through Mo, V or Nb is moderate. However, the β phase may be significantly strengthened by fine coherent ω particles which homogeneously precipitate in heavily β stabilised alloys. Unfortunately the high strength is often associated with poor ductility. For oxidation resistance Nb is the preferred β stabilizer while minor additions of Pd are known to improve corrosion resistance.

The microstructures of Ti alloys tell how the two phases α and β are distributed. They are a result of thermomechanical treatments (TMT), which are schematically outlined in Figure 2.16. TMT essentially consist of combinations of forming and heat treatment procedures and can be roughly separated into three processing steps: deformation, solution-heat treatment (SHT) and aging (stresss relief). These three steps determine the microstructure of Ti alloys, which is furthermore influenced by the starting structure of the as-received condition (AR).

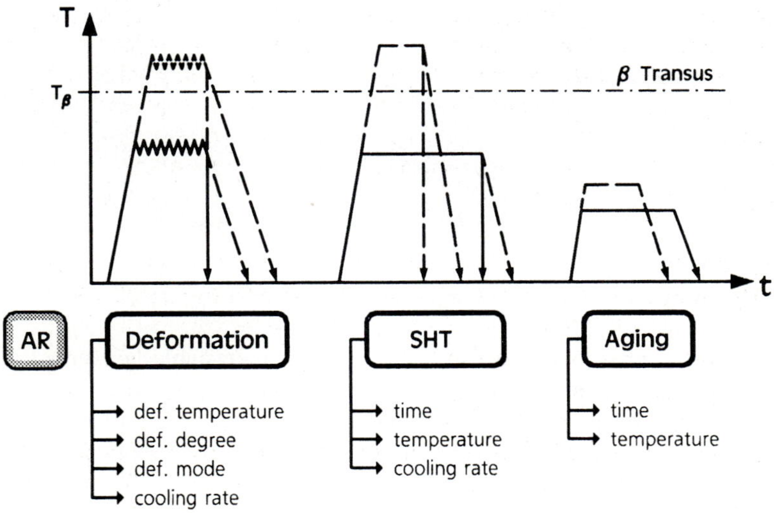

Figure 2.16 Thermomechanical treatment of Ti alloys

A key element in the TMT process is the β-transus temperature, T_β. The microstructures of Ti alloys can be roughly divided into lamellar and equiaxed structures [134]. Lamellar structures (β annealed, β processed, transformed β) are a result of cooling from the β phase field. Here the α phase nucleates at prior β grain boundaries and grows to a lamellar morphology into the former β grain. Depending on the cooling rate, the final structure can either be fine (Widmanstätten structure, basketweave microstructure) or coarse (colony structure) (Figure 2.17c).

Conversely, equiaxed structures are a result of a recrystallization process [134]. Therefore they require sufficiant stored cold work as a result of deformation at temperatures in the $(\alpha+\beta)$ phase field, followed by SHT in the two-phase field. Extended SHT after recrystallization leads to grain growth, thus allowing fine and coarse equiaxed structures (Figure 2.17a). The SHT temperature determines the volume fraction of the primary α phase. Temperatures close to the β transus lead to a bi-modal type of microstructure (duplex structure), consisting of equiaxed primary α in a lamellar $(\alpha+\beta)$ matrix (Figure 2.17b). This structure is known for its excellent balance of engineering properties [135, 136] and is also recommended for some of the latest elevated temperature Ti alloys [128].

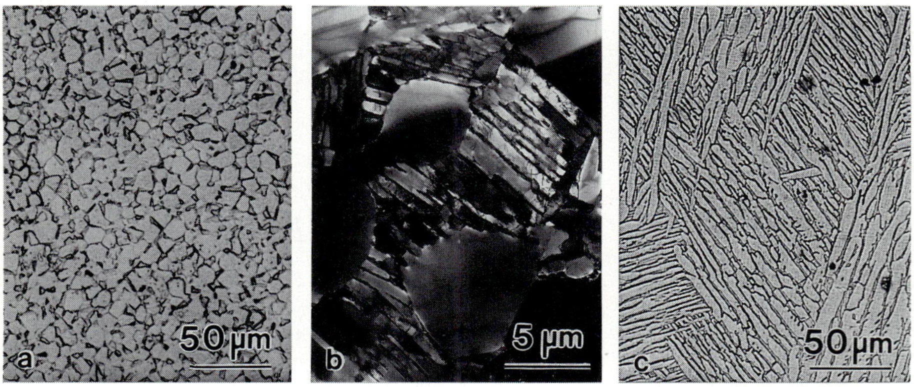

Figure 2.17 Microstructures of Ti alloys: equiaxed (a), bi-modal (b), lamellar (c)

Upon quenching, the β phase can (partly) transform into metastable hexagonal α' or orthorhombic α'' martensite, which can further decompose upon aging into $(\alpha+\beta)$. β phase decomposition products include fine precipitates of ω or α phase, while fine ordered α_2 can precipitate out of supersaturated α, provided the Al concentration is sufficiently high.

The various microstructures can substantially influence the mechanical properties of Ti alloys [137]. Fine structures tend to increase both strength and ductility, inhibit crack nucleation and are a prerequisite for superplasticity. Conversely, coarse structures reveal improved resistance against crack growth and creep [138, 139]. Equiaxed structures have a potential for high strength and ductility, fatigue strength and superplasticity, while lamellar type structures are more creep and fatigue crack growth resistant and reveal higher fracture toughness [140]. As a compromise between lamellar and equiaxed microstructures, the bimodal structure has particularly gained interest since it has a well balanced set of properties of both basic microstructure types [141, 142].

Unlike Al alloys or steels, Ti alloys can develop a strong crystallographic texture. This is a result of the anisotropic nature of the hexagonal α phase. Upon hot working a deformation texture develops in the α phase which may change into an even more pronounced recrystallization texture after annealing. Depending on the processing mode and temperature, a basal or a transverse type texture may result with the basal {0002} planes either parallel or perpendicular to the rolling plane. A strong texture can result in marked anisotropy of the mechanical behaviour [143, 144].

Corrosion of Ti Alloys

Regarding standard potentials titanium is a very active metal, but nevertheless it exhibits excellent corrosion resistance due to the presence of a stable oxide film. A further advantage of this film is its rapid formation on active Ti interfaces in oxidizing environments, including oxidizing acids [145]. Spontaneous activation with uniform corrosion is only observed in reducing acids, e.g. HCl ($c_{HCl} > 8{,}5\%$ at RT) or H_2SO_4 ($c_{H_2SO_4} > 5\%$ at RT).

In aerospace structural applications of Ti alloys, uniform corrosion is negligible, but localized types of corrosion like pitting corrosion, crevice corrosion or stress corrosion cracking (SCC) and corrosion fatigue (CF) gain increasing importance and are characterized by the simultaneous existence of active and passive interfaces; in terms of electrochemistry the behaviour of a heterogeneous electrode has to be considered. Other localized corrosion phenomena like intergranular and galvanic corrosion are of less importance for aerospace applications.

Crevice Corrosion: This type of corrosion is observed particularly in halide and sulfate solutions at elevated temperatures and becomes more severe with increasing temperature and decreasing pH. In deaerated or alkaline solutions crevice corrosion is not to be expected. In chloride containing media ($c_{Cl} = 1.0\ \mathrm{Mol} \cdot l^{-1}$) titanium activates sponaneously at pH ≤ 2.3 [146]. At passive titanium interfaces in aqueous environments anodic reactions occur which may be illustrated as overall reactions (2.3), producing passive oxide and dissolved titanium hydroxide (2.4).

$$Ti + 2H_2O \rightarrow TiO_2 + 4H^+ + 4e^- \tag{2.3}$$

$$Ti + 2H_2O \rightarrow [Ti(OH)_2]^{2+} + 2H^+ + 4e^- \tag{2.4}$$

which are in balance with cathodic reactions, generally oxygen reduction (2.5)

$$O_2 + 4H^+ + 4e^- \rightarrow 2H_2O \tag{2.5}$$

These electrochemical reactions are coupled with the chemical hydrolysis reaction of titanium dissolution products (2.6)

$$Ti^{4+} + 2H_2O \rightarrow [Ti(OH)_2]^{2+} + 2H^+ \tag{2.6}$$

Both first reactions (2.3) and (2.4) and the hydrolysis reaction (2.6) produce protons, whereas reaction (2.5) consumes protons. This reaction (2.5) depends upon the diffusion of oxygen to the interface, which is strongly slowed down within the crevice. On the other hand the diffusion of protons from the base region of the crevice is also limited, resulting in an increasing concentration of protons. Both limitations of diffusion processes cause a spatial separation of the above mentioned anodic and cathodic reactions, with the anodic reactions in the base region and the cathodic reaction near the crevice opening. Within the crevice, pH values below 1.0 were observed in electrolytes having near neutral bulk pH [147], significantly less then the above stated value of 2.3 below which titanium interfaces are activated. In this way a fairly stable active-passive system is established with the active walls in the base region (dissolution region) of the crevice serving as the anode and the crack entrance as the cathode. The mechanism of crevice corrosion was discussed in detail for a model system by E.J. Kelly [146].

Generally crevice corrosion of Ti alloys in neutral aqueous chloride solutions does not appear at temperatures below 70 °C. Impurities of Fe increase the attack; by alloying with small amounts of Pt, Pd, Ru, and Ir resistance to crevice corrosion is significantly increased. Ni and particularly Mo increase the temperature and pH range where the material is not susceptible [148], as shown in Table 2.13, whereas Cr and Sn deteriorate the crevice corrosion behaviour.

Alloy	pH_{krit}
Ti-6Al-4V	10,0-10,5
unalloyed Ti	9,5-10,0
IMI 550	3,0
Ti Grade 12	2,5
Ti-6Al-2Sn-4Zr-6Mo	2,0
Ti-3Al-8V-6Cr-4Zr-4Mo	1,0
Ti Grade 7 or Ti11	0,7-0,8
Ti-15Mo-5Zr	<1,0

Table 2.13 Threshold pH values below which crevice corrosion in hot (>90 °C), concentrated NaCl solutions may be expected [149].

Welding does not affect, and cold working improves, crevice corrosion resistance. Acicular β structures, obtained by annealing above the β transus, deteriorate the resistance compared to equiaxed microstructures [150].

Pitting Corrosion: Contrary to crevice corrosion, pitting can initiate at the complete metal interface, at heterogeneities like nonmetallic inclusions, microdefects and precipitates, at grain boundaries and dislocations, and at weak points of the passive layer. Similar to crevice corrosion the composition of solution in the pit is changed and pH is decreased; however, the extent of these changes is much less, even in immobile solutions, because of better accessibility by diffusion of solution components, particularly oxygen. This in combination with the stable and dense oxide film makes pitting corrosion for titanium materials a negligible problem in most media.

Only in halide containing aqueous solutions at elevated temperatures (except F^-) is pitting corrosion observed. In chloride solutions with a concentration of 0.53 $Mol \cdot l^{-1}$ the pit initiation potential at RT is about 11.4 V_{SCE}, decreasing with increasing temperature to about 2.0 V_{SCE} at 150 °C. Decreasing pH from 7 to 1 shifts this potential to about 1.3 V_{SCE} at 200 °C [151]. Alloying with aluminium significantly reduces the pit initiation potential at RT significantly to 1.5 V_{SCE} for an aluminium content of 8%. This influence is most pronounced between 40 to 80 °C [152]. Also Fe contents above 0.07% deteriorate the resistance against pitting [153], while Mo improves it.

Stress Corrosion Cracking: The high expectations of aerospace engineers regarding SCC properties of Ti alloys were considerably damped by the results of B.F. Brown et al. in 1964 [154] who found SCC susceptibility on precracked Ti-7Al-2Nb-1Ta specimens in seawater. This resulted in a complete reevaluation of the potential usefulness of Ti alloys and caused a huge amount of research work which was not at all justified by the number of service failures [155].

Present understanding of SCC mechanisms are founded upon the assumption of a hydrogen embrittlement mechanism. The following observations support this model:

- The cleavage-like SCC fracture surfaces through the α phase are very similar to those observed in hydrogen containing specimens which failed by delayed fracture under sustained load or by slow strain rate. Cracking occurs on $\{10\bar{1}7\}_\alpha$ and $\{10\bar{1}0\}_\alpha$, both hydride habit planes [156].
- Identical cleavage-like fractures are observed on specimens pre-exposed in HCl-containing methanol and subsequently tested with slow strain rate [157].
- The electrochemical conditions in the crack are favourable for development and absorption of hydrogen, e.g. active crack tip and low pH.
- By electron-diffraction measurements titanium hydride could be identified on SCC fracture surfaces [156]. It has to be mentioned that this result was not confirmed by other authors [158], who postulated the dissolution of these hydrides after stress relieving caused by crack propagation [159].
- By audioradiographic investigations tritium was found on SCC fracture surfaces of specimens tested in tritium-containing aqueous solutions [160].

- Acoustic emission studies identified SCC crack growth as discontinuous, confirmed by crack arrest markings on fracture surface [161].

In spite of these results the existence of other observations which are not consistent with the hydrogen model must not be overlooked:

- The diffusion rate of hydrogen in α phase is too slow to follow the measured crack propagation rate (see Table 2.14).
- The occurrence of SCC in solvents free of water like CCl_4, where $TiCl_2$ was observed on fracture surface [166].

Material	Phase	$D^H_{77\,°C}$ $cm^2 \cdot s^{-1}$	$D^H_{677\,°C}$ $cm^2 \cdot s^{-1}$	Ref.
Electr.-Ti	α	$2{,}0 \cdot 10^{-11}$	$1{,}2 \cdot 10^{-5}$	[162]
I-Ti	α	$3{,}3 \cdot 10^{-10}$	$2{,}6 \cdot 10^{-5}$	[163]
	β	$1{,}4 \cdot 10^{-7}$	$5{,}8 \cdot 10^{-5}$	[163]
Beta 120 VCA	β	$9{,}6 \cdot 10^{-7}$	$1{,}04 \cdot 10^{-4}$	[164]
Ti-88-23	β	$8{,}0 \cdot 10^{-7}$		[165]

Table 2.14 Diffusion coefficient of hydrogen in α and β titanium alloys.

Basing on hydrogen embrittlement and the formation of brittle hydrides, a model was developed (Figure 2.18) to explain qualitatively the mechanism of SCC as well as slow strain rate embrittlement of hydrogen containing Ti-8Al-1Mo-1V specimens (SSRHE) [156].

1. Cathodically generated hydrogen is absorbed at the active crack tip and accumulates in the triaxial stressed region in front of the crack tip. In the case of SCC this is the main source of hydrogen. Residual hydrogen plays a minor role. In the case of SSRHE, residual hydrogen diffuses from the bulk material into the region ahead of the crack tip because of the triaxial stress state there.
2. Within the α plates brittle hydrides are formed and
3. crack under the influence of tensile stress. At the side of these plates the crack front is pinned at the ductile β phase and shows a semi-circular profile, observed by TEM investigations [167].
4. The crack propagates by an increment, forming a crack-arrest marking. Because of the low diffusivity of hydrogen in α phase the single increments are extremely small and their detection by SEM is difficult.

It has to be mentioned that this hydride rupture model is not able to explain all observations, but the controlling influence of hydrogen via hydrides on SCC seems to be unequivocal.

Most titanium alloys are not susceptible to SCC in aqueous halide containing solutions if not pre-cracked. This may be due to the special electrochemical conditions in the crack (see section "Crevice corrosion") but strain rate is also a

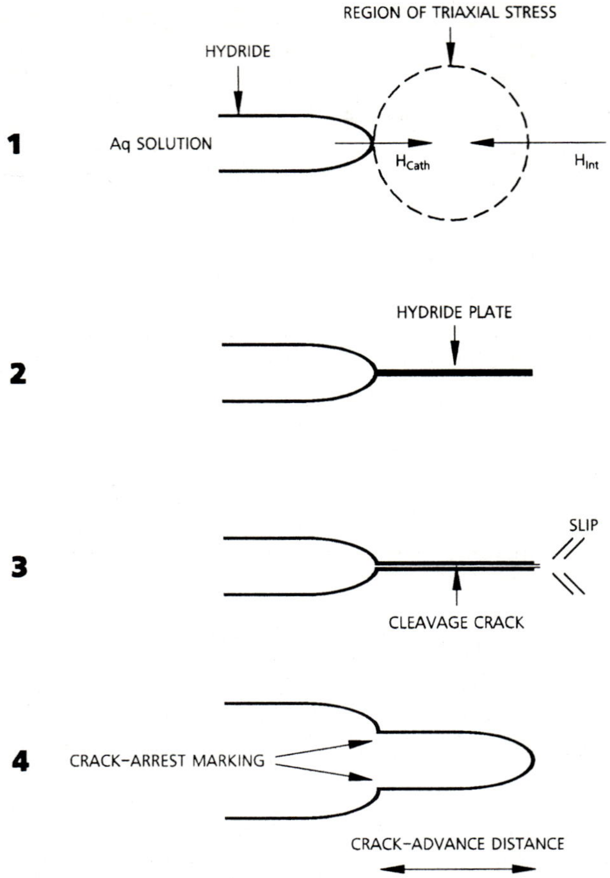

Figure 2.18 Schematic of the hydride-rupture model for SCC and SSRHE [156]

controlling parameter. Because of the extremely high repassivation rate, shown by H.-J. Rätzer-Scheibe and H. Buhl [145] a high strain rate is necessary to keep a propagating crack tip active. Such a strain rate cannot be realized by constant load or deformation at the surface of a smooth specimen, but only at the tip of a crack. That smooth specimens also become susceptible to SCC if only the strain rate is drastically increased is illustrated in Figure 2.19.

The dependence of relative fracture energy upon strain rate indicates strain-rate induced SCC (characterized by the minimum) for the precracked as well as the smooth specimens. For smooth specimens the required strain rate to release SCC (minimum) is two orders of magnitude higher than for the precracked specimen,

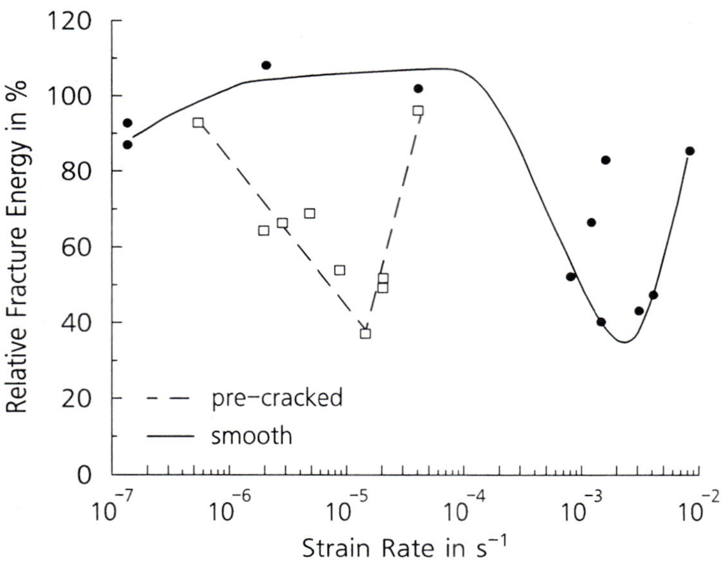

Figure 2.19 SCC susceptibility of Ti-6Al-4V in 3.5% NaCl in precracked and smooth condition evaluated by the SSR method [168]. The scale on the abscissa is only valid for the smooth specimen, because all the strain of the precracked specimen is focussed on the crack tip.

which can be correlated to the difference between the strain rate at the surface of a smooth specimen and at the crack tip of a precracked specimen.

Corrosion Fatigue: CF does not need a specific system to occur, like SCC. Every environment different from vacuum or dry inert gases deteriorates the CF behaviour within the boundaries given by materials susceptibility. Frequently CF is released at local corrosion points serving as notches. The high resistance of most titanium materials against local corrosion has the consequence that CF resistance is also high, at least in neutral aqueous salt solutions and seawater [169]. CF behaviour of Ti-6Al-4V in hot concentrated NaCl solution is nearly identical with that in air [170, 171]. In contrast to this observation CF crack propagation is markedly accelerated in solution-treated Ti-6Al-4V. For the SCC-susceptible alloy Ti-8Al-1Mo-1V this effect is even more pronounced. Generally, Ti alloys which are susceptible to SCC in a distinct environment without notches or precracks are also susceptible to CF at any frequency. The improvement of the fatigue behaviour by β annealing of (α+β) alloys is transferable to their CF behaviour. The combination of CF and fretting corrosion deteriorates the CF behaviour drastically. Coatings and lubrication films are remedial measures.

With decreasing frequency and increasing ΔK, CF crack propagation is accelerated. At low growth rates this effect is reversed, understandable by a superposition model of SCC and CF [172], regarding repassivation behaviour.

Oxidation of Ti Alloys

In oxygen-containing gaseous environments at elevated temperatures, Ti and Ti alloys are more or less susceptible to oxidation, which leads to structural instabilities, influencing the mechanical properties of the material. This oxidation includes two processes:

1. Dissolution of oxygen in the bulk material, which depends upon phase composition and temperature. Below the β transformation temperature, oxygen dissolves in α phase, and above it in β phase. In the α phase the solubility is about 34 at% with weak influence of temperature. In β phase solubility is much less and strongly dependent upon temperature. It reaches 8 at% at 1700 °C.
2. Growth of TiO_2 with rutile structure. At temperatures above 1000 °C and reduced oxygen pressure lower oxide phases are also formed [173].

The oxidation kinetic between 300 and 900 °C may be described by equation (2.7)

$$\Delta m_A^n = k \cdot t \quad \text{with} \quad k = k_0 \cdot e^{-\frac{Q}{RT}} \tag{2.7}$$

In this temperature range the predominant growth law is the parabolic one, characterized by n=2. Below 300 - 400 °C oxidation follows a logarithmic law and above 900 - 1000 °C oxidation behaviour becomes linear. These threshold temperatures are shifted to lower values with extended reaction times. Within the scope of linear growth the oxide scale becomes lamellar and less protective [174, 175].

The growth rate of the oxide as well as the diffusion coefficient of oxygen in the bulk material are controlled by alloying elements like Al, Zr, Nb, Mo and particularly Si. Carefully directed application of these elements led to the actually most oxidation resistant Ti alloys IMI 834 and Ti-1100 (see Table 2.12) with application temperatures up to 600 °C. However, compared to Ni-base superalloys, the oxidation behaviour is still unsatisfactory. Improved oxidation behaviour is expected from the Ti aluminides. Furthermore, reliable coatings have to be developed.

Applications

First applications of Ti alloys date back to the 1950s when they paved the way for the first fan-type gas turbine engines from Pratt & Whitney and Rolls Royce, first as compressor blades, then disks. Still today most of the Ti alloys used on an aircraft go into the engine. Ti alloys represent one quarter to one third of the weight of modern turbine engines and - following Ni-base superalloys - are the second most used engine material.

The large front fan of modern high by-pass engines as well as most of the stator and rotor blades and discs of the low and intermediate sections of the compressor withstanding temperatures approaching 600 °C are made of Ti alloys. But also less critical components like compressor casings, fan frames, bearing housings or ducts and vanes - often as castings - are made out of titanium.

Usually the weight of titanium parts in commercial aircraft structures does not exceed 4 to 6% while for some military aircraft - e.g. where the wing box or bulkhead is totally made of Ti alloys - the level of use reported for engines can be exceeded. Typical airframe applications are hot parts like firewalls, exhaust shrouds or wing leading edges of high Mach planes. High strength forgings are employed for critical components like flap and slat tracks in wings, nacelles, engine pylons, spars and beams or undercarriage parts. Further applications include speed-brake doors, fuselage frames serving as crack stoppers, seamless tubing for hydraulic lines, fasteners, rivets and springs. In helicoptors titanium is used for rotor hubs and blades. In rockets large amounts of Ti alloys go into propulsion and tankage [117, 176].

Outlook

A major drawback of Ti alloys is their high cost. Furthermore, for a lot of components the material waste is substantial. Therefore, efforts have been devoted to increase the fly-to-buy ratio by employing (near) net-shape processes. Casting technology in combination with hot-isostatic pressing has matured and led to both engine and airframe applications [177]. The development of precision forgings - starting in the 1970s - led to new technologies like isothermal/hot-die and closed-die forging. A high amount of cost savings and weight reductions is possible with superplastic forming/diffusion bonding. This technology allows the manufacture of complex parts with improved materials utilisation at reduced assembly cost [178, 179]. Powder metallurgy has been considered for quite a while as an effective tool to reduce component cost [180]. However, the expensive production and compaction processes have meanwhile focussed the PM activities on novel alloy compositions and microstructures which can not be produced by ingot metallurgy. Dispersion strengthening by rare earth additions is one example [181].

Apart from these net-shape processes, further developments stress high strength/stiffness components produced via fibre reinforcement (see section "Metal Matrix Composites"). Substantial efforts are put into increasing the upper use temperature of Ti alloys, with titanium aluminides being most promising (see section "Titanium Aluminides"). Their success is linked to overcoming the poor oxidation behaviour of Ti alloys. Apart from chemistry optimisation, efforts to develop improved oxidation resistant coatings have to be substantially increased.

2.1.2.2 Titanium Aluminides

J. Kumpfert and C.H. Ward

Conventional high-temperature Ti-based alloys have reached a temperature barrier set by their creep and oxidation resistance close to 600 °C. The needs for advanced aerospace systems are primarily higher service temperatures and higher specific strength, at the same time. Currently, the materials most suited to these demanding needs are alloys based on titanium aluminides [182].

The most recent Ti-Al phase diagram is shown in Figure 2.20 [183, 184]. The central portion of this phase diagram was controversial for nearly 40 years before the currently accepted version was constructed. It shows the line compound Al_3Ti as well as the compounds Al_2Ti, TiAl and Ti_3Al which are stable in a more or less narrow range. Besides these ordered intermetallic phases there are two disordered phases on the Ti rich side: the hexagonal α-Ti and the high temperature β phase with a bcc structure known from conventional Ti alloys (see section "Titanium Alloys").

In the following, a brief overview will be given for the Ti_3Al and TiAl based alloys.

Ti_3Al Based Alloys

In the pursuit of ever stronger, hotter and lighter aerospace alloys, titanium alloys with high concentrations of aluminum have been developed. As a group, these alloys are termed the titanium aluminides; they have concentrations of aluminum which exceed roughly 20 at%. With such a high aluminum content, the resulting phases are intermetallic compounds - having an ordered structure. The system based on the compound *Ti_3Al* is the oldest and most comprehensively studied of the titanium aluminides. The first major research effort to develop an alloy based on this compound occurred in the 1950s when higher temperature materials were required for the rapidly developing turbine engine industry [185]. While some promising mechanical properties were reported, interest waned as nickel-based superalloys were developed more rapidly. As turbine engine designs matured, more focus was placed on developing materials which enhanced engine performance through lighter weight and higher operating temperatures; the oil crises of the 1970s spurred this interest even further. Large research efforts were undertaken to develop *Ti_3Al* alloys during this time. By the end of the 1970s the reference alloy for several years to come had been developed, Ti-24Al-11Nb (at%) [186]. This alloy was later improved in its high temperature mechanical properties with the alloy Ti-25Al-10Nb-3V-1Mo in 1982 [187]. Further research efforts have lead to the development of Ti-24.5Al-12.5Nb-1.5Mo, Ti-25Al-17Nb-1Mo and alloys near Ti-25Al-25Nb [188-190]. Numerous reviews are available which discuss the details of the various aspects of α_2 alloy development [182, 190-194].

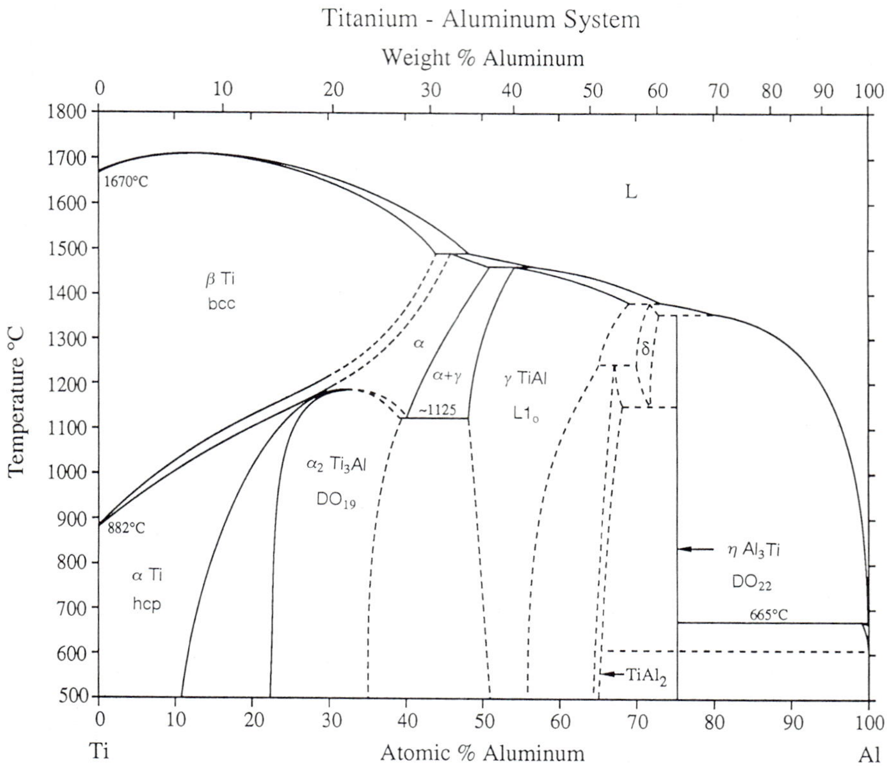

Figure 2.20 The Ti-Al phase diagram [184]

Physical Metallurgy: The room temperature equilibrium phase of Ti_3Al is referred to as the α_2 phase. This phase has an ordered DO_{19} hexagonal crystal structure, $P6_3/mmc$ space group. The single α_2 phase has a relatively large room temperature solubility region for aluminum, extending from approximately 22 to 35 at%, Figure 2.20 [183]. Due to the similarities in microstructure and the alloying response of Ti_3Al alloys to conventional titanium alloys, classification of alloying elements to Ti_3Al by α- and β-stabilizing and neutral elements also remains the same. Niobium has become the most widely used and important element in the development of Ti_3Al alloys. It has proven to be the most effective addition for increasing the inherent low room temperature ductility of binary Ti_3Al. It achieves this by increasing the number of active slip systems and occurrence of cross slip [195]. Once a niobium content of 10 at% is reached, the high temperature β phase is retained to room temperature. The equilibration of the β phase has also been an important step in increasing ductility by relieving strain incompatibilities at α_2 interfaces [196]; in addition, it has also increased the fracture toughness of these alloys.

Molybdenum is the second most widely used alloying element, though it is used in much smaller quantities than niobium - rarely exceeding 2 at%. Molybdenum has two important benefits, one is as a solid solution strengthener, probably within the β phase, which leads to increased high temperature strength and creep resistance. Molybdenum also slows transformation kinetics, allowing increased processing flexibility and microstructural control [197]. Vanadium has also been used in lieu of density penalizing niobium to stabilize the β phase; this, however, reduces the oxidation resistance of the resulting alloy, making it an undesirable addition [187]. Other elements have been added from time to time, the most common being chromium and tantalum, but mixed mechanical property results have kept interest in them from becoming widespread [198].

The phase transformations, equilibrium phases, and morphological stabilities in these alloys are still not completely understood. The requirement that large amounts of alloying elements be used to arrive at alloys with a balance of mechanical properties has made phase characterization much more difficult. Currently, it appears that eleven phases are required to completely describe the phase transformations of this system. The most important of these phases are the four equilibrium phases found in the alloys of interest. These include α_2, β, B2 (an ordered bcc derivative), and the ordered orthorhombic (O) phases. The O phase can be described as a distortion of the α_2 phase's crystal structure in which niobium occupies its own sublattice in addition to those of titanium and aluminum [199, 200]. On a microscopic level of deformation, dislocation behavior in O appears to be similar to that of α_2, and perhaps even more capable of carrying plastic deformation [201]. In alloys containing approximately 10 to 11 at% niobium, the equilibrium phases present are α_2 and β [202]. However, with only small additions of niobium the equilibrium phases become α_2, B2 and O. Further increasing levels of niobium stabilize the O phase at the expense of α_2 [203-205]. By the time 20 at% niobium is added, these alloys are probably more appropriately called O alloys versus α_2 alloys.

Microstructure and Mechanical Properties: Like the conventional titanium alloys, the mechanical properties of alloys based on Ti_3Al are strongly influenced by both alloy composition and microstructure. In fact, as mentioned above, the microstructures of Ti_3Al alloys are completely analogous to those found in conventional titanium alloys and are arrived at through the same processing control concepts.

Strengthening of Ti_3Al alloys is most readily controlled through thermomechanical processing (TMP). The proper combinations of working and heat treatment can increase the strength of a particular alloy by more than 50%, more than can be accomplished through minor alloying changes [206, 207]. An increase in strength is accomplished through the refinement of microstructural scale, leading to a commensurate increase in boundary strengthening mechanisms. As compared to conventionally based high temperature titanium alloys such as IMI 834 and Ti-1100, the Ti_3Al based alloys have the potential for improved strength from room temperature through the anticipated use temperature of approximately 600 °C (Figure 2.21).

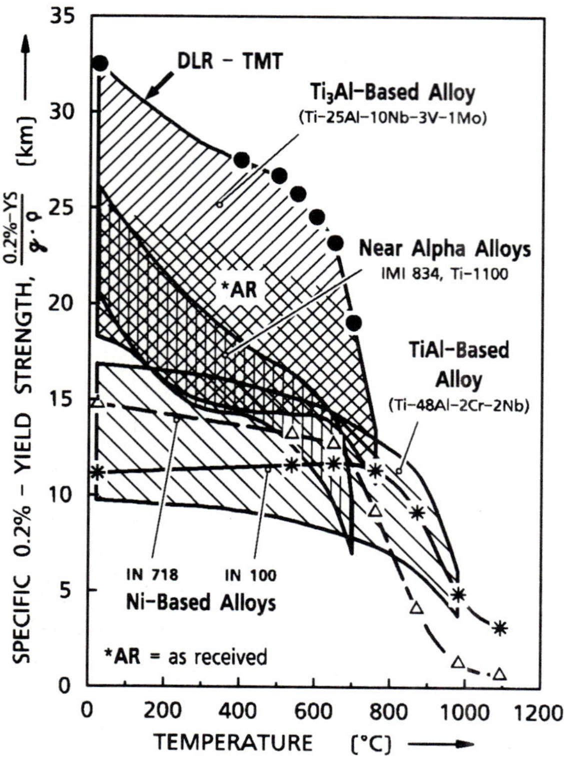

Figure 2.21 Specific yield strength values of the titanium aluminides (Ti-25Al-10Nb-3V-1Mo, Ti-48Al-2Cr-2Nb) in comparison to the near alpha alloys (IMI 834, Ti-1100) and Ni based alloys (IN 718, IN 100)

Additions of niobium, molybdenum and oxygen, among other additives, are known to increase the strength of these alloys [187, 208]. Molybdenum and oxygen are particularly potent strengtheners. However, it is difficult to assign the amount of strengthening due to alloying, as these elements also affect phase stability and transformation kinetics.

For quite some time low room temperature ductility was considered to be the most important barrier to overcome before application. However, this has been largely solved through extensive alloying additions, as described above, and through microstructural manipulation. The most important point to keep in mind when considering room temperature ductility is that these materials fail in a brittle manner. As such, the normal approach of refining grain size to reduce slip length can only be taken so far. On extreme refinement of the microstructure it is possible to embrittle the material, as the yield strength approaches the cleavage strength of the microstructural components. There are a number of factors which are important to the retention of high room temperature ductility. Perhaps foremost is slip dispersal through a basketweave morphology of secondary α_2. This is a much more important factor than the size of the individual laths, Figure 2.22a and b [206]. A basketweave morphology avoids the slip accumulation along a common slip plane that can occur in aligned colony microstructures. A second consideration is the presence of α_2 sideplate formation at prior β grain boundaries. This morphological feature can lead to strain incompatibilities across the prior grain boundaries due to the limitation of active slip planes, Figure 2.22c [209]. Usually connected with this feature is grain boundary α_2 which acts as a fast path for crack propagation [206, 209]. Another benefit to ductility is a recrystallized grain structure, which provides a microstructural unit of deformation at a much smaller scale.

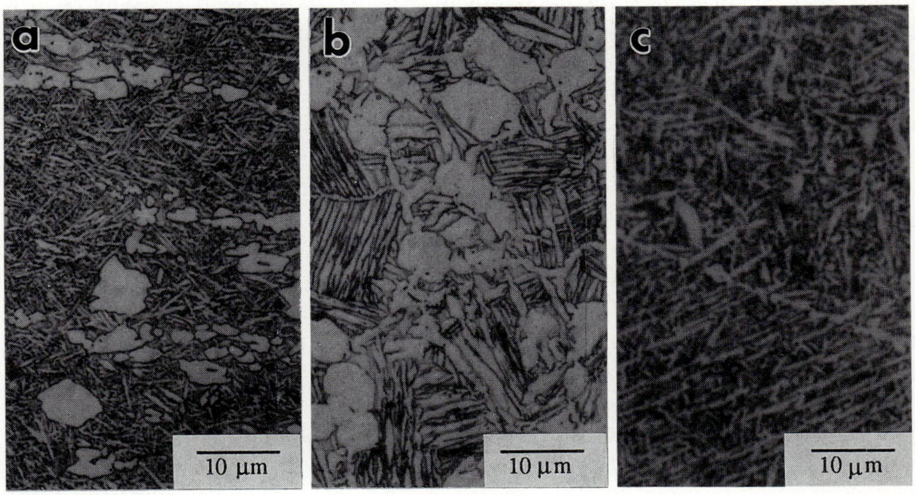

Figure 2.22 a) α_2+ β forged plus direct age at 815 °C. Yield strength 942 MPa, ultimate tensile strength 1097 MPa, plastic elongation at failure 2.7%.
b) α_2+β forged, α_2+β solution treated plus aged at 815 °C. Yield strength 703 MPa, ultimate tensile strength 907 MPa, plastic elogation at failure 1.6%.
c) α_2+β forged, solution treated, air cooled, aged at 815 °C. Yield strength 823 MPa, ultimate tensile strength 950 MPa, plastic elongation at failure 0.8 %.

Attempts have been made to increase the fracture toughness of α_2 alloys by attempting to either modify the plastic deformation capacity of the α_2 phase or increase the volume fraction of the ductile β phase [189, 198]. Of these approaches, the latter has been more successful in maintaining a balance of mechanical properties. This was accomplished although very little additional β phase was stabilized; instead it was found that the O phase was equilibrated. This has led to a number of interesting high niobium content alloys based on the O phase [190]. In contrast to conventional titanium alloys, the fracture toughness of α_2 alloys appears to be less sensitive to microstructure. Values for both Ti-24Al-11Nb and Ti-25Al-10Nb-3V-1Mo are typically at or below $20 MPa\sqrt{m}$ [187, 210-213]. However, there is some evidence that a β processed colony microstructure may significantly increase fracture toughness through crack tortuosity, while other factors such as primary α_2 volume fraction and microstructural texture may also have an influence [214]. The impact resistance of the alloys is extremely low at room temperature, 2-4 J [189, 210]. This value has not been significantly raised by alloying or microstructural manipulation. Although this value rises with temperature, it remains an impediment to the application of these materials.

The creep behavior of α_2 alloys is again similar to that of conventional titanium alloys, with a coarse colony microstructure being most creep resistant at low stresses and all temperatures [215, 216]. The steady state creep rates are lower for the α_2 alloys than for conventional titanium alloys [216]. However, the extent to which α_2 alloys undergo primary creep can be rather large, often approaching and even surpassing 1% strain [215, 216].

The fatigue properties of α_2 alloys have undergone a number of studies which examine behaviour as a function of test parameters [210, 217, 218]. Studies to characterize behaviour as a function of microstructure have shown that similar to conventional titanium alloys, fatigue crack growth rates are lower in coarser Widmanstätten and colony microstructures [213, 219]. As in fracture toughness, it is once again presumed that crack tortuosity leads to the decreased crack growth rates associated with the coarser microstructures.

Oxidation Behavior: While α_2 alloys have a lower oxidation rate than conventional titanium alloys, as measured by thermogravimetic analysis, Figure 2.23, they have a much greater sensitivity to environment in terms of mechanical properties [220, 221]. Like the conventional titanium alloys, α_2 alloys form an "α case" at the surface of a specimen when exposed to air at elevated temperatures. Tensile elongation, creep-rupture life, fatigue crack growth rates, and probably fracture toughness can all be adversely effected by the presence of this brittle phase [218, 220, 221]. While certain additions to α_2 alloys such as niobium appear to improve oxidation resistance, vanadium in particular is very detrimental.

Application: Original applications envisioned for the α_2 class of materials included compressor vanes, blades and disks in gas turbine engines. More recently that list has included impellers and nozzle components. Other applications in airframe components such as skins and ducts have also been considered. To date, this alloy class has not been used due in part to a reluctance by aerospace designers to use

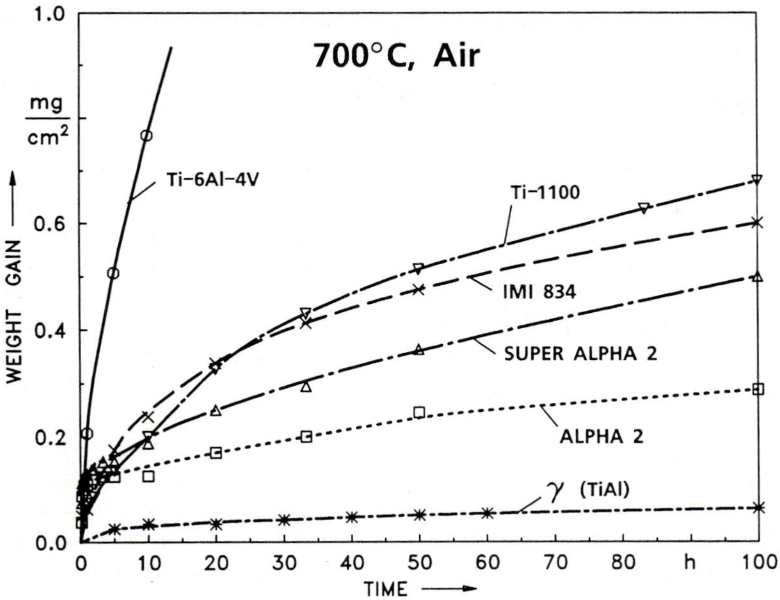

Figure 2.23 Weight-gain measurements of γ TiAl in comparison to Alpha 2 (Ti-24Al-11Nb), Super Alpha 2 (Ti-25Al-10Nb-3V-1Mo), the conventional high temperature Ti alloys IMI 834, Ti-1100 and the most used Ti alloy, Ti-6Al-4V.

a material which does not have the same balance of mechanical properties as traditional aerospace materials. Of the problems yet to overcome, burn resistance, environmental sensitivity and low temperature impact resistance are the most important. To a lesser extent, issues of primary creep response and low temperature fracture toughness also need to be resolved before this material will see large scale use in gas turbine engine components.

TiAl Based Alloys

The development of TiAl based alloys is much more recent than that of those based on Ti_3Al. This is because the initial TiAl base alloys had a much lower ductility at RT than Ti_3Al, causing great difficulty in even casting TiAl ingots. However, the higher oxidation resistance (Figure 2.23), elastic moduli, thermal stability of the microstructure and the lower density were the motives in attracting attention to TiAl [222].

Physical Metallurgy: TiAl, also called the γ phase, has the $L1_0$ superlattice (ordered face centered tetragonal) with an axial ratio of the stoichiometric compound of c/a=1.015. This tetragonality ratio increases to 1.03 with increasing aluminum concentration and decreases to 1.01 with decreasing aluminum. The γ-TiAl phase has a composition range that varies between 48.5 and 66 at% Al depending on temperature, and remains ordered up to its melting point of 1450 °C [223].

Binary TiAl alloys show a direct relation between mechanical properties and aluminum content. The highest elongation is observed at 48 at% Al and the lowest strength values at 51 at% Al. The most important factor, of course, is that these alloys are single phase alloys above 51 at% Al, and two phase alloys below this value. An optimum volume ratio of α_2/γ occurs in the range of 5 to 15% depending on processing and heat treatments [224]. Additionally, it is supposed that decreasing aluminum content leads to increased metallic bonding [225] and to decreased tetragonality of the unit cell [226].

Elongation is also strongly affected by interstitial impurities such as oxygen [227], carbon, nitrogen and boron [228]. For this reason low oxygen contents (≤600 ppm) are required for γ-TiAl alloys. Higher oxygen contents drastically reduce the ductility of the γ phase by decreasing the number of active slip systems. In a two phase alloy ($\alpha_2 + \gamma$) the higher solubility of the α_2 phase is thought to remove the interstitial elements from the γ phase, thereby increasing the ductility of this phase.

At all temperatures the solubility for hydrogen in γ-TiAl is distinctly lower than for Ti_3Al based alloys. However, under extreme conditions such as high temperature and high hydrogen pressure, hydrogen also diffuses into the γ phase, which can contribute to lower room temperature ductility [229]. A conclusive influence of hydrogen on the mechanical properties has not been seen so far. Nevertheless, this is a major point of interest, because these alloys are considered for application in hydrogen-fueled hypersonic aerospace vehicles.

Alloy Development and Microstructures: The multicomponent alloys that have been developed so far can be grouped into (1) single phase γ and (2) two phase α_2+γ alloys [230],

1. Ti - (50-52) Al - (0-2) n; n = Nb, Ta, W
2. Ti - (44-49) Al - (1-3) m - (1-4) n - (0-1) p
   m = Cr, V, Mn, Si; n = Nb, Ta, W, Mo; p = Si, C, B, N, Ni, Fe

In early development attention was focused on the single phase alloys due to their high oxidation resistance and their low hydrogen absorption. However, their low room temperature ductility led the attention to the two phase alloys. Additions of the m elements have been reported to increase the ductility of the γ alloys. Cr additions in particular substantially increase ductility in the range of 46-50 at% Al [231]. On the other hand, m elements in general reduce the oxidation resistance of γ alloys. To compensate for this, n elements are added [226]. p elements have various effects, wherein C and N are known to improve creep resistance [232],

Si, B, Ni and Fe to decrease the melt viscosity [233], which is important for near net shape considerations.

In addition to alloying, a second metallurgical treatment has been applied to improve the critical mechanical properties of this family of intermetallic alloys - microstructural control by various heat treatments or thermomechanical processing (TMP) [234]. The cast microstructures in 46-50 at% Al alloys are normally fully lamellar, containing alternate layers of γ and α_2 plates of a defined crystallographic orientation [235] (Figure 2.24).

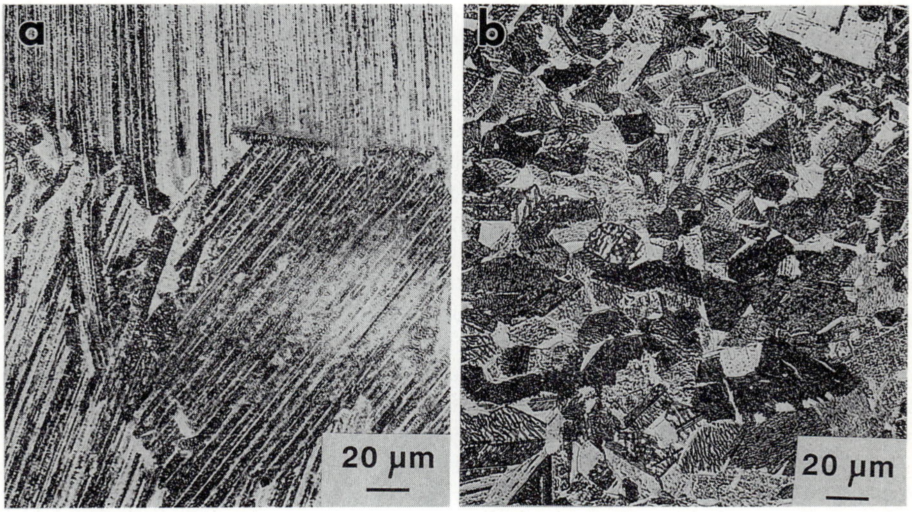

Figure 2.24 Microstructure of Ti-48Al-2Cr-2Nb (a) fully lamellar, (b) duplex

This microstructure is obtained after heat treatments above the α transus and is referred to as the fully transformed structure. The grains here are typically larger than 500 µm in diameter due the high diffusion and grain growth rates in the single α phase field [226].

Further microstructural control is achieved by hot working followed by a two step heat treatment (TMP). Hot working by forging is usually carried out isothermally, at a low strain rate and near the eutectoid temperature T_e (Figure 2.20). The first heat treatment employed is to solutionize and/or recrystallize the hot worked material in either the α or $\alpha + \gamma$ phase field. The second step is to stabilize the microstructure and phase compositions by heat treating at temperatures in the $\alpha_2 + \gamma$ phase field. TMP refines the entire grain size if the process is controlled such that recrystallization nucleation events are maximized while the growth rate is minimized. By varying the solution heat treatment temperature it is possible to produce a wide range of grain morphologies [236]. Decreasing solution heat treatment temperatures from just below the α transus to temperatures 50 °C above T_c results in nearly-lamellar, duplex and near gamma structures. The nearly-lamellar microstructure consists of finely dispersed gamma grains in an intermediate size lamellar-grain matrix, while the near gamma microstructure consists of predominantly equiaxed γ grains with finely dispersed α_2 structures. The finest microstructure is represented by the duplex structure with a fine mixture of equiaxed γ grains and lamellar grains [222]. The pronounced differences between the fully lamellar and the duplex microstructure have a substantial influence on mechanical behavior.

Mechanical Properties: Microstructure is the result of the alloying composition as well as the processing route. The mechanical properties vary over a wide range depending on the microstructure. The yield strength values, represented in Figure 2.21 as specific yield strength values, indicate the potential of the γ alloys. It can be concluded from these results and from results of oxidation resistance in Ti_3Al and TiAl based alloys [237], that the useful range of γ alloys starts at temperatures above 650 °C due to their higher thermal stability. The highest strength values and reasonable ductility are obtained from nearly-lamellar microstructures with intermediate size grains, resulting from solution heat treatment temperatures just below the α transus [230]. The much larger grain size of the fully lamellar microstructure leads to markedly lower ductility and thus ultimate tensile strength. For further reductions in the grain size, it is necessary to perform the solution heat treatment at lower temperatures (in the α + γ field) resulting in higher volume fractions of equiaxed γ grains. These fine microstructures, even duplex or near gamma, show increased fracture elongation but lower strength values as compared to the intermediate size lamellar microstructures [222]. However, single phase γ alloys with a fine microstructure comparable to those of the fine duplex alloys show very little fracture elongation [238]. Thus grain size is not the only factor in enhancing ductility. Rather, a certain volume fraction of the α_2 phase is advantageous for the mechanical properties, probably due to its beneficial influence on interstitial elements in the γ phase, as mentioned above.

In the fully lamellar microstructure fracture elongation also appears to depend on the orientation of the lamellae. The yield strength is high for orientations of the lamellae perpendicular or parallel to the load axis, while it shows a minimum for intermediate orientations. A room temperature tensile elongation of 12.8% has been observed by a Japanese group [239] for a specimen with the tensile axis inclined at 51° to the lamellar boundaries. A German group found for a similar test only fracture strain values as low as 1.05% [240].

A positive temperature dependence of the yield strength has been observed in single-phase γ alloys with a large grain size (650 μm) and a high aluminum content (≥52 at%) [241]. However, no anomaly was observed in polycrystalline materials with a fine microstructure processed by rapid solidification [242]. The ductile-brittle transition temperature (DBTT) shifts for γ alloys in the range between 550 °C and 750 °C, depending on the microstructure. A fine duplex microstructure shows a lower DBTT than a coarse fully transformed structure [243].

Creep properties of fully lamellar structures show significant improvements over duplex microstructures at temperatures higher than 650 °C. This is believed to be the result of reduced grain boundary diffusion of this coarse microstructure and the effect of the "hard" α_2 phase which acts as a reinforcing phase, while the deformation takes place, in almost all cases, in the "soft" γ phase [244]. The effect of grain size on the creep behavior of the single phase alloys, however, is very sensitive to the Al concentration. This effect only becomes noticeable at Al concentrations below 52 at% Al [245]. The highest creep resistance of binary two phase alloys was observed in Ti-48Al, which had serrated grain boundaries. Creep resistance decreases with aluminum content in the range from 48 at% Al to 44 at% Al, which is due to a reduction in the serrated nature of the lamellar grain boundaries [246]. Moreover, it has been demonstrated that the creep rupture stress in a two phase alloy is very sensitive to oxygen at 900 °C [244]. This has been attributed to initiation and growth of cracks from the surface (creep crack growth) as well as the possibility that oxygen diffuses during this long term test toward the crack tip, accelerating crack growth. Thus, a protective coating needs to be found for high temperature applications.

Fracture toughness is also very sensitive to microstructure. Again, the fully transformed microstructure shows significantly better values than a fine duplex microstructure [222]. In as-cast structures, Ti-48Al shows the highest fracture toughness values [247]. Lower Al content leads to well defined, smooth grain boundaries, as described above, which are less fracture tolerant. At higher Al content, interdendritic γ forms, which provides an easy transgranular fracture path.

To date very little information is available on fatigue. However, a fine duplex microstructure seems to be favorable for high cycle fatigue at temperatures lower than 800 °C and higher amplitude-load conditions, while for the fully lamellar microstructure the fatigue life appears to be lower at higher temperatures [248].

To summarize, the fully lamellar microstructure has the potential to develop a balanced set of properties if its grain size can be reduced. However, the grain growth rates above the α transus are currently too fast to allow grain refinements by thermomechanical processing.

2.1.3 Superalloys and Coatings

K. Fritscher, M. Peters, H.-J. Rätzer-Scheibe and U. Schulz

2.1.3.1 Superalloys

More than 50% of the weight of modern aircraft gas turbine engines comes from superalloys. Superalloys represent a class of metallic materials which exhibit an excellent combination of long-time strength at temperatures above 650 °C and resistance to corrosion and erosion. Since there are no other alloys with equally high temperature capability, the term "superalloys" - first used in the late forties - is more than justified. Depending on the prime alloying element, superalloys can be divided into three main classes: Fe-base, Ni-base and Co-base alloys. From these, the Ni-base superalloys are by far the most important class, so that they are often referred to simply as superalloys. There is no other alloy system, where the upper use temperature of components is so close to the melting point of the base metal as for Ni-base superalloys. Therefore, their domains are the hottest and most highly stressed parts in gas turbine engines [249].

Microstructure and Properties

The key microstructural feature responsible for the outstanding high temperature properties of Ni-base superalloys is the ordered face-centered cubic γ' intermetallic phase Ni_3 (Al,Ti). (For a few alloys Nb substitutes for some of the Al and Ti to form the γ'' intermetallic phase Ni_3Nb.) Due to its strong coherency with the γ matrix, γ' precipitates homogeneously, in high volume fractions often exceeding 60%. Thus being stable even after long-time exposure at high temperatures allows using these alloys at temperatures up to 85% of the melting point. Furthermore, Ni-base superalloys become stronger with temperature, peaking in strength at about 600 to 800 °C. This is due to the ordered nature of γ', forcing dislocations to move in pairs, leading to extra antiphase boundary hardening. The γ matrix is furthermore solid-solution strengthened by Ta, W and Mo. Carbides, mostly formed by Cr, Mo, Ti, Ta and Nb, but also minor additions of B and Zr, strengthen the grain boundaries.

Alloys and New Developments

Table 2.15 shows the chemical compositions of commercial and newly developed Ni-base superalloys. First there were the wrought superalloys. They were, however, limited in alloying additions and thus strength since forging becomes difficult once the γ' volume fraction exceeds 45%. Therefore, wrought alloys reveal moderate elevated-temperature strength and are used up to about 650 °C. Consequently, cast alloys were developed with higher elevated temperature strength due to increased levels of Al and Ti, raising the γ' volume fraction, and reduced Cr or increased Co contents, raising the γ' solvus temperature. Furthermore, improved creep resistance is inherent to the coarser cast structure, which is, however, also more brittle. Often small amounts of Hf are added to regain ductility. Hot-isostatic pressing (HIP) was routinely employed to reduce porosity of the cast structure and increase creep-rupture and fatigue strength [250]. The reduced Cr content deteriorates the hot corrosion resistance. Therefore, protective coatings have to be developed to make ultimate use of the high temperature capability of cast Ni-base superalloys (see section "Coatings").

	Ni	Cr	Co	Mo	W	Nb	Ti	Al	Fe	C	Zr	Ta	Hf	B	Others
Wrought Ni-base superalloys															
Astroloy	Bal	15	17	5	-	-	3.5	4	-	0.03	0.04	-	-	0.02	
Hastelloy X	Bal	22	1.5	9	0.6	-	-	2	15.8	0.15	-	-	-	-	
Inconel 718	Bal	19	-	3	-	5.1	0.9	0.5	18.5	0.08	-	-	-	-	0.15 Cu
Nimonic 80A	Bal	19.5	1	-	-	-	2.25	1.4	1.5	0.05	-	-	-	-	0.1 Cu
René 41	Bal	19	11	10	-	-	3.1	1.5	<0.3	0.09	-	-	-	0.01	
René 95	Bal	14	8	3.5	3.5	3.5	2.5	3.5	<0.3	0.16	0.05	-	-	0.01	
Udimet 500	Bal	19	19	4	-	-	3	3	4	0.08	-	-	-	0.01	
Udimet 700	Bal	15	18.5	5	-	-	3.4	4.3	<1	0.07	-	-	-	0.03	
Waspaloy	Bal	19.5	13.5	4.3	-	-	3	1.4	2	0.07	0.09	-	-	0.01	
Cast Ni-base superalloys															
B 1900	Bal	8	10	6	-	-	1	6	-	-	0.05	4.3	1.3	0.01	
IN-100	Bal	10	15	3	-	-	4.7	5.5	<0.6	0.15	0.06	-	-	0.01	1 V
IN-713LC	Bal	12	-	4.5	-	2	0.6	6	-	0.05	0.1	-	-	0.01	
IN-738LC	Bal	16	8.5	1.75	2.6	0.9	3.4	3.4	-	0.11	0.05	1.75	-	0.01	
DS/SX - Directionally solidified/Single crystal Ni-base superalloys															
AM1	Bal	7.5	6.5	2	5.5	-	1.2	5.3	-	-	-	8	-	-	
CMSX-2	Bal	8	4.6	0.6	7.9	-	0.9	5.6	-	-	-	5.8	-	-	
CMSX-3	Bal	7.5	4	0.5	7.5	-	0.9	5.5	-	-	-	6	0.1	-	
CMSX-4	Bal	6.5	9.5	0.6	6.4	-	1	5.6	-	-	-	6.5	0.1	-	3 Re
CMSX-6	Bal	10	5	3	-	-	4.7	4.8	-	-	-	2	0.1	-	
MC2	Bal	8	5	2	8	-	2	5	-	-	-	6	-	-	
MERL 76	Bal	12	19	2.8	-	1.2	4.2	4.9	-	0.01	0.04	-	0.3	0.02	
NASA IR 100	Bal	9	-	1	10.5	-	1.2	5.75	-	-	-	3.3	-	-	
René N4	Bal	9	7.5	1.5	6	0.5	4.2	3.7	-	-	-	4	-	-	
SRR 99	Bal	8.5	5	-	9.5	-	2.2	5.5	-	-	-	2.8	-	-	
MAR-M 002	Bal	8	10	-	10	-	1.5	5.5	-	0.15	0.03	2.6	1.5	0.02	
MAR-M 200	Bal	9	10	-	12.5	1	2	5	-	-	0.05	-	2	0.01	
MAR-M 246	Bal	9	10	2.5	10	-	1.5	5.5	-	-	0.05	1.5	-	0.01	
MAR-M 247	Bal	8.3	10	0.7	10	-	1.0	5.5	-	0.15	0.05	3.0	1.5	0.02	
PWA 1444	Bal	10	5	-	4	-	1.5	5.0	-	-	-	12	-	-	
PWA 1480	Bal	10	5	-	4	-	1.5	5	-	-	-	12	-	-	
PWA 1484	Bal	5	10	2	6	-	-	5.6	-	-	-	8.7	0.1	-	3 Re
PM/ODS - Powder metallurgy/Oxide dispersion strengthened Ni-base superalloys															
N 18	Bal	11.5	15.7	6.5	-	-	4.35	4.35	-	0.02	0.03	-	0.5	0.02	
MA 754	Bal	20	-	-	-	-	0.5	0.3	-	0.05	-	-	-	-	0.6 Y_2O_3
MA 6000	Bal	15	-	2	4	-	2.5	4.5	-	0.05	0.15	2	-	0.01	1.1 Y_2O_3

Table 2.15 Chemical composition of Ni-base superalloys

Since failure at high-temperature service usually occurred at grain boundaries, carbides were added to strengthen grain boundaries. However, long-term exposure led to the formation of embrittling grain boundary phases with a deteriorating effect on stress-rupture properties. These observations initiated two parallel developments. Firstly, air melting changed to vacuum (re)melting in the late fifties to reduce the amount of chemical segregations, brittle intermetallics or low melting point eutectics, all known to deteriorate alloy properties. Still today novel vacuum melting techniques like electron-beam cold hearth refining or plasma arc melting are being evaluated to further improve the cleanness of the superalloys. Secondly, directional solidification (DS) was used to produce columnar-grained superalloys without weak transverse grain boundaries. From here it was only a logical step to single crystal (SC) technology in the early eighties [251]. This development not only led to further increased creep strength; the elimination of the unnecessary grain boundary strengtheners such as C, Hf, Zr or B resulted in an appreciable increase of the incipient melting point of the alloy, thus increasing the solvus temperatures of strengthening phases, and simultaneously the useful alloy strength limit. Today, the most advanced SC alloys can be used at temperatures up to 1100 °C [252].

Contrary to Ti alloys, powder metallurgy (PM) processing techniques are meanwhile routinely employed to produce superalloy components. Compared to conventional ingot metallurgy, the PM route offers several benefits. Apart from the net-shape capability with its potential for materials and cost savings, the PM route also allows an increase in the amounts of alloying elements, leading to higher-strength alloys. Furthermore, the increased microstructural and chemical homogeneity inherent in this technology avoids severe macrosegregation with the risk of ingot cracking. However, care has to be taken to produce clean powders, since defects and inclusions are potential crack nucleation sites, which might lead to catastrophic failures. Efforts include the use of advanced melting techniques to start with cleaner material. Furthermore, emphasis is placed on finer and cleaner powders as well as new consolidation techniques like metal injection molding and spray deposition [253].

A lot of efforts have been made to use PM for dispersion hardening of superalloys by introducing fine dispersions of refractory oxides, nitrides or carbides, which can be stable up to the melting point of the alloy. The general problem was to generate a sufficiently fine and homogeneous distribution of these ceramic particles. Mechanical alloying (MA) turned out to be a successful approach. In a solid-state process the powdered constituents are mixed in high-energy ball mills in a way that they not only fracture, but with help of process control agents also reweld, and true alloying occurs. For dispersion fine inert oxides - usually Y_2O_3 - are included to provide oxide-dispersion strengthening (ODS). The compacted MA superalloys achieve higher strength at temperatures above 1000 °C than DS and SC alloys [254].

Ni aluminides based on the intermetallic compounds Ni_3Al [255] and NiAl [256] have recently become the subject of large development programmes. These were initiated by the findings of Aoki and Isumi [257], who were able to ductilize the essentially brittle γ' phase by adding small amounts of B to Ni_3Al. Brittleness is due to the ordered nature of γ' Ni_3Al ($L1_2$ crystal structure) and β NiAl (B2 crystal structure; ordered bcc) and has been the reason for disregarding the intermetallics for use as structural materials. Overcoming the low temperature brittleness offers significant advantages like reduced density and improved oxidation resistance compared to conventional superalloys. Intermetallic alloys have been intensively investigated, but so far no semi-commercial alloy has been announced with the required balance of ductility, creep resistance, toughness and strength.

Applications

The gas turbine aeroengine is the domain of Ni-base superalloys. They are used at temperatures above about 500 °C, where poor oxidation resistance limits the use of the lighter Ti alloys (see "Titanium Alloys"). Main applications are turbine discs, blades and vanes. Furthermore, they are selected as material for high-temperature fasteners, shafts, casings, burner cans, afterburners and thrust reversers. The high-strength, fatigue resistant forging alloys are used up to temperatures of 650 °C. More highly alloyed cast polycrystalline superalloys are used up to about 1000 °C, while DS and SC technology pushed this limit by another 100 °C. At the maximum use temperature, ODS superalloys show the highest overall strength levels.

2.1.3.2 High Temperature Corrosion

There are several reviews dealing with the high temperature corrosion occuring in boilers, combustion turbine engines and other types of directly fired energy conversion systems which is caused by the combustion gases [175, 258, 259]. The following chapters will concentrate on high temperature corrosion of superalloys and coatings used in aeroengines [260].

Corrosive Conditions in Aircraft Engines

Corrosion is normally not expected to occur in aircraft engines since both intake air on cruise and combustion fuel (nominally less than 0.01% S) seem to be polluted to an uncritically low extent. However, turbines on aircraft operating near the seacoast or close to the ocean surface can ingest from it sea salt (main constituents: 68% NaCl, 14% $MgSO_4$, 11% Na_2SO_4) and aerosols. Potential corrodants can also be transmitted by continental atmospheres if industrial emissions, such as sulphur dioxide and nitrogen oxides, are present. All pollutants interact with residual fuel contaminants (sulphur and sodium compounds) during combustion, whereby gaseous species such as NaCl, NaOH, HCl, and Na_2SO_4 are formed. In addition to the sodium sulphate directly ingested as constituent of sea salt with the turbine intake air, Na_2SO_4 can be produced by a reaction between NaCl or Na, O_2 and SO_2. If molten Na_2SO_4 is deposited on turbine components, it will cause an accelerated corrosion called 'hot corrosion'.

Condensation onset temperatures (dew point) of Na_2SO_4 in flame environments were calculated as a function of sulphur content in fuel and sea salt concentration in intake air, on the assumption that reactions forming Na_2SO_4 are in chemical equilibrium [261]. The results are plotted in Figure 2.25 for various sea salt concentrations. For a given fuel/oxidant ratio the condensation temperature varies strongly as a function of sulphur content at low sulphur concentrations, but is practically independent at high levels.

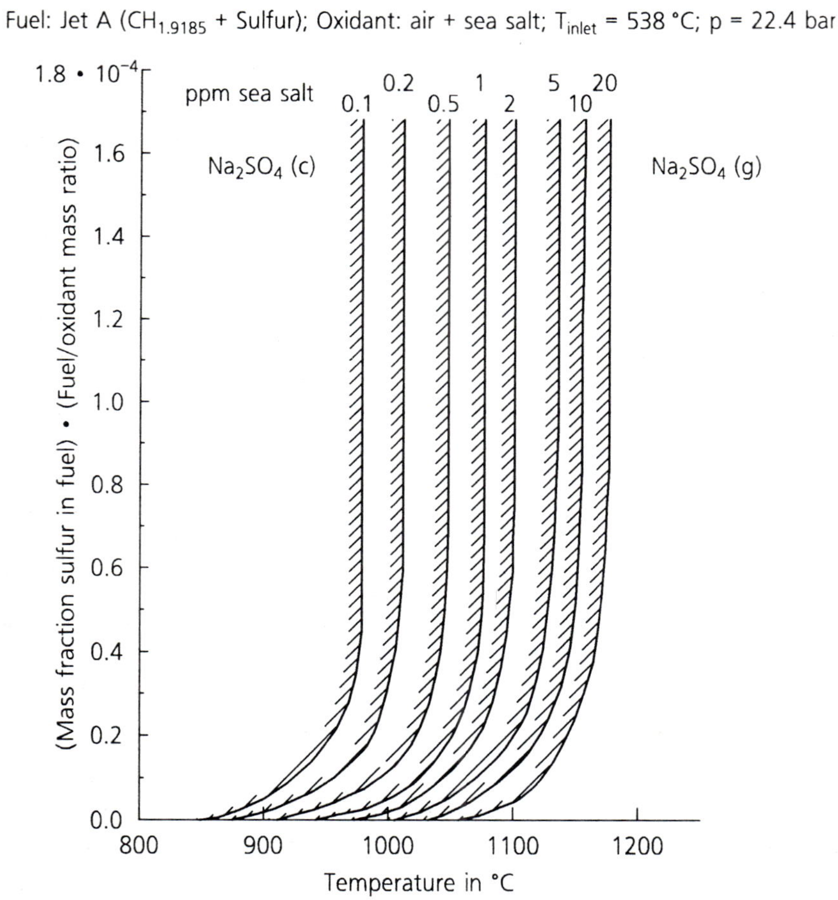

Figure 2.25 Condensation temperature of Na_2SO_4 as a function of sulphur content in fuel and of sea salt concentration in combustion gas for an aircraft turbine engine flame (g=gaseous, c=condensed) [261]

Condensation of Na_2SO_4 can only occur at conditions which correspond to the range on the left of the equilibrium dew point curves. Conditions for hot corrosion are really satisfied only above the melting point (MP) of Na_2SO_4 at 884 °C where the deposited salt is molten. For instance, thermodynamic calculations indicate that Na_2SO_4 is not expected to condense on turbine components exposed at 1100 °C to flames combusting fuel with 0.04% S at an air/fuel ratio of 20 (corresponding to a value on the ordinate of $0.2 \cdot 10^{-4}$) and contaminated with 0.5 ppm sea salt. If the sea salt concentration increases to 10 ppm at otherwise the same conditions, condensation of sodium sulphate could occur. This example demonstrates how sensitively hot corrosion depends on minute variations in the sea salt contamination of air.

Oxidation and Corrosion Mechanisms

High Temperature Oxidation: The degradation of alloys is caused by the reaction of metals with the environment. Heat-resistant alloys are however protected by the formation of dense oxide layers having extremely low growth rates and excellent adhesion. Three main categories of passive oxide forming alloys have commonly been established: chromia, alumina, and silica formers. Ni-base chromia formers contain at least 15% Cr and less than 3% Al, alumina formers more than 5% Al and less than 15% Cr. Ni-base silica formers are too brittle to be applicable for structural components. However, for coating applications they perform well. In essence, high temperature oxidation of protective oxide-forming alloys is not a degradation process, but a highly appreciated process protecting against surface decay.

Among these oxides, alumina, the most stable one, is provided for high temperature protection above 1000 °C, whereas chromia protection is limited to 1000 °C due to progressing evaporative loss of CrO_3. Oxides of other alloying elements are unstable at high temperatures and melt, e.g. certain vanadium oxides, or evaporate, e.g. molybdenum and tungsten oxide. Silica protection may be applicable to high temperatures exceeding 1200 °C but usage of silica-forming coatings may be restricted to 900 °C due to excessive coating/substrate interactions.

In addition to the vaporization of oxides, degradation of the oxide scale can also occur by spallation, which usually results from a mismatch in the coefficients of thermal expansion of the metal and the oxide during thermal cycling. Improvement in scale adherence of oxides may be obtained by small additions of reactive elements, such as Y and Hf.

Chloride-Induced Corrosion: The mechanism by which chlorine contamination accelerates oxidation rates has only partially been established. Figure 2.26 shows a possible mechanism of attack for an oxide scale exposed to gaseous NaCl in an oxygen-nitrogen atmosphere [262]. Alumina scales are more resistant to NaCl attack than chromia scales. The NaCl vapour can react with chromia in an oxygen environment to form volatile products, such as Na_2CrO_4 and CrO_2Cl_2, which disrupt the protective scales.

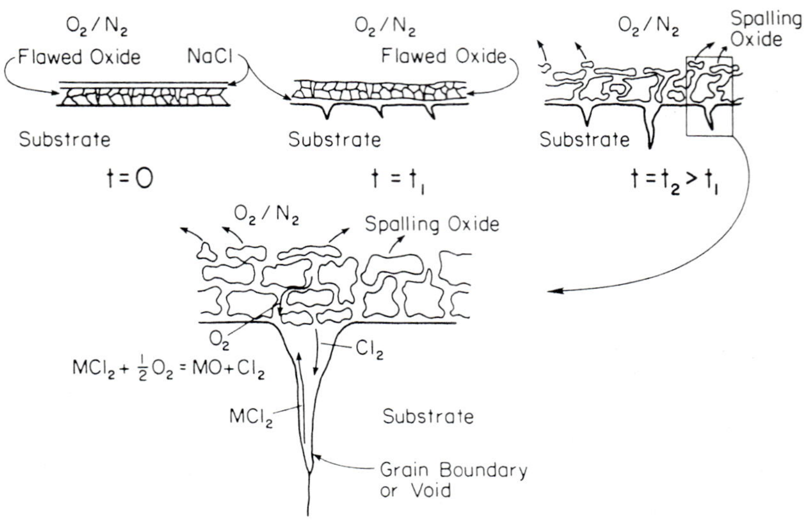

Figure 2.26 Schematic illustration of the mechanism of NaCl attack [262]

The flawed oxide accelerates the transport of gaseous species to the scale-metal interface. Severe disruption and uplifting of the oxide scale occur due to the formation of volatile metal chlorides at the scale-metal interface, whereby formation of $CrCl_2$, $FeCl_2$, and $CoCl_2$ is specially favoured. The volatile MCl_2 can react at higher partial pressures of O_2, producing MO and recycling Cl_2 back to the substrate surface. These processes favour oxide spallation, void formation, and intergranular attack.

Hot Corrosion: If Na_2SO_4 is deposited above its melting point (884 °C) on the surface of materials, the molten sulphate can cause hot corrosion. Corrosion conditions are modified by NaCl-Na_2SO_4 deposits, because, compared with pure Na_2SO_4, the melting point of the salt mixture is lowered by the content of NaCl (MP 742 °C). Some authors [263, 264] proposed a mechanism of high temperature (type I) hot corrosion based on dissolution of the protective oxide on the metal, and a reprecipitation of the oxide as nonprotective particles in the fused salt film (Figure 2.27). Metal oxide dissolves by acidic or basic fluxing at the oxide/salt interface and migrates down a concentration gradient to a region of lower solubility, where MO precipitates. The solubility gradient is a consequence of a Na_2O activity gradient. The result of these processes is a porous layer which no longer effectively protects the alloy.

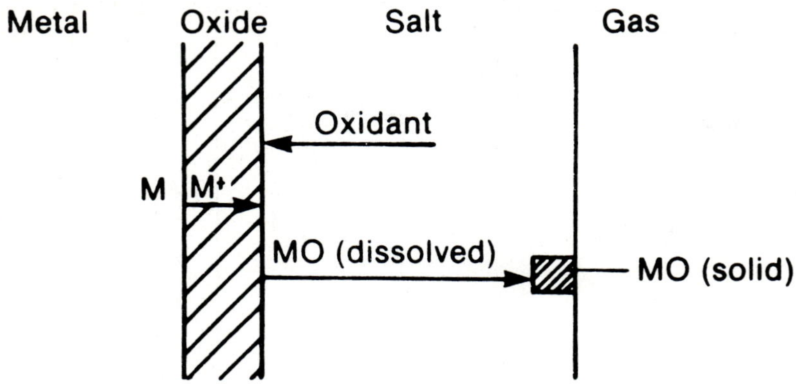

Figure 2.27 Schematic diagram for fluxing mechanism of protective oxides by a fused salt film [265]
Reprinted with permission from the NACE. Copyright NACE.

In the case of sulphate-induced hot corrosion an additional attack usually occurs by sulphidation. In this process Na_2SO_4 is a source of sulphur which diffuses into the metal to form sulphides beneath the oxide scale. Thus a zone of metal sulphides is introduced in the alloy and advances in front of accelerated oxidation (Figure 2.30).

The solubility of an oxide depends on the acidity or basicity of the Na_2SO_4 melt (Figure 2.28). The acidic or basic character of the Na_2SO_4 melt is usually determined by the concentrations of minority species such as O^{2-}, SO_3 and Na_2O. In basic dissolution, solubility decreases with decreasing O^{2-} activity (increasing p_{SO_3}), whereas in acidic dissolution, solubility increases with decreasing O^{2-} activity (increasing p_{SO_3}). According to the conditions of the melt/gas system, the oxide solubility curves predict whether basic or acidic fluxing can occur. In particular, the solubility minima characterise the range where the oxides are most corrosion resistant.

Molten Na_2SO_4 deposits on turbine components usually have an acid character (pO^{2-}-values between about 12 and 14) where the chromia solubility has a minimum. This coincidence may explain the superior type I hot corrosion resistance of chromia forming alloys in gas turbines.

Under special conditions so-called low temperature (type II) hot corrosion can occur in gas turbines between about 600 to 800 °C. This type of corrosion, however, is more prevalent in marine and industrial gas turbines. Here, at temperatures below the melting point of Na_2SO_4, rapid attack is only possible when a liquid sulphate mixture is formed. The mechanism of attack has been described for some Co-base alloys, because particularly CoCrAlY coatings exhibited rapid degrada-

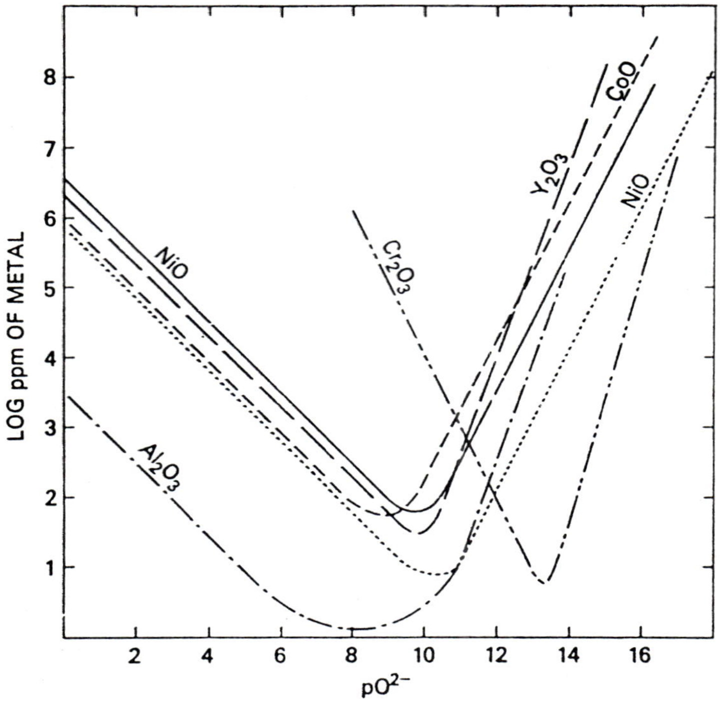

Figure 2.28 Solubility diagrams of some metal oxides as a function of acidity (defined as pO^{2-} = - log $a_{O^{2-}}$) in Na_2SO_4 at p_{O_2} = 1 bar and 927 °C [266]

tion in gas turbines operating in marine environments at lower temperatures. This corrosion behaviour has been attributed to the formation of a low melting Na_2SO_4-$CoSO_4$ eutectic (MP 565 °C) [267].

Testing Methods

Testing in engines or in high pressure burner rigs is expensive and thus a need exists to establish simple and reliable methods of determining hot temperature corrosion resistance of candidate materials for gas turbine applications.

Kinetics of oxidation and corrosion of metals exposed to the reactive atmospheres can be investigated by thermogravimetry [175]. The accelerated hot corrosion attack, e.g. induced by fused Na_2SO_4, can be tested with the same procedure when the specimen is coated with a Na_2SO_4 deposit before it is inserted in the thermobalance.

A simple way to investigate hot corrosion is the so-called crucible test. Thereby specimens are completely or partially submerged in fused salt, such as Na_2SO_4 or Na_2SO_4-NaCl mixtures, at elevated temperatures. Gases may be bubbled through the melt or passed over it to provide oxidizing or reducing environments. Further variations may include alternately dipping and removing the specimen from the melt. A similar procedure is packing test pieces in synthetic ash. Because of their simplicity and because the attack is rapid and severe, the crucible test as well as testing in ash have found use for screening a series of candidate alloys.

Since molten salts are electrolytes, hot corrosion is really an electrochemical phenomenon. In electrochemical tests samples are exposed to an environment similar to that of the crucible test. The experimental arrangement involves an electrochemical cell comprised of a fused salt as the electrolyte, a reference electrode, a working electrode (specimen), and auxiliary electrodes [268]. The purpose of the experiment may be, besides determining corrosion resistance and incubation period, to investigate corrosion reactions and mechanisms more fundamentally.

Burner rigs are frequently used to evaluate the corrosion resistance of alloys for application in gas turbines because they permit materials testing at high gas temperatures (550 °C to 1650 °C) and velocities (Mach 0.3 to 0.8) [260]. Hence, ignoring the pressure conditions of the gas stream, service conditions in gas turbines can be simulated more realistically. High velocity burner rigs can be applied to oxidation or corrosion testing under isothermal as well as cyclic conditions [268]. For testing hot corrosion, a salt solution can be injected into the combustion chamber. To accelerate corrosive attack, contaminations such as S, V, Na, and Cl can also be added to the fuel as soluble compounds.

The extent of corrosion is characterized by determination of weight loss and depth of penetration of the reaction products. Since hot corrosion often involves highly localised attack and because of the frequent occurrence of oxide and sulphide precipitates beneath the metal-scale interface, penetration measurements on a metallographically prepared cross section are usually preferred.

Base Alloys and Their Problems

Strength increases of nickel base superalloys required for modern aircraft gas turbines caused decreases in chromium contents and increases in the contents of aluminium, molybdenum, and tungsten (Table 2.15). The good oxidation resistance in this class of alloys is attributed to a scale mainly consisting of Al_2O_3 based on approximately 5.5% aluminium and less than 10% chromium present in the alloy.

The oxidation resistance of some Al_2O_3-forming Ni-base superalloys was investigated in cyclic burner rig tests at 1100 °C, including the single crystal alloys SRR99, CMSX-2, CMSX-6 and the directionally solidified alloy MAR-M002 [269]. All of these high-strength alloys have preferentially been used as blade and vane materials in aircraft gas turbines. MAR-M002 has the best oxidation resistance due to 1.5% hafnium which forms a good adherent oxide layer. The alloys SRR99, CMSX-2, and CMSX-6 show similar moderate oxidation characteristics apparently due to similar chromium, cobalt, and aluminium concentrations. The different, but also relatively low, molybdenum and tungsten contents of these alloys have no significant effect on oxidation resistance.

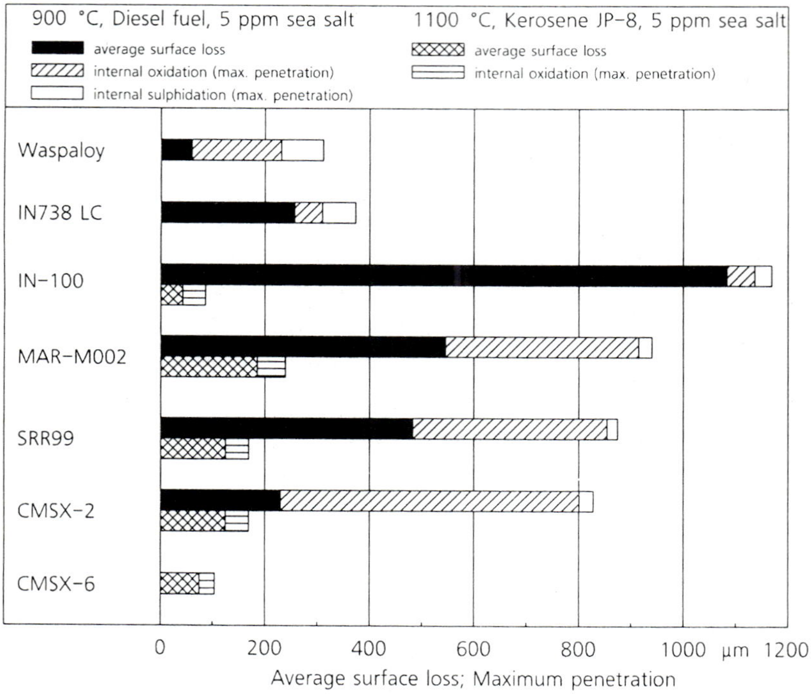

Figure 2.29 Results of metallographic examination for corrosion testing in a burner rig at 900 °C and 1100 °C for 100 h under sea salt addition

The typical airfoil materials usually forming Al_2O_3 as surface oxide are much less affected by gaseous chlorides at temperatures of 700 to 900 °C than Cr_2O_3-forming superalloys which are particularly susceptible to attack in the presence of NaCl [262]. But above 900 °C alumina formers are also more heavily attacked because reactions with NaCl may cause the restructuring of the Al_2O_3 film.

The effect of chloride-induced corrosion could be confirmed in burner rig tests performed with sea salt contaminated combustion gas [269]. Data of the corrosion attack are shown in Figure 2.29 for some Ni-base superalloys cyclically tested at 1100 °C for 100 hours. The cast alloy IN-100 and the single crystal alloy

CMSX-6 proved to be the most corrosion resistant alloys. Apart from the cobalt concentration, the composition of these two alloys is nearly identical. The directionally solidified MAR-M002 and the two single crystal alloys SRR99 and CMSX-2 containing comparable chromium and tungsten concentrations show a similar corrosive attack. Regarding the MAR-M002, the benefits of hafnium observed in the oxidation test are obviously lost in the presence of salt contamination.

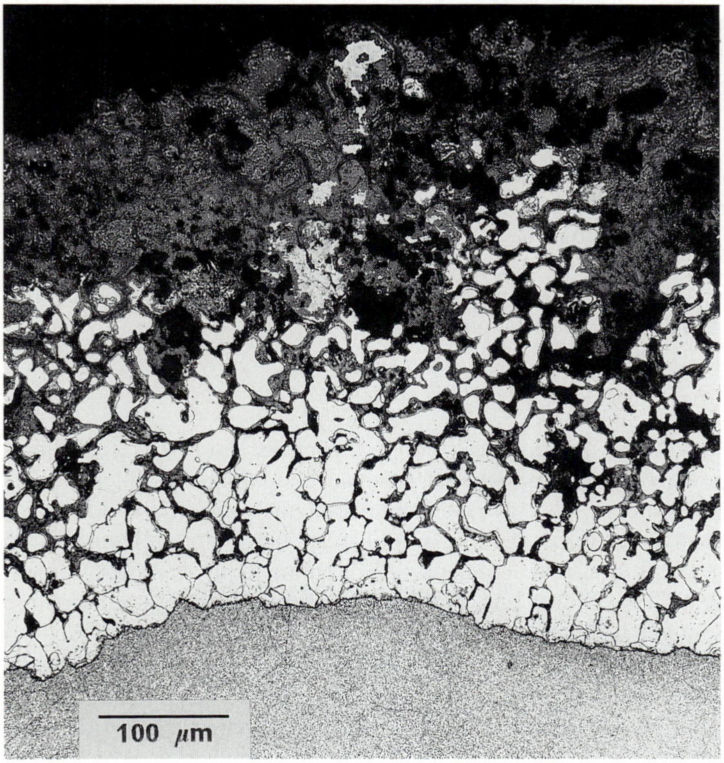

Figure 2.30 Micrograph of the scale formed on the alloy CMSX-2 after cyclic burner rig testing at 900 °C for 100 h in sea salt contaminated combustion gas

If Na_2SO_4 or NaCl-Na_2SO_4 mixtures condense on turbine components exposed to low temperatures, type I hot corrosion will also be possible in aircraft engines. The typical morphology of scale and subscale on Ni-base alloys formed under hot corrosion conditions is shown in Figure 2.30 for CMSX-2. These alloys form a thick and retentive oxide scale with fine precipitates of chromium sulphides within the alloy ahead of the advancing oxide.

The attack of hot corrosion enhanced by sulphidation is usually more severe than that induced by gaseous species such as oxygen or chlorides. This was proved for some Ni-base superalloys tested at 900 °C as well as 1100 °C in combustion gas with 5 ppm sea salt addition (Figure 2.29) [270]. Especially, for the IN-100 it is evident that high oxidation and corrosion resistance under aircraft conditions above 1000 °C is not compatible with high hot corrosion resistance.

In general, Cr_2O_3-forming alloys are more hot corrosion resistant than are Al_2O_3 formers. Cr_2O_3-forming alloys ordinarily contain less than 2-3% Al and 15% Cr or more depending on the alloy base. The high hot corrosion resistance of Cr_2O_3-forming alloys could be proved with the high chromium alloys Waspaloy and IN738 LC (Figure 2.29). The results of metallographic examination demonstrate the superiority of both alloys to IN-100, MAR-M002, SRR99, and CMSX-2.

Development of superalloys involves a compromise between satisfactory mechanical properties, oxidation resistance, and corrosion behaviour at elevated temperature. This aim is difficult to achieve, because the selection of the alloys for a given application is mainly determined by the creep behaviour which is often not compatible with good oxidation and corrosion resistance. Protective coatings can, however, partially overcome this problem.

2.1.3.3 Coatings

Although superalloys show fairly good high temperature corrosion resistance, extreme environmental conditions require protective coatings [272-274]. The principal role of metallic coatings is the build-up of reliable protective oxide scales which have to control the ingress of oxygen and other detrimental elements into substrates, whereas thermal barrier coatings (TBCs) aim to increase the turbine gas inlet temperature (TIT). By the late 1950s the application of high-temperature protective coatings to gas turbine aerofoils was introduced, first in the USA.

Dip aluminizing of simple solid blades was first done in order to preclude shorter inspection intervals for descaling these components. Pack aluminizing of first stage vanes and blades has successively been adopted by aero industries as a reliable coating route to produce diffusion coatings. Since about 1970 almost all first-stage vanes and blades have been coated by a pack cementation process. Many other coating procedures have been developed and investigated [275-277]. There are now production-scale supplementary coating techniques for the protection of internal cooling passages in aerofoils, e.g. gas phase aluminizing in the USA [278] and pulsed-pressure aluminizing in the UK [277]. Coatings have become more and more a substantial part of superalloy components where they contribute their share to unique versatility in a great many hostile environments.

Hot corrosion resistant coatings have been applied in the USA in a chromizing process preceeding aluminizing [273]. Coatings of superior resistance to hot corrosion and high-temperature oxidation were simultaneously verified in Germany by pack aluminizing platinum precoated substrates [279]. All coatings mentioned above are of the diffusion type. This fact implies a reduction of the effective load bearing cross sectional area, as coatings do not contribute any relevant mechanical properties to substrates. Repeated stripping of coatings from aerofoils and refurbishing them by a diffusion coat cannot be tolerated since they will become too thin. The deposition of overlay coatings as a primary coat is a prerequisite processing route for rejuvenating aerofoils unless reduced mechanical properties of recoated substrates are acceptable.

The pack aluminide process yields 50 μm thick coatings containing about 30% aluminum on substrates having only 6% aluminum at maximum. High operational temperatures of coated aerofoils will bring about an unacceptable flux of aluminum from the coating into the substrate. Higher operating temperatures, however, necessitate diffusionally stable coatings. Steadily increasing TIT which are now reaching up to 1700 °C primarily initiated the development of overlay coatings and deposition processes in the late 1960s.

Overlay Coatings of the MCrAlY type (M = Ni; Co; Ni+Co) with a thickness of about 100 μm have been deposited in fully automatic machines by utilizing the processes of argon shrouded plasma spraying (ASPS), low pressure plasma spraying (LPPS), and electron-beam physical vapor deposition (EB-PVD). The processes are listed here in the order of increasing costs for investment, operation, and maintainance, whereas they show a decreasing order as to their respective versatilities. These considerations seem to account for a worldwide distribution of plasma spraying facilities but also for a monopolized operation of plants in the USA which use EB-PVD processes for coating aerofoils. The fuel shortage in 1973 emphasized the need to save fossil fuels by improving the overall efficiency of heat engines. Cycle efficiencies of aero gas turbines have been improved inter alia by developing advanced high bypass engines. Potential benefits in component efficiency have been envisaged by a reduction of internal cooling of aerofoils with redirected compressor air. This fact attracted attention to TBCs. TBCs on critically heat loaded parts might also allow TIT increases thus improving the Carnot efficiency. A 200 μm thick TBC on a cooled blade lowers its metal temperature by some 125 °C [280]. In 1976 zirconia based TBCs were applied on blade platforms by LPPS where they proved beneficial on thermal fatigue accessible superalloy parts. The performance of TBCs on aerofoils, however, proved to be not sufficiently reliable. Thus no increase in TIT could be realized. Only in 1989 were fully TBC coated vanes successfully subjected to open-field real engine tests [281]. The TBCs were deposited by EB-PVD.

The Chemical Design of Coatings

The metallic coatings act simply as oxide formers. The selection of the appropriate oxide forming coating composition depends on the environment in which a substrate component should be operated, and also on the substrate material. Aircraft gas turbines, e.g., operate in extremely high temperature oxidizing environments which contain only minor pollutants. Thus they need coatings which are most resistant to high temperature oxidation. Alumina forming coatings may be the best choice, while chromia and silica formers might suffer from evaporative losses of suboxides, e.g. CrO_3 or SiO. The tip and the root areas of blades, however, operate at lower temperatures than does the middle part. Condensation of harmful compounds may occur there, especially on concave surfaces. These locations are susceptible to hot corrosive attack. Here a different type of coating, e.g. a chromia former, might be better. One has to bear in mind that no superior coating can be composed which can control all decay mechanisms equally well. The choice of a coating always reflects a compromise between a number of implications.

Applying coatings with low interdiffusion to aerofoil substrates is becoming more and more essential, since as 1100 °C substrate temperature during operation of modern engines causes appreciable atomic movement. The diffusive escape of elements from coating to substrate can be many times as high as the loss by selective oxidation. This is why within coating/substrate systems the gradients of elemental activity, which are the driving forces for interdiffusion, have to be minimized [282]. The activity of aluminum, e.g., has to be adjusted very carefully: The diffusivities of the three basic elements Al, Cr, and Ni, present in most coatings, differ widely. They are given here in decreasing order, with aluminum the most mobile element. Hence most aluminide coatings are inferior to overlayers on aerofoils of advanced gas turbines. Generally speaking, different superalloys may require different coatings under identical operational histories as well.

Thermal fatigue (LCF) is a serious problem of superalloys in aeroengines. They are the more susceptible to LCF failure the more they are inherently hardened by γ' phase. This is a situation which cannot be markedly ameliorated by use of the "right" coating. The best choice would be to utilize a component which has a low Young's modulus as well as a low thermal expansion coefficient (TEC) along the axis of greatest restraint in order to minimize stresses which arise on thermal cycling. A good choice - and an elegant one, too - is the use of (001) oriented single crystal blades. The (100) crystallographic axis exhibits a favorable 40% reduction in Young's modulus. LCF failure is then no longer a problem. This prominent advantage should also be maintained by careful handling and intelligent coat processing. Any minute plastic deformation must be precluded in order to prevent formation of new grains. Transferred arc cleaning, which is common procedure before plasma spray deposition, bears the risk of local melting, and subsequent plasma deposition on thin sections can locally superheat them as well.

The alloying elements in metallic coatings are briefly introduced in the context of their respective effect on the coatings [273].

Nickel — Generally taken as base element for overlay coatings on Ni base blades to minimize interdiffusion; depresses chemical activity of Al.

Cobalt — Base element for overlay coatings on Co base vane alloys to minimize interdiffusion; raises chemical activity of Al and raises γ' solvus temperature in Ni base alloys; slightly increases TEC.

Aluminum — Alumina forming element in Ni and Co base alloys; suitable against HT oxidation up to about 1200C; aluminide forming constituent in diffusion coatings; γ' phase hardener in γ/γ' Ni base alloys.

Chromium — Chromia forming element; essential constituent in Ni and Co base coatings against hot corrosion (see section "High Temperature Corrosion") and against oxidation up to about 900 °C; reduces critical level of Al needed to form protective Al_2O_3 on coatings.

Silicon — Silica forming element; most effective against "low temperature hot corrosion", e.g. common in marine environments; promotes formation of protective Al_2O_3 in Ni-Al alloys [283].

Yttrium and Rare Earth Elements: Impart adherence to protective Al_2O_3 and Cr_2O_3 on Ni and Co based alloys; convert the growth characteristics of these oxides from predominantly cation to predominantly anion diffusion; reduce the oxidation rate of chromia formers.

Hafnium — Improves adherence of in-situ grown Al_2O_3 and Cr_2O_3 scales and of artificially deposited zirconia based layers on coatings.

Platinum — Improves adherence of Al_2O_3 on diffusion type coatings; lowers chemical activity of Al in Ni base alloys and thereby reduces its interdiffusion; effective element against acidic fluxing (see section "High Temperature Corrosion").

Tantalum — Solid solution strengthener of γ and γ' phases; reduces interdiffusion of Ni; beneficial against hot corrosion [284]; improves oxidation resistance if <3%; promotes early establishment of α-Al_2O_3; reduces TEC.

Titanium — Accelerates formation of Cr_2O_3 at metal/oxide interface in Ni-Cr alloys; reduces TEC if >6% in Ni base alloys [285].

The role of alloying elements in zirconia based ceramic TBCs is mainly to stabilize the cubic or nontransformable tetragonal T' phase of ZrO_2 Their stabilizing effect in respective environments is of vital importance for proper service. Alkaline earth oxides (CaO, MgO) are taken for lower temperature ranges <1000 °C whereas yttrium and some rare earth oxides (Y_2O_3, CeO_2, Yb_2O_3) are also applicable at higher temperatures.

Processing of Coatings

The three most commonly used coating procedures - Pack Cementation, Electron-Beam Physical Vapor Deposition (EB-PVD), and Low Pressure Plasma Spraying (LPPS) - are described in some detail.

Pack Cementation: The most common chemical vapor deposition processes (CVD) for siliconizing, chromizing, and particularly aluminizing relate to the pack process [286]. The workpieces are located or "packed" into an airtight vessel which is filled with a powder mixture. In the case of aluminization, e.g., it consists of (i) a donor, e.g. Al or Al-Cr alloy powder, (ii) an inert filler, e.g. corundum powder, and (iii) a small amount of activator, e.g. 1% NaF or NH_4Cl. The vessel is then argon flooded and subjected to a reactive temperature regime. Gaseous AlF_3 or $AlCl_3$, which represents a high chemical activity regime of aluminium, will form and is attracted by the superalloy workpieces due to the low chemical activity of aluminium there. The gases decompose on the workpiece surface and thus in-diffuse to form aluminide "diffusion" coatings.

Two variant of aluminide coatings can be produced this way

- Low activity aluminides. By using a "low activity" prealloyed powder of, e.g., Cr-15Al and an appropriate cementation treatment at 1000 °C to 1100 °C external β NiAl will form via predominant Ni diffusion. The coating, although brittle up to about 700 °C, is fairly damage tolerant due to its moderately graded Al content.
- High activity aluminides. By using a "high activity" powder of Al or Cr-30Al and a cementation treatment at 700 °C to 800 °C, a brittle layer of Ni_2Al_3 + β NiAl will form via predominant Al diffusion. A subsequent homogenization anneal (1050 °C to 1200 °C) will suppress the Ni_2Al_3 phase but leave a high Al content at the surface.

The advantages of pack cementation are (i) simple installation and good process reproducibility, and (ii) uniform coating of the entire surface of complex shapes. The limitations are: (i) limited repairability of thin and critical parts since the subsurface of components is affected, (ii) lack of compositional flexibility, (iii) brittleness of layers, (iv) interdiffusion, which might be critical.

Electron-Beam Physical Vapor Deposition: The first commercial overlay coatings for gas turbine aerofoils were applied by EB-PVD. This process makes use of multielemental evaporation from a single pool onto real foil shapes [276, 287]. Although not new, this process had to be modified in many engineering details in order to satisfy the physical, chemical, and mechanical requirements which otherwise are uncommon with PVD applications. The differences concern

- coating deposition on heated substrates which are preheated in an extra heating chamber interlocked to the main chamber to give good adherence and growth conditions;
- a tight valve between preheater and main chamber which prevents the entry of any contaminant during the heating sequence in order to preclude growth defects;
- geometrical preconditions which concern pool/substrate distance, pool area, and e-gun power during evaporation to maintain appropriate substrate temperature;
- an oversource heater for additional temperature adjustment;
- high deposition rates on rotating parts of 10 to 20 μm/min to be competitive to other coating processes;

- the proper movement of aerofoil substrates in the evaporant by simultaneously rotating and slewing them to attain a fairly uniform coating thickness on foil and platform sections;
- multielement evaporation from one pool.

The last item addresses a most complex relationship between starting ingot (identical to the liquid-pool average composition), the composition of the ingot being bottom-fed into the evaporation pool, the composition of the vapor cloud, and the condensing overlay coat on the substrate. For a MCrAlY coating having nominally 0.3% Y, the pool composition has been found to be about 20% Y. Other implications arise from pool depth, substrate temperature (reevaporation) and movement, electron beam focusing and scanning characteristics. In practice, EB-PVD coatings often show a columnar structure which exhibits weak bonding of grains. For metallic coatings this structure is subsequently densified by using glass bead peening and vacuum heat treatment near 1080 °C for recrystallizing. Otherwise, however, if oxide ceramic TBCs have to be processed via reactive evaporation, a microcolumnar structure is desired which imparts, e.g., unique erosion tolerance [281] to components. The PVD growth of TBCs on EB-PVD bond coatings will first bring about a thin equiaxed zirconia layer on which nucleation and textured columnar growth occurs. Substrate and coating are thermally equilibrated to about 1000 °C. The columns show weak bonding and the overall density is about 0.65 to 0.8 of compact yttria partially stabilized zirconia (YPSZ) [288]. Hereby a favorable low internal stress TBC is accomplished which persists even under severe thermocyclic histories.

The advantages of EB-PVD coatings are

- controlled interdiffusion for good bonding; controllable temperature histories of substrates in order to preclude initiation of temperature related failure modes;
- excellent surface finish and roughness; these depend on previous surface conditions which can be approximately replicated;
- preservation of free cross-section of cooling holes;
- tailored compositional coating distribution with regard to Cr content [289, 290];
- ease of stripping and recoatability,
- possible manufacture of TBCs via reactive evaporation.

The limitations concern

- cooling passages which cannot internally be coated as well as multiple components which are partly overshadowed by neighbouring sections;
- specified compositional ranges which should not be too narrow;
- elements which differ excessively in vapor pressure. In case of need they have to be evaporated simultaneously from an additional source.

The last item adresses future needs which have to be met. The coatability of more-complex shaped turbine blades may initiate further process variations, e.g. plasma assisted evaporation [291] as is common with ion plating.

Low Pressure Plasma Spraying: The LPPS process [276, 292, 293] is based on atmospheric plasma spraying. It attained commercial status following the EB-PVD process. It offers production yields comparable to EB-PVD processing at lower capital costs and small batch operation. The argon shrouded spray process is similarly effective and qualified. Although every component is singly processed, the reproducibility in chemical composition is excellent, if guaranteeing a high standard of operational control.

A main precondition for proper processing is the availability (strategic restrictions can be harmful!) of reproducible powder qualities, meeting all criteria concerning powder feeding and size distribution. A uniform alloy-powder particle size is required (10 to 35 μm). Smaller sizes give denser coatings whereas bigger sizes entrap less adsorbed gases. Careful attention must be paid to powder storage and conveyance because of the danger of explosion or adsorption of water vapor and gases. Prior to coating the components are commonly shot-peened by corundum or silicon carbide particles. After loading in prechamber the fully-automated coating cycle involves (i) vacuum preheating of the component to about 700 °C, (ii) electric-arc surface activation leading to an additional 150 °C, and (iii) spray metallizing, with the first layers produced using negative arc polarity thus eliminating residual oxide inclusions. A subsequent vacuum anneal generates a fine recrystallized and diffusion bonded coating. A multitude of different coating procedures exist concerning equipment and process parameter control. Apart from customization, the advantages of LPPS include (i) verification of nearly unrestricted compositional ranges, (ii) high product yield, (iii) ease of processing large and small number batches, (iv) good reproducibility, and (v) processability of TBC ceramics. The limitations are (i) low quality surface finish, (ii) higher gas content than in PVD processed coatings, (iii) no temperature control on thin parts, (iv) possible blocking of cooling holes, and (v) the fact that prepeening processing can affect monocrystallinity of single crystal blades.

In summary the performance of coatings on special shapes depends on chemistry as well as on processing. The processing route brings about distinctive coverage, microstructures, and topographies. A modern trend is apparently to crosslink the respective process-related advantages in order to tailor an optimum coating route for individual purposes. Although the distinction between different processes may become rather blurred, this is a promising option for making the best coatings.

Corrosion of Metallic Coatings

The mechanisms of high temperature corrosion of the superficial and subsurface regions of coatings are qualitatively identical to those operative on base alloys (see section "High Temperature Corrosion"). The differences refer to more or less quantitative features. They concern (i) the gain in time between overhauls (TBO) which necessitate refurbishment of coatings, (ii) the gain (or loss) of incubation time before the onset of corrosive decay compared with unprotected or differently coated substrates, and (iii)the reduction of corrosion rate and depth.

An internal decay mechanism exclusively operative in coatings is interdiffusion between substrate and coating. It is not really corrosion but can promote corrosion processes by critically impoverishing subsurface regions of essential elements. E.g. the β-NiAl which represents the main aluminum reservoir in aluminide diffusion coatings can be consumed by inward diffusion, as already mentioned above. Chemical activity gradients, temperature, and respective diffusivities of the elements are the rate-controlling factors. Interdiffusion can result in the formation of brittle topologically close packed (TCP) phases (risk of internal crack initiation) and in Kirkendall porosity (liable of decohesion). Modifications in subsurface chemistry - either by in-situ oxidation/corrosion or by an artificial treatment, e.g. aluminizing - will in turn affect the quantity of interdiffusion. Internal and superficial compositions are interlinked: the more external Al depletion the less internal diffusion, and vice versa, e.g. enhanced growth of Al depleted zones occurs by action of sea salt (chloride induced oxide spallation), whereas slow growth is attained by additionally aluminizing the overcoat [269].

Compared to overlay coatings the degradation of diffusion coatings via selective oxidation can be more rapid due to their relatively higher chemical Al gradients. Burner rig tests at 1050 °C demonstrate the consumption rate of Al rich phases from simple and Pt-modified (RT22) substrate aluminized substrates (Figure 2.31) [271]. The RT22 coatings perform substantially better because of the reduced Al activity imparted by Pt. A favourable operational regime for diffusion coatings is of course at lower temperatures, where the rate of degradation decreases.

Although coatings are primarily designed for high temperature protection they also protect effectively against type I (800 °C to 1000 °C) and type II (600 °C to 800 °C) hot corrosion. The most reliable coating against type I hot corrosion was found to be aluminized CoNiCrAlY overcoat (LCO22 + Al), whereas NiCrAlSi (8% Al, 8% Si, 35% Cr) overcoats will have their domain in the type II hot corrosion regime, where interdiffusion is less detrimental [270, 282].

Degradation of Thermal Barrier Coatings

The TBC System: The destruction of TBCs during service is evidence of the breakdown of a whole system consisting of at least four elements which should be in long-lasting accord. These are mainly: substrate, bond coat, TBC, and environment. Models for oxidation-based degradation of TBCs are referenced [294]. The interplay of the elements, however, is complex, and - as TBC processing is a young technique - not yet fully understood. Failure mechanisms may not originate from one single precondition, but from a combination of different causes.

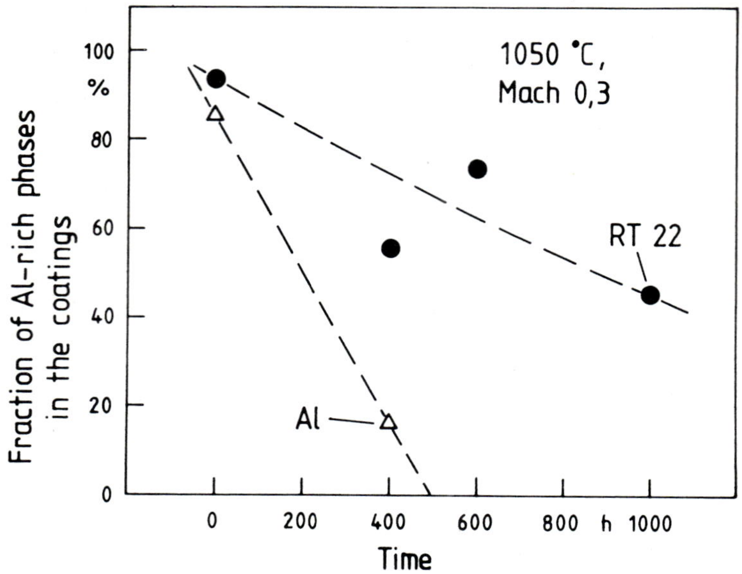

Figure 2.31 Consumption of aluminide and platinum aluminide (RT22) coatings on the single crystal base material SRR99 as a function of time during cyclic oxidation in a kerosene combustion gas stream of Mach 0.3 at 1050 °C [271]

In spite of the caution expressed above, the most probable aspects of failure modes of TBCs are summarized below.

Microstructural Phenomenology:

1. If a superalloy blade aerofoil has to be coated with a TBC, its crystalline state - conventionally cast polycrystalline, DS, or single crystalline - is of concern. For single crystals, e.g., (and in analogy DS substrates) Young's modulus is reduced by about 40% in the (001) direction. The response to bending stresses of single crystalline blades will therefore be higher than that of conventionally cast blades. Thus TBCs on single crystal blades will be subjected to more severe mechanical fatigue than commonly expected. Furthermore, single crystal blades have to operate at higher service temperatures, and so there will be higher thermal expansion mismatches between substrate and TBC. Thus higher internal stresses from thermal fatigue will occur. Diffusion-related processes will dominate.

2. At present only two TBC deposition processes are available: Plasma spraying (PS, either APS or LPPS) and PVD. PS bondcoats need a rough surface finish for good bonding of TBCs. Otherwise PVD bond coats have a flat surface finish and will form on it an adherent alumina film on annealing. The alumina film provides adhesion between TBC and bondcoat [295]. No mechanical bonding is provided.
3. On alumina forming MCrAlX bond coats, alumina is expected to grow under oxidizing "equilibrium" conditions whereas in the initial state of oxidation, Ni-rich spinels are formed [296]. They are also present in APS bond coats. These oxides show poorer mechanical properties than α-alumina. They may induce premature cracking in TBC systems. Al_2O_3 and ZrO_2 are considered superior to NiO and spinels on the basis of minimal interface energy arguments [297]. Thus NiO and spinels have to be avoided for the sake of good bonding, as is unequivocally accepted. The type of bonding, however, is still controversial.
4. APS processing of TBCs on bond coats will bring about inherent segregation phenonena. The substrate temperature has to be controlled with utmost care in order to preclude compressive stresses in TBCs, which generate multiple perpendicular microcracks on cooling which provide strain tolerance to TBCs.
5. Sprayed YPSZ coatings as well as PVD coatings contain nontransformable tetragonal T' which may dissociate to cubic and to transformable tetragonal T zirconia. On cooling T may transform to the monoclinic M phase, which will expand by about 9% [298]. In some cases small amounts of transformable T phase may be helpful to lower internal stresses, but normally the volume change on T ↔ M transus is detrimental.
6. *Crack Initiation and Propagation in APS TBCs* occurs intraceramically near the ceramic/bond coat interface and partly along spinel and Ni/Cr rich oxides. Initiation sites for cracks are these oxides. The crack deflection into the TBC is accomplished by the action of moderate cooling cycles at the interface: they initiate compressive stresses which favor latitudinal crack propagation [299]. On the other hand ASPS bond coats (= formation of detrimental oxide possible) were found to be superior to LPPS bond coats (= low oxide formation expected) [300] which may contradict the hypothesis given above.
7. *Crack Initiation and Propagation in EB-PVD TBCs* was identified at bondcoat oxidation product/metal interfaces and within the bond coat oxidation products (α-alumina and spinel) [301], but not within the thin equiaxed zirconia layer nearby. It was noted that Zr/Y affects ionic transport in α-alumina which causes porosity there [296, 302]. It has not yet been decided whether holes in α-alumina or spinels on top of it, or both, are crack initiators.

The microstructural details are given in order to explain those failure mechanisms in TBCs which are activated by accumulation of harmful stresses in and close to the ceramic layer. They will all result in spallation. Some phenomena which contribute to stress generation are given below.

Mechanisms for TBC Degradation

Bond Coating Oxidation/Hot Corrosion: Ionic transportation of oxygen in zirconia as well as gaseous oxygen transport along microchannels allows bond coat oxidation. PVD processed TBC systems have a flat metal/ceramic interface which allows the oxide growth in normal direction without generation of any stress. Curved shapes (leading and trailing edges of blades), however, will induce stresses into TBCs on oxidation. Oxide protrusions which occur on irregularly oxidizing interfaces will produce stresses too. PS processed components will also generate internal stresses on oxidation due to their corrugated interface, and microcracks in TBCs may favor irregular oxidation and subsequent stress accumulation. A "critical" oxide thickness value of about 5 to 10 µm for terminating safe operation of TBC systems is so far assumed to be reasonable. Salt deposits can penetrate the porous ceramic layer and can attack the bond coat. "Bulky" corrosion products will form and hence induce stresses into TBCs. Entrapped salts and deposits can also lessen the strain tolerance of TBCs on cooling. At temperatures >1000 °C chloride induced corrosion may affect the adhesion of α-alumina to the metal base.

Chemical Reactions on TBCs: Hot corrosive decay of TBCs by Na_2SO_4 and vanadates is reported which involves leaching out of stabilizers from the parent zirconia. The degradation occurs by transformation of high temperature phases on cooling to the monoclinic crystal lattice. This failure mode is not likely to occur on components of clean-fuel operated engines. It may be found in heavy duty engines and off-shore service as referenced elsewhere [303]. An analogous reaction at low temperatures, however, which is provoked by water vapor, may be harmful [304]. The "open" microstructure of PVD TBCs may be assumed to be more susceptible to this moisture decay than are PS structures.

Sintering of TBCs bridges those gaps otherwise needed for stress relief. The thermal shock resistance will then be reduced. Sintering phenomena became apparent for PS TBCs at 1350 °C and higher [299], whereas PVD TBCs show a sintering affected change of appearence (corn-cob structure) after 1150 °C cyclic exposure [288].

Bond Coating Strength: Bond coats have to smooth down critical stresses near the TBCs which may be partly introduced by thermal expansion mismatches. Thus they need some ductility. On the other hand the interface may be deformed due to creep and fatigue loading if it is not strong enough. Apparently the provision of creep resistant coatings which are TEC adapted to substrate and zirconia is essential for safe performance. Selective oxidation of Al leads to loss of strength within the Al depleted zone (Figure 2.32), which may be a TBC life-limiting process.

Erosion/Microspallation: A smooth surface finish on aerofoils has to be guaranteed at first in order to maintain satisfactory aerodynamic efficiencies. Increasing surface roughness will initiate gas flow turbulences which in turn favor increased mixing with the main stream hot gases. Thus hotter gases cause higher

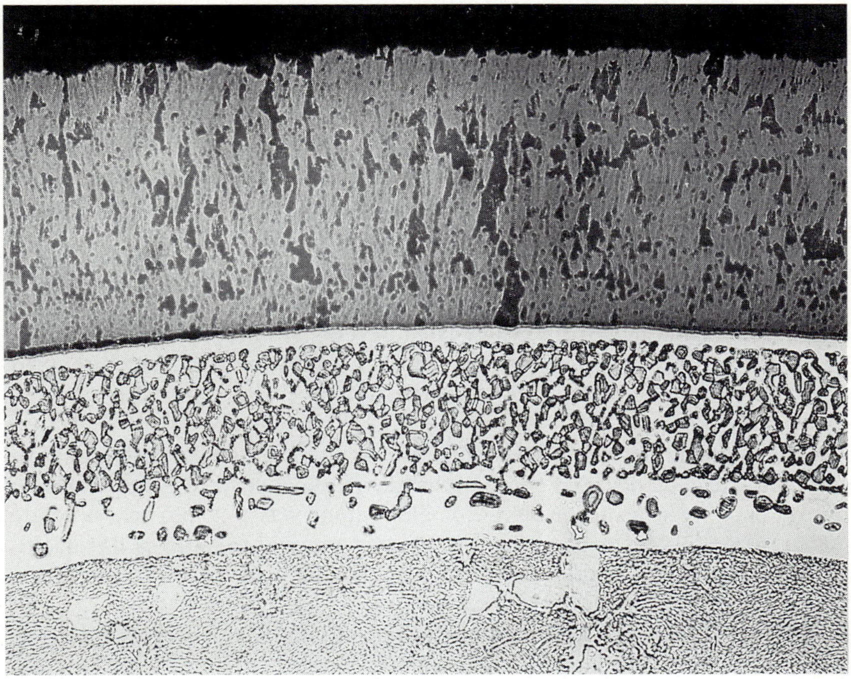

Figure 2.32 Cross section of a EB-PVD processed, 210 μm thick, YPSZ TBC/NiCoCrAlY bond coat system on IN100 cyclic-burner rig tested at 1150 °C for 130 h [305]

material temperatures. Higher temperatures will in turn shorten the expected life time. The surface finish and performance of EB-PVD TBCs is so far superior to PS TBCs. On real-engine tested blade platforms they outperformed different PS TBCs by about five times [300]. A perpetual minute material loss from EB-PVD TBCs during service, however, has to be noted. It results from frictional movements of the microcolumns introduced during severe heating in a cyclic burner rig program at 1150 °C [288]. Some minute loosened zirconia particles become entrapped between the parent microcolumns and preclude there their contact on heating. They may contribute a pitching moment to the columns which initiates their "microspallation".

2.2 Metal-Matrix Composites

2.2.1 Metal-Matrix Composites with Aluminium Matrix

K. Schulte

Metals and alloys are useful and versatile materials which have been developed during mankind's long history to near perfection in such properties as strength, toughness, deformability, thermal stability, fatigue and creep.

However, it seems that metals have now been developed in their major properties close to the ultimate achievable so that any further improvement will remain rather limited. But still, some properties can be further improved.

In the past, mainly in the late sixties and early seventies, metal-matrix composites already attracted interest and some major developments were achieved. As a result, the boron fibre used to reinforce of aluminium was put into the structure of the space shuttle. However, the cost of this composite was and remains extremely high. Therefore, hardly any further applications are known.

In the last 10 to 15 years advances in the development of fibres capable of withstanding very high temperatures offer the possibility of reinforcing metals. Most of the fibres presently available are based on silicon carbide and aluminium oxide, although other variations are being studied. It is this, together with the comparatively low price, which is exciting so much interest in the reinforcement of metals.

Why Metal-Matrix Composites?

In many technical areas extreme material properties are needed. High strength and stiffness should be combined with high temperature retention and low specific gravity. MMC can fulfill these needs. In Figure 1.7 on page 9 the specific strength of metals, various fibre reinforced plastics and fibre reinforced metals is plotted vs. the specific modulus. With fibre reinforcement, improved stiffness and strength can be achieved when compared to metals. If, additionally, higher temperature retention is needed, then fibre reinforced plastics reach their limit at about 200 °C. However, fibre reinforced metals can be used beyond this point. Their application temperature is mainly dependent on the matrix material used. Only beyond 1000 °C do MMC reach the maximum temperature loads which limit their application. Higher temperatures can only be achieved if a ceramic matrix is used. Figure 2.33 shows a comparison of the temperatures up to which polymer matrix, metal matrix and ceramic matrix composites can be used. In general, fibre reinforced metals can be taken into consideration for structural design [306, 307] because:

- they have improved temperature retention, when compared to the neat matrix metal.

- the high ductility and strength of the metal matrix improves the mechanical properties when compared to fibre reinforced polymers, specifically interlaminar shear strength or transverse tensile strength, as well as fracture toughness.
- the high elastic modulus of the fibres increases the stiffness of the composite. Compared to fibre reinforced polymers, the higher elastic modulus of the metal matrix leads, at load levels under the matrix yield strength, to a further increase in MMC stiffness.
- the superior physical properties, such as electrical and thermal conductivity, or the magnetic properties, are advantageous for many technical applications.
- because of the low thermal expansion coefficient of the matrix, MMC remain dimension stable under temperature variations.
- they can resist environmental attack from ultraviolet radiation, moisture or many chemicals. Besides, they are nearly indiffusible for gases and liquids.

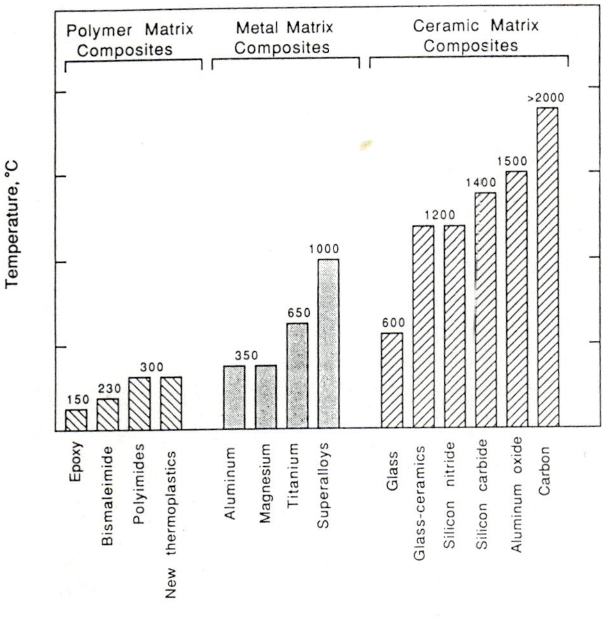

Figure 2.33 Maximum service temperatures of various polymer, metal and ceramic matrix composites

In the case of particulate reinforced composites in general, no reinforcement, but an increase in fracture toughness when compared to fibre reinforced composites, and related properties can be achieved. Their influence in improving high temperature properties or compressive strength seems to be rather limited. A substantial improvement is observed in the elastic modulus and the thermal expansion coefficient. However, it is less than in the case of fibre reinforcement and does not at all reach the values predicted with the rule of mixtures [308, 309].

Positive characteristics are the relatively small anisotropy in the case of particulate reinforcement and the comparably uncomplicated fabrication based on casting or powder metallurgy techniques. Both fabrication technique and particulates are relatively cheap; therefore the cost for a component can remain within acceptable margins. Machining this type of composite seems to be relatively uncomplicated.

Besides all these advantages, MMC have some drawbacks: Even light metals such as Al, Mg or Ti - the most often used matrix materials - have a higher specific gravity as the polymers and the fabrication processes for MMC are in general more complicated and costly than for polymer matrix composites.

In general, the properties of MMC are dependent on the type and distribution of fibres (whiskers, short or continuous fibres) or particulates, the fibre or particle content, the orientation of fibres and the matrix alloy. The fibre-matrix bonding, which includes cohesion, wettability and chemical interaction, is of further importance.

The relatively high cost of MMC will limit their application to primary structures in a variety of application fields, such as helicopter parts (parts of the body, support for rotor blades, drive shafts) rotor vanes of compressors, airplane parts, and car parts (gear box, pistons, transmission parts, connecting rods, etc.). One prominent example for application in the automotive industry in Europe is the use of a selectively reinforced piston for Diesel engines, where Al_2O_3 fibres are used as reinforcement elements. While this part is now in production, others are still under development, such as connecting rods. However, an increasing acceptance to use these parts can be observed in the automotive industry, which in the future needs to reduce fuel consumption and pollution, as well as weight. This makes the shift to advanced materials an obligation.

While fibre reinforced parts have the drawback of being rather costly, particle reinforced MMC promise to be relatively inexpensive. One typical part to be manufactured is a driving shaft. It is reported that with a driving shaft produced from particle reinforced MMC a weight reduction of about 14% can be achieved, while the critical number of revolutions can be increased by about 25%.

When preferably used for primary structures, the reliability and durability of MMC have to be, from early development stages on, at a high level. It is, therefore, important to know and understand the mechanisms and development (initiation and propagation) of damage.

The Reinforcement Fibres

The principal fibres which are currently available and of potential interest for use in MMC and some physical and mechanical properties are listed in Table 2.16 [310].

Fibre Type	Trade name	Density g/cm³	Diam. μm	Tensile strength MPa	Elastic Modulus GPa	Strain to failure %
Bor	Avco	2.27	140	3500	400	1
C-fibre	T300	1.75	7	3430	230	1.5
	HM35	1.79	6.7	2350	360	0.7
	FT700	1.8	10	3300	700	0.5
SiC	SCS6	3.05	140	3400	400	1
	Tyranno	2.4	10	2800	200	1.5
Al_2O_3	FP	3.9	20	1500	380	0.4

Table 2.16 Mechanical properties of various types of fibres [310]

In the sixties boron fibres were used to reinforce aluminium alloys. The extremely high goals with respect to mechanical properties could be reached. However, their extremely high cost reduced applicability such that they were only used in spacecraft. The high diameter of the boron fibres of 140 μm allows their use only in relatively plane parts.

The ceramic non-oxide SiC fibres have, because of their low reactivity to the metal matrix, a high potential for technical applications. The oxide fibres, mainly the Al_2O_3 based fibres, are already used in technical parts, because they are relatively cheap, have suitable mechanical properties, and their reactivity to the molten metal is low.

The high variability in the mechanical properties of carbon fibres, along with their high thermal stability, makes them the ideal reinforcement for metals. However, the reactivity of carbon fibres with the metallic matrix very often reduces the superior fibre properties in the composite. Carbon fibres, therefore, generally need a protective coating, which increases fibre cost.

Of the ceramic oxide fibres, the Al_2O_3 fibres have focussed the greatest attention, while the ZrO_2 and SiO_2 fibres have not found the same interest. The mechanical properties of the Al_2O_3 fibres do not reach the same values as the latter, but their thermal stability, their low reactivity to the matrix materials and especially their low price have made the Al_2O_3 fibre the favourite reinforcement fibre for light metals.

The Matrix

The main matrix materials which found attention are the light metals magnesium, aluminium and titanium. For high temperature applications, nickel and cobalt, as well as tungsten, molybdenum, niobium or tantalum are also considered (see Table 2.17). At present scientific work is mostly concentrated on aluminium and magnesium as matrix materials. For the compressor blades of aircraft turbines fibre reinforced titanium is also of interest. The main aim in science and technology is to increase the temperature retention of the matrix metal and to load it as close to the material solid temperature (about 0.8 Ts) as possible.

	Density in g/cm³	Melting point in °C
	Low density metals	
Aluminium	2.7	660
Magnesium	1.8	650
Titanium	4.6	1660
	High melting point metals	
Nickel	8.9	1453
Superalloys	~8	~1400

Table 2.17 Matrix metals for metal-matrix composites

The idea of MMC is to strengthen the matrix by adding a second phase, the fibres or particulates. Strengthening of the matrix can easily be achieved if continuous long fibres are used. The mechanical properties follow the rule of mixtures. The possibility of matrix strengthening is indeed an advantage of metal-matrix composites; however, it has the drawback that intensive internal stresses are built up which have to be considered in the design.

Fibre-Matrix Compatibility

Fibre-matrix combinations are seldom in a state of equilibrium; thus in most systems fibre and matrix will tend to interact, usually with detrimental effects on the composite properties. The main interaction phenomena are [311]:

- fibre dissolution in the matrix
- fibre dissolution and reprecipitation (coarsening processes)
- chemical reaction between fibre and matrix
- segregation and/or precipitation of matrix constituents at the interface
- poisoning of fibres by matrix atoms
- thermal expansion mismatch

Apart from the last, the above phenomena are diffusion-dependent and, therefore, increase in importance with increasing temperature and time of exposure. Thus for most composite systems production methods are sought with as low temperatures and exposure times as possible. Provided they survive manufacture, composites with potentially incompatible constituents can be used to certain temperature limits. To achieve a strong fibre reinforcement, the fibres must be protected from extensive chemical reactions between fibre and matrix. This can mainly be achieved by coating the fibre surface. The main techniques used are chemical vapour deposition (CVD) and physical vapour deposition (PVD).

To achieve a strong and tough fibre reinforced material, the interface also has to meet conflicting requirements, which are [312]:

- a weak interface, to achieve longitudinal strength and toughness by fibre debonding;
- a strong interface for good transverse properties.
- Further, ease of fabrication, including good wetting of the fibres if the matrix is combined in the liquid state, depends critically on interface chemistry for most economical processing methods.

The criteria to be satisfied will vary depending on the fibre, the matrix and the application. There is, therefore, no given set of rules dictating what chemical engineering of the interface will achieve optimized properties. Contradictory demands have to be fulfilled, which shows that interface chemistry has to be further studied.

Experimental Results

Results discussed in this paper are from experiments performed on a number of materials which contain various high strength fibres from different manufacturers. The fibres chosen for investigation were the HM35 carbon fibre from Toho Beslon, the SiC fibre (Tyranno) from Ube, and the α Al_2O_3 FP fibre from Dupont. Common to all materials was the aluminium matrix, either pure as Al 1070 or with the FP fibre as an Al-2.5Li alloy.

	E in GPa	σ_B in MPa	ε_B in %
Al 1070	69	59	40
Al-2.5Li	80	320	~25

Table 2.18 Properties of neat metal matrices

The mechanical properties of the neat matrix alloys are given in Table 2.18. The pure aluminium matrix (Al 1070) has a comparatively low fracture stress, but a high strain to failure. When using the Al-2.5Li matrix, coherent δ' particles form leading to a precipitation hardening. Therefore, a relatively high fracture stress σ_B can be observed. Using lithium as an alloying element also increases Young's modulus.

Manufacturing of the composites was accomplished by various squeeze casting techniques; only in the case of the C-Al system were the composites were produced by hot pressing aluminium coated carbon fibres, following the prepreg route [313]. During consolidation at temperatures above 580 °C, fibre matrix interactions can occur. The most prominent is the formation of reaction products, such as Al_4C_3 carbides in the C-Al system. Masson et al. [314] show that the carbides have a needle-like structure and grow both into the fibre and into the matrix. They harm the fibres by forming cavities which reduce their strength.

a)	[0 °], SiC-Al		[90 °], SiC-Al	
	Tension	Compres.	Tension	Compres.
σ_B in MPa	920	1605	168	204
ε_B in %	0.89	1.77	1.30	7.80
E in GPa	125	118	102	90
b)	[0 °], Al_2O_3-Al-2.5Li		[90 °], Al_2O_3-Al-2.5Li	
σ_B in MPa	583	1460	177	225
ε_B in %	0.36	1.09	1.20	7.40
E in GPa	181	166	126	117

Table 2.19 Mechanical properties of SiC-Al and Al_2O_3-AlLi after tensile and compressive testing

Tensile and compressive tests were performed on the unidirectional composite with fibres in the 0 ° and 90 ° directions. In addition, fatigue tests were made at stress ratios of R=0.1 and R=-1, respectively.

Material Characterization

Static Loading: In Table 2.19 the mechanical properties after tensile and compressive testing for both 0 ° and 90 ° fibre orientation are summarized [310, 315]. The strain-to-failure is controlled by the fibre failure strain. However, under compressive loading the strength is considerably higher than in tension, which is in agreement with a higher strain-to-failure under compressive loading. The generally good compressive behaviour can mainly be related to the facts that:

- The metal matrix itself has relatively high Young's and shear moduli, which allows better fibre support and avoids their short wave kinking.
- The ceramic fibres themselves can carry a higher compressive than tensile load, because of their microstructure.

In a test transverse to the fibre direction, the fracture stress of the composite exceeds the fracture stress of the matrix. Due to the fibre reinforcement, an increase of the Young's modulus can also be observed.

Although the matrix material has a high strain-to-failure, it cannot be realized in the composite. Only about 1.3% strain-to-failure can be achieved. Under compressive loading Young's modulus is not as high as in tension; but the fracture stress is again essentially higher. Therefore, the strain-to-failure is by far higher under compression loading than under tensile loading.

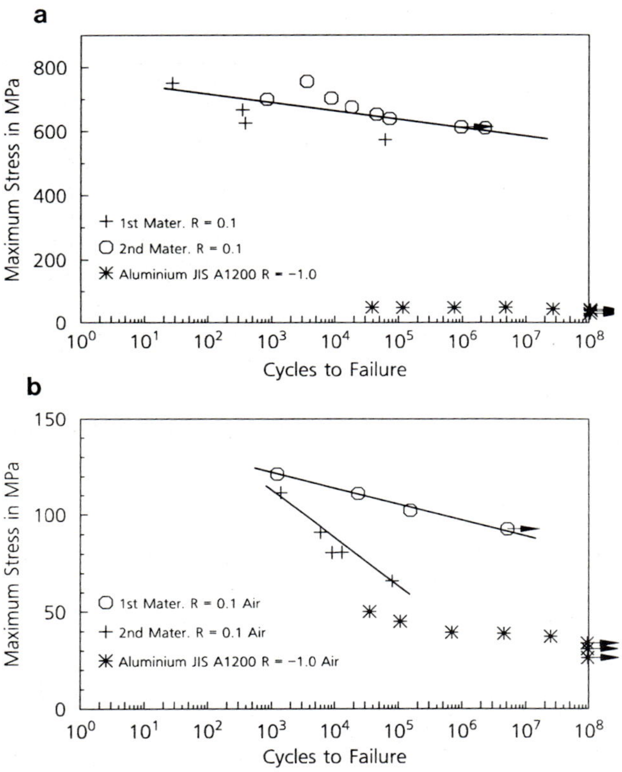

Figure 2.34 S-N curves of SiC-Al composites
a) Fatigue behaviour in fibre direction [0 °]
b) Fatigue behaviour transverse to fibre direction [90 °]

The strength of a metal matrix composite strongly depends on the fibre-matrix interaction. In Table 2.20 the fracture stress of a carbon fibre reinforced aluminium is given [313]. Due to the processing temperature the fibre matrix interface is influenced. For longitudinal specimens a low processing temperature (580 °C, 15 min) leads to only few fibre-matrix interactions with a weak interface and pronounced fibre pull-out [310], but a relatively high fracture stress. At a high processing temperature (600 °C, 30 min) with extensive fibre-matrix interactions [314], the strong interface results in fast crack propagation with a relatively plane interface and a reduced fracture stress. The longitudinal strength depends on the temperature and the time of C-Al processing. Transverse to the loading direction just the opposite can be observed. At high processing temperatures, due to good fibre-matrix bonding because of fibre-matrix interactions, higher fracture stresses can be realized, as in the case of poor bonding (at low processing temperatures). A sufficiently good C-Al composite can probably only be produced when a protective fibre coating is developed which suppresses fibre-matrix interactions.

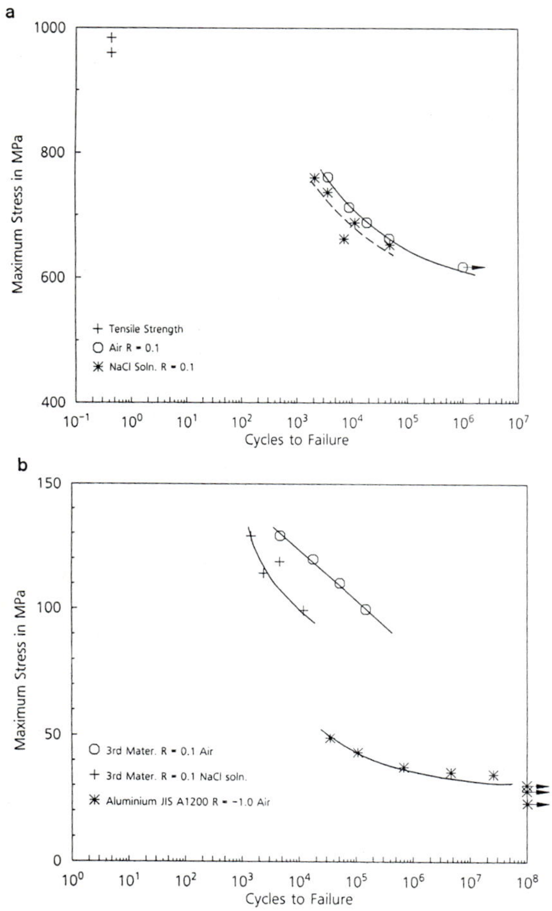

Figure 2.35 Fatigue behaviour in 3.5% NaCl solution.
a) Fatigue behaviour in fibre direction [0 °].
b) Fatigue behaviour transverse to fibre direction [90 °].

The type of specimen used in a tensile test essentially influences the fracture stress achieved, while the Young's modulus remains uneffected [316]. Table 2.21 shows a comparison of the tensile fracture stresses achieved with flat and cylindrical specimens. The material properties obtained using flat specimens were inferior to those obtained with round specimens. This can be explained by differences in strain constraint between specimens, accuracy of area assessment, and circumference-to-area ratio [317].

	[0 °], C-Al		[90 °], C-Al	
	600 °C 30 min	580 °C 15 min	600 °C 30 min	580 °C 15 min
σ_B in MPa	445	638	33	8
ε_B in %	0.38	0.45	0.11	0.03
E in GPa	154	175	34	30

Table 2.20 Tensile properties of C-Al dependent on processing temperature

Fracture stress MPa	Flat specimen [0 °]	Cylindrical specimen [0 °]
SiC-Al	594	920
Al_2O_3-Al	333	538

Table 2.21 Comparison of tensile fracture stresses achieved at flat and cylindrical specimens

Dynamic Loading: In fatigue tests on unidirectional fibre reinforced composites, superior fatigue properties can be achieved compared to bulk aluminium alloys. Figure 2.34a shows the S-N curve for SiC fibre reinforced aluminium. When the same material is fatigue loaded (R = 0.1) transverse to the fibre direction, its fatigue properties remain comparatively good (Figure 2.34b). However, at a fatigue load level of R = -1 a pronounced influence on the fatigue behaviour is observed and with increasing number of load cycles the composite reaches the fatigue level of the matrix alloy.

Environmental Influence: The environmental influence on the mechanical properties of metal matrix composites has only scarcely been studied. Therefore, tests were made in a 3.5% NaCl solution. The investigation has shown that under static mechanical loading no influence of the environment can be expected. Under fatigue loading parallel to fibre direction, the fatigue strength was slightly smaller than in air (Figure 2.35a). However, when the fibres were oriented transverse to the loading direction, a pronounced reduction in fatigue life could be observed (Figure 2.35b) The results show that metal matrix composites with an aluminium matrix have, on the present development stage, already sufficiently good mechanical properties, so that their transfer into composite parts is of great interest.

They seem to be especially suitable for fatigue loaded parts. Designers and engineers seem not to be aware that metal matrix composites have their greatest potential when loaded in compression. No other engineering material is known to the author having similar properties.

2.2.2 Fibre Reinforced Aluminium Laminates

K.-H. Trautmann and H. Döker

An essential step with respect to an improved crack propagation behaviour of lightweight materials was reached in the 1970s at Fokker by adding high strength fibres between the layers of adhesive bonded aluminium structures [318-320]. After extensive investigations together with Delft University of Technology the new material ARALL (Aramid Reinforced Aluminium Laminates) was developed. This material is a compound of thin sheets of high strength aluminium alloys (thickness 0.3 mm - 0.5 mm) and layers of a structural adhesive reinforced with fibres of an aromatic polyamid (aramid fibres) [321-323]. Aramid fibres combine high specific tensile strength with low fatigue sensitivity. Figure 2.36 shows the laminating order of two types of ARALL.

The aluminium sheets have to be pretreated for the bonding process (cleaning, pickling, anodizing, and applying of a primer). In the ARALL type on the left side of Figure 2.36 the aluminium sheets are bonded together by fibre-adhesive prepreg layers consisting of about 50% undirectionally oriented fibres and 50% partially cured adhesive. A prepreg of this type has been developed by the 3M company especially for its application in ARALL. The thickness of the prepreg is about 0.2 mm. In the ARALL type on the right side of Figure 2.36 the fibre-adhesive layer is built up by an aramid fabric with 90% of the fibres in the loading direction, which is embedded in two thin layers of a structural adhesive. The laminate is hot cured under pressure. For typical adhesives the curing temperature is 120 °C and the curing time between 30 min and 1 h. Pressure between 6 and 10 bar is necessary for the fabric material.

Meanwhile some other types of fibre reinforced aluminium laminates have been developed. In GLARE the reinforcement consists of high strength glass fibres. Whereas the reinforcement for ARALL is mainly undirectional, for GLARE undirectional and cross-ply prepregs with different fibre volume ratios for both axes are used. This extends the range of application of GLARE to biaxially loaded structures.

In a report by C.T.Lin et al. first investigations of carbon fibre reinforced aluminium laminates CARALL are described [324]. The CARALL material consisted of 1 mm thick 2024-T3 aluminium sheets and a unidirectional Toray T300 pre-

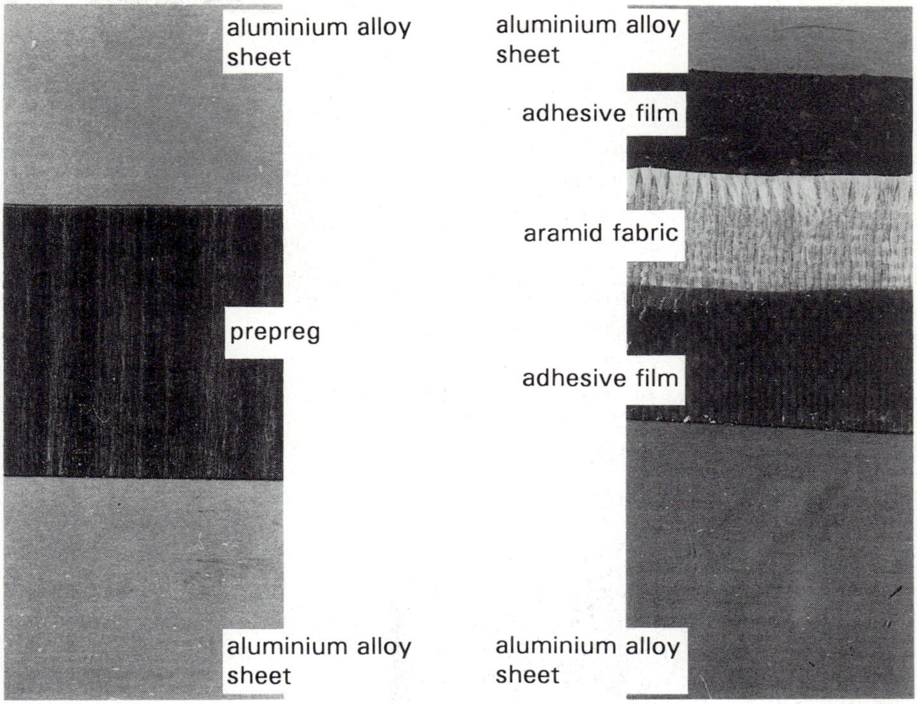

Figure 2.36 Laminating order of two types of ARALL

preg. Because of the different thermal expansion coefficients between the aluminium sheets and the carbon layers a glass cloth/epoxy prepreg is placed between these layers to reduce the production-introduced residual tensile stresses in the aluminium layers.

Fatigue Behaviour of Fibre Reinforced Aluminium Laminates

The outstanding mechanical property of these laminated composites is their excellent resistance to fatigue crack growth. Figure 2.37 shows the principal failure mechanisms of ARALL. The good fatigue properties are caused by intact fibres which bridge the fatigue cracks in the metal sheets, thus limiting the crack opening displacement and the stress intensity at the crack tip. Parallel to the crack in the metal sheets a delamination between the fibre layers and the aluminium sheets develops. The bridging forces in the fibres are transferred into the metal sheet by shear deformations in the adhesive.

The crack propagation process in the aluminium sheets is additionally affected by residual stresses. Marissen et al. showed already in 1980 that the curing process causes residual tensile stresses in the aluminium sheets and compression stresses

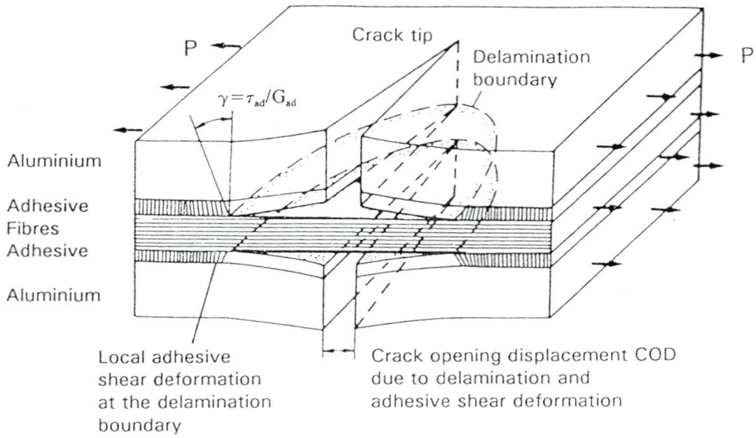

Figure 2.37 Schematic illustration of the failure mechanisms of ARALL during fatigue crack growth

in the fibres [325]. This is due to the low thermal expansion coefficients of the fibres. Tensile stresses in the metal sheets basically have an unfavourable influence on crack growth behaviour. By prestressing the aramid fibres during the curing process or by prestraining the finished laminate before manufacturing structural parts, the residual curing stresses can be eliminated or even reversed &lbracket324-327] Figure 2.38 shows the favourable crack growth behaviour of ARALL as an example of these laminated composites. It is compared to the behaviour of pure monolithic 7075-T6 sheets and to laminated but not reinforced 7075-T6 sheets. The influence of different amounts of residual stress is also shown. The tests were performed with centre notched specimens and flight-by-flight loading. The applied civil aircraft loading spectrum TWIST was truncated at a level of 1.3 S_{mf}, and the maximum stress level was 230 MPa. It can be seen that ARALL's crack propagation characteristics are completely different from those of 7075-T6. The curves show for increasing crack lengths increasing crack propagation rates for 7075-T6 specimens, however decreasing crack propagation rates for ARALL.

The influence of post stretch is shown for the three different compression residual stress levels in the aluminium sheets (-95 MPa, -120 MPa, -145 MPa). These curves show that the higher the stress level, the better the crack propagation behaviour.

Fatigue crack growth in the metallic part of fibre reinforced laminates is accompanied by delamination growth between the fibre layers and the aluminium

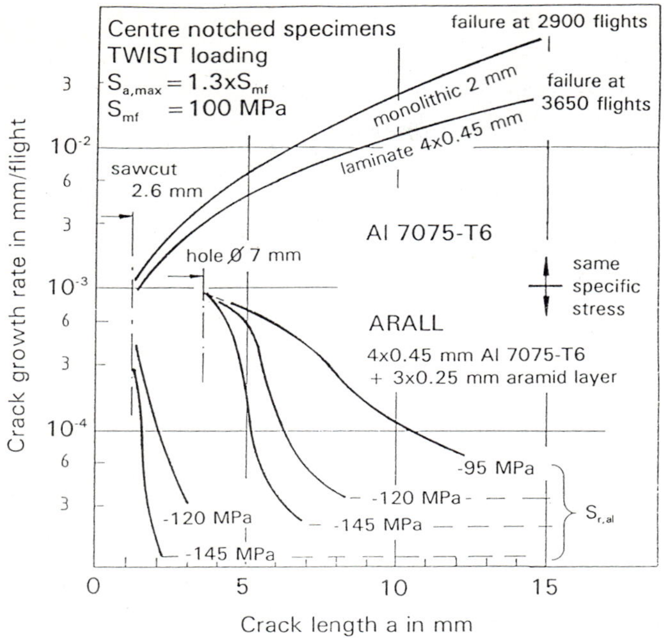

Figure 2.38 Crack propagation behaviour of ARALL and Al 7075-T6

sheets. This delamination growth as a consequence of cyclic shear stresses in the adhesive between the different layers has been studied thoroughly by Marissen (see Figure 2.37). In this study special ARALL specimens were used, where the loads introduced in the metallic parts were transferred only by shear stresses to the aramid fibres. The investigated delamination growth curves were described analytically by an equation of the type $db/dN = C \cdot \Delta G^m$, with b as the delamination size and G as the energy release rate. With this equation for delamination growth and the basic da/dN vs ΔK_{eff} behaviour of the metallic part of ARALL a prediction model for crack extension in fibre reinforced aluminium laminates has been developed [328]. Delamination growth can be monitored by ultrasonic inspection (C-scan). Figure 2.39 shows a C-scan from the middle part of a fatigued GLARE 2 specimen as an example. The shape and size of the delamination zone can clearly be recognized and the extension in crack propagation direction is comparable to the measured crack length.

Beside the excellent crack propagation behaviour in the macrocrack stage, microcrack growth in fibre reinforced aluminium laminates is often also improved. Cracks normally initiate at locations with high stress levels (notches, rivets, etc.) and propagate in monolithic material with one crack front through the whole

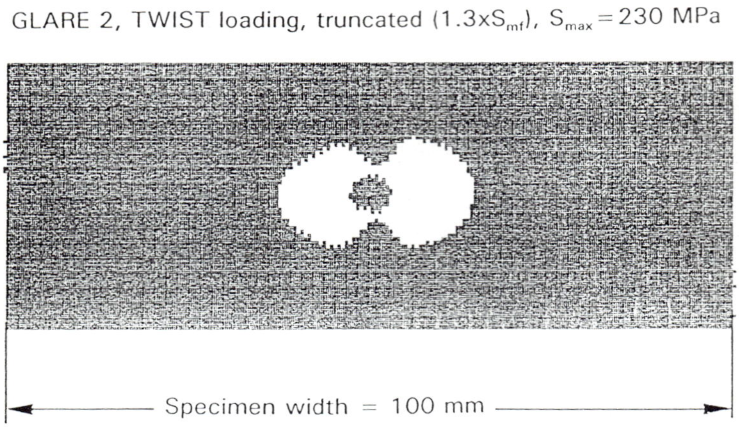

Figure 2.39 Delamination zone size (white area), taken after 200,000 TWIST flights by ultrasonic scan. The black region in the middle is the centre notch of the specimen.

specimen or structural part. In the reinforced material, however, the reinforced adhesive layers act as a barrier and cracks usually do not penetrate these layers. Therefore, individual cracks have to initiate in each aluminium sheet and this effect can improve short crack behaviour, especially in the case of compressive residual stresses in the Al sheets [329].

Figure 2.40 shows the results of a comparison of an ARALL type and different aluminium alloys [330]. The test series was performed with double-lug specimens, representing a highly loaded structural part, under constant amplitude loading and the standard flight-by-flight spectrum FALSTAFF. The specimens were pin loaded. Crack initiation was influenced by the contact pressure and the friction between pin and specimens.

The thickness of all specimens was in the range between 5 mm and 6.3 mm. Two types of constant amplitude loading were applied to the specimens, one without compression and one with a compression part in the load cycle, and the flight-by-flight loading with a higher maximum net section stress. The parameters evaluated were crack initiation, crack propagation and total fatigue life. Crack initiation life was defined as the number of cycles, after which the first crack reached a crack length of 1.0 mm.

For constant amplitude loading without compression, ARALL shows nearly the same crack initiation life as the metallic specimens, but a more than ten times

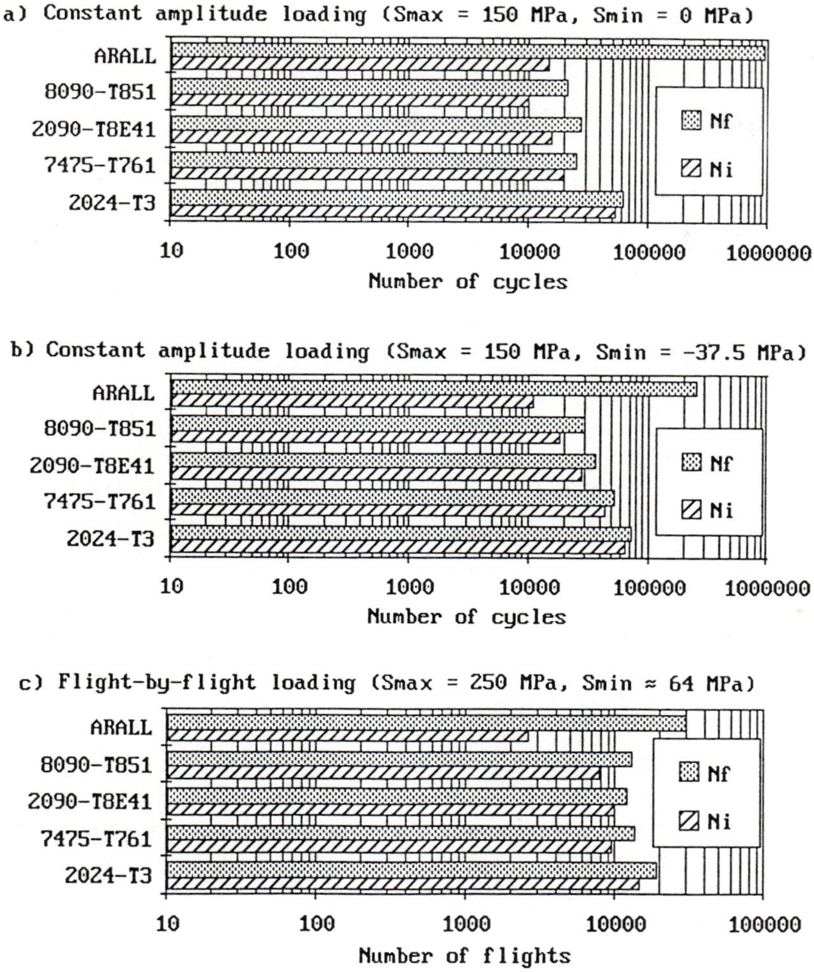

Figure 2.40 Total fatigue life Nf and crack initiation life Ni for ARALL and different advanced aluminium alloys

higher total life. Under constant amplitude loading with compression the pure metallic specimens showed nearly the same results as for R=0 conditions, but the crack initiation life of ARALL was now the shortest and the total fatigue life also decreased significantly. For the test with flight-by-flight spectrum, where the maximum net section stress was relatively high, a further decrease of Ni and Nf was found. Under these circumstances ARALL showed a much shorter life to crack initiation, but was slightly superior in total fatigue life to the other materials investigated.

Application of Fibre Reinforced Laminates

Different types of ARALL and GLARE are now available from material suppliers. Because of their superior crack growth behaviour these materials are ideally suited to fatigue critical sheet structures like the pressure cabin and the lower wing skin of aircraft. With ARALL instead of conventional Al alloys for the lower wing skin weight savings between 20% and 30% have been calculated. If a damage tolerant material like ARALL is applied, the fail safe requirements can more easily be fulfilled, because fatigue cracks grow very slowly. Moreover, certain parts previously considered to be safe life items, can now be considered to be fail safe. A typical example are lug type joints.

Material properties other than strength are important for the manufacturing of aircraft. In contrast, for instance, to carbon fibre reinforced plastics several technological properties of fibre reinforced aluminium laminates are similar to those of monolithic aluminium sheets. Hence similar manufacturing processes can be used, even for sheet forming. A comprehensive survey of the basic properties of ARALL, the best characterized fibre reinforced aluminium laminate, is given in [328].

2.2.3 Titanium Matrix Composites

H.J. Dudek and R. Leucht

A short review of the present state of the development of SiC fibre reinforced Ti-6Al-4V alloys is given, with special emphasis on composites processed by applying fibre coating with matrix and hot isostatic pressing (HIPing). Composites processed by hot pressing and by fibre coating and HIPing are characterized and compared.

SiC fibre reinforced titanium alloys are presently mainly processed by hot pressing of alternating layers of titanium alloy foils and fibres [331-335]. The fibres are fixed into position by suitable methods, such as adhesives, weaving with titanium or molybdenum wires [335] or by deposition of a layer of the matrix using, e.g., plasma spraying [336]. Beside hot pressing, hot isostatic pressing of the foil-fibre layers can also be applied [332, 333].

For the processing of Ti-SiC composites another method is also suitable however, in which the fibres are coated with matrix and bundles of coated fibres are hot isostatically pressed (or hot pressed) [337-344]. For simplicity in the following discussion the first processing procedure will be called the "hot pressing", the second one the "fibre coating and HIPing" method. In the present paper a general discussion of the properties of Ti-SiC composites is given and the properties of composites obtained by fibre coating and HIPing are compared with those obtained by hot pressing of metal-fibre layers.

Experimental Procedures

The processing of composites by fibre coating and HIPing is described in some detail in [344]; therefore only a short summary is given here. Approximately 300 fibre pieces of 80 mm length are put into a cylindrical holder which is introduced into a deposition chamber. During matrix deposition the holder is rotated about the cylinder axis. Three different methods for matrix deposition were applied: electron beam evaporation (EB-PVD), magnetron sputtering and random arc. The coated fibres are put into tubes made of the matrix material, encapsulated in stainless steel capsulae, outgassed at 450 °C for several hours, closed by welding and hot isostatically pressed at 900 °C/0.5h/1900 bar. Samples with a round cross section of 3.5 mm in diameter and a test length of 30 mm were machined. Single fibre testing, metallographic examination, tensile and fatigue investigations and fractography were performed.

In the present report only composites reinforced with protected SiC fibres (SCS-6-fibres from Textron Speciality Materials, Lowell, MA) are discussed. (The microstructure of these fibres and their protecting coating is investigated in [345-347]). Earlier publications on unprotected SiC, or boron fibres are not included in the present discussion, because strong fibre strength degradation due to fibre-matrix interaction during processing was observed in these composites [339, 348, 349].

Composite Characterization

Coating of Fibres: The three deposition methods applied result in matrix coatings on the fibre with different properties. For the random arc and the EB-PVD deposition methods the chemical composition of the coatings on the fibres is different from that of the target material. Only by using magnetron sputter deposition in the coating the same element concentration is obtained as in the target.

More important than identity of the chemical composition with the target is however the incorporation of oxygen from the residual gas pressure of the vacuum chamber into the coating during deposition. For the three deposition methods the oxygen content of the coating could be decreased below 2000 ppm by choosing suitable deposition conditions. As will be discussed later, a higher oxygen content results in a brittle fracture surface, lower tensile strength and reduced fatigue properties.

Considering an ideal hexagonal fibre arrangement in the cross-section of the composite, Figure 2.41a, all fibres have equal distances B to each other. Each fibre can be related to a parallelogram with height $\sqrt{3/2}$ B, as shown in

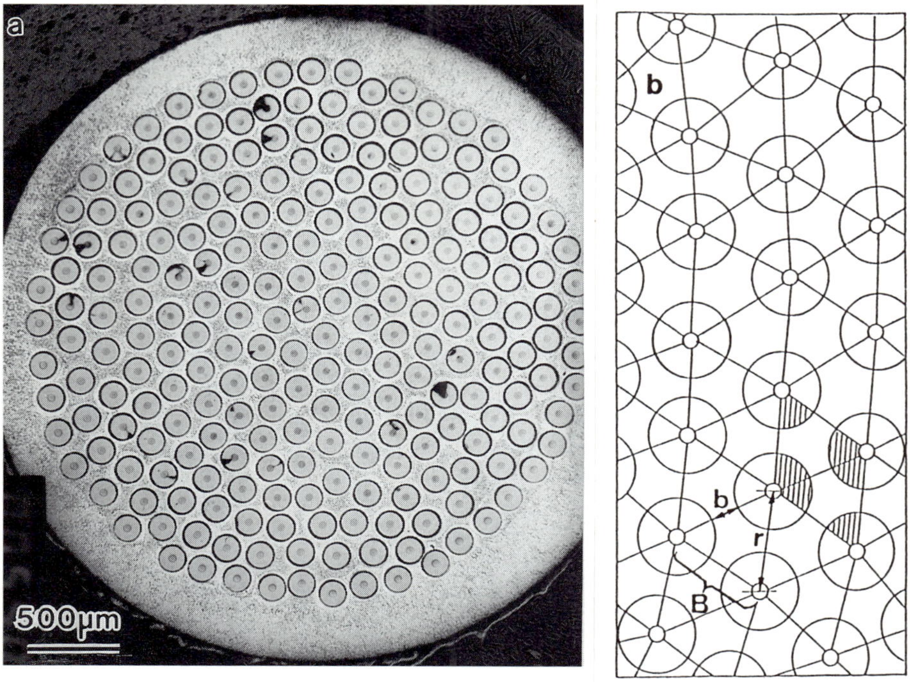

Figure 2.41 a) Typical metallographical polish of a transverse section of the SiC-Ti-6Al-4V-composite processed by fibre coating and HIPing.
b) Enlarged cutting from a) showing the hexagonal fibre distribution and the determination of fibre distance b and fibre volume fraction v_f.

Figure 2.41b. The area of the fibre in the cross section $F_f = \pi \cdot r^2$ (r fibre radius) in relation to the area of the parallelogram $F_p = \sqrt{3/2} \cdot B^2$, results in the fibre volume fraction $v_f = F_f/F_p$. Using this relation the distance b between the fibre surfaces can be computed to

$$b = B - 2r = r \cdot \left[\sqrt{\frac{2\pi}{v_f \cdot \sqrt{3}}} - 2 \right] \tag{2.8}$$

In the reversal case using the mean value of the fibre distance, the fibre volume fraction can be determined (see later). For fibre distances b≥0 the fibre volume fraction has to be

$$v_f \leq \frac{\pi}{2\sqrt{3}} = 0.907 \tag{2.9}$$

In Figure 2.42a the ratios b/r and b for two kinds of fibres with r=50 μm and r=70 μm are shown as a function of the fibre volume fraction v_f. In a composite with a fibre diameter 2r=140 μm and a volume fraction of $v_f = 0.2$ the fibre distance is b=160 μm (with 2r=100 μm, b=115 μm) and for $v_f = 0.6$ a fibre distance of b=32 μm results (with 2r=100 μm, b=23 μm).

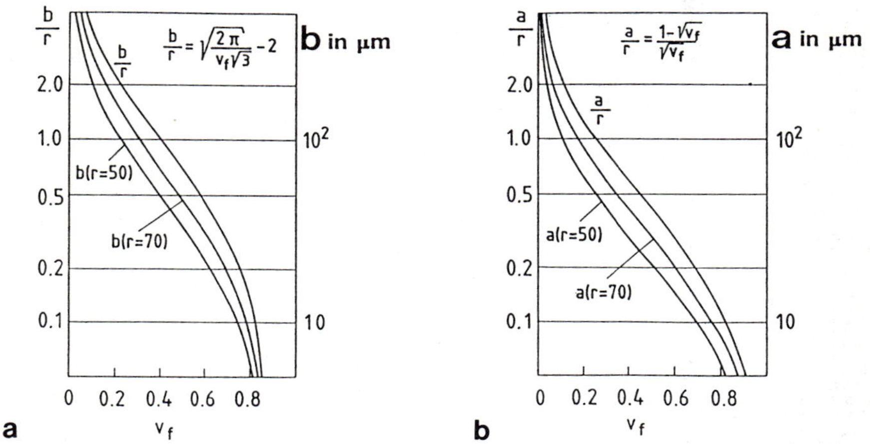

Figure 2.42 a) Fibre distance b in relation to fibre radius and fibre distances b for two fibre diameters 2r=100 μm and 140 μm as a function of the fibre volume fraction v_f.
b) Fibre coating thickness a in relation to the fibre radius r and a values for two fibre diameters 2r=100 μm and 140 μm needed to obtain a composite with a volume fraction v_f.

For a fibre coating thickness a, the area of the cross section of the matrix on the fibre is $F_m = \pi \cdot (2ar + a^2)$. Supposing the material is compressed completely, then the fibre volume fraction is

$$v_f = \frac{F_f}{F_f + F_m} = \frac{r^2}{(r+a)^2}$$

which results in a coating thickness for a volume fraction v_f of

$$\frac{a}{r} = \frac{1 - \sqrt{v_f}}{\sqrt{v_f}} \tag{2.11}$$

In Figure 2.42b the ratio a/r and a values for r = 50 μm and r = 70 μm are shown. The effect of v_f on a is similar to its effect on b: a decreases sharply with v_f. For $v_f = 0.2$ and for a fibre diameter 2r = 140 μm a coating thickness of a = 88 μm is needed, and for $v_f = 0.6$ a thickness of a = 20 μm.

Composite Consolidation: The quality of the composite also results from the fibre distribution in the cross section. Using the cross section of Figure 2.41a, the frequency of fibre distances is determined and the results are shown in Figure 2.43a. In Figure 2.43b, a frequency of fibre distances obtained from a polish of a hot pressed sample processed by Textron is shown. The polish of the Figure 2.41a from [350] was evaluated for these results. In samples processed by fibre coating and HIPing a sharp maximum in the fibre distance distribution is obtained, Figure 2.43a, with a probable fibre distance of approximately 35 μm and a FWHM of 40 μm. In composites processed by hot pressing two maxima are obtained: one sharp maximum at roughly 110 μm with a FWHM of 20 μm and a second maximum at 50 μm with a FWHM of 80 to 100 μm. The maximum at 110 μm has to be assigned to the thickness of the foils, and the broad maximum at 50 μm to the distances of the fibres in the fibre layers. The broad distribution of the maximum at 50 μm results from the difficulty of fixing the fibres in the layers during hot pressing.

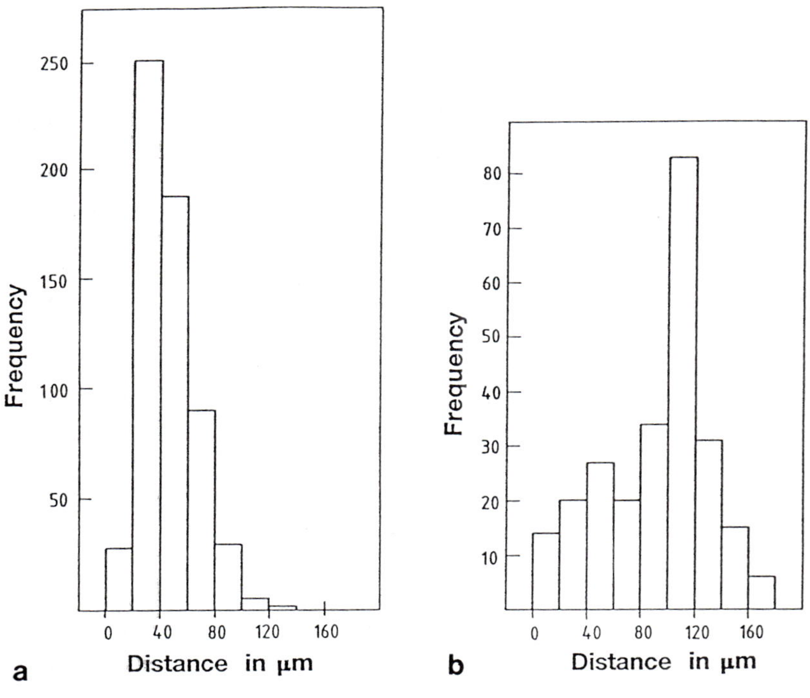

Figure 2.43 a) Frequency of fibre distances for a SiC fibre reinforced Ti-6Al-4V-alloy processed by fibre coating and HIPing.
b) Frequency of fibre distances for a SiC fibre reinforced Ti-6Al-4V-alloy processed by hot pressing using a polish published in [350].

A further criterion of the quality of composite consolidation is short fibre distances. For the fibre coating and HIPing process of Figure 2.43a the number of short distances between 0 and 20 μm is 27/618=0.044 (ratio of the number of distances in the region 0 to 20 μm to the number of all computed distances). A similar result of 14/240=0.058 is obtained for the hot pressed composite in Figure 2.43b, however for a much smaller volume fraction (see below).

From all distances of Figure 2.43a the mean value in composites processed by fibre coating and HIPing, is computed to be $b^*=49.3$ μm. From the hot pressed sample according to Figure 2.43b, $b^*=90.9$ μm. Using equation (1) and b^* the fibre volume fraction v_f can be computed. For the fibre coating and HIPing process v_f $(b^*=49.3) =0.496$ and for the hot pressing process $v_f(b^*=90.9)=0.334$. This is in fair agreement with the results obtained by integrating the areas of all fibres and the polish in the cross section. (Textron supplies the customers with the value 0.32 ± 0.02 [351].)

Grain Structure: The matrix properties of the composites cannot be determined easily. For composites processed by hot pressing one can assume that the properties of the matrix are similar to those of the unreinforced alloy with the same thermal treatment (compare [352] where the properties of the matrix are determined after processing). This assumption cannot be made for the composites processed by fibre coating and HIPing, because the matrix is transferred onto the fibres by an evaporation process and, as discussed, the chemical composition as well as the grain structure is changed. The grain structure of the material in the coating is very fine and different for the three deposition methods investigated. Applying HIPing on this grain structure favours superplastic deformation [353] (using HIPing a deformation speed of approximately 0.1 μm/s is realized) and a very fine grained structure with grains of 3 μm for sputter deposition, 5 μm for EB-PVD and 10 μm for the random arc deposition results [344].

A further advantage of the fibre coating and HIPing process is that each material which can be deposited on the fibre can also be used as matrix for the composites [354]. The method is not restricted to matrices for which foils can be produced (compare [332]).

Interfaces: The main components of the fibre-matrix interface in the SiC-SCS-6-fibre reinforced titanium alloys are [355-357]: a) the interface between the SiC fibre and the protecting coating, b) the protecting coating, which consists mainly of isotropic pyrolytic carbon with varying concentrations of SiC-particles, some ten nanometres in diameter, d) the interface between the protecting coating and the reaction zone, e) the reaction zone which develops during composite processing, consisting mainly of titanium carbide, titanium silicides and carbon enriched matrix, and d) the interface between the reaction zone and the unreacted matrix.

During thermal loading of the composite the reaction initiated during processing progresses, the protecting coating dissolves in the matrix and the reaction zone thickness increases. The thickness of the reaction zone increases according to a parabolic law [358-360]. However, the remaining coating protects the fibre as long as no titanium from the matrix can reach the fibre surface [355]. The reaction speed between the coating and the matrix is decisively determined by the chemical composition of the matrix, with high reaction rates for pure titanium and the Ti-6Al-4V alloy and low rates for alloys such as Ti-14Al-21Nb and Ti-1100 [360, 361].

As will be discussed later the fibre-matrix interface is one of the main components influencing the mechanical properties of SiC-Ti-composites. The adhesion between fibre and matrix determines transverse strength, toughness and thermal strength. The fatigue life of the composites depends on the strength of the reaction zone, which decreases with its increasing thickness [362]. The optimization of the interface is therefore one of the major subjects in the development of Ti-SiC-composites.

Mechanical properties of the composites

Stress-Strain Behaviour: For most Ti-SiC-composites processed by hot pressing, tensile test experiments result in a stress-strain behaviour with two linear regions and a nonlinear region [335, 362]. The transition from the first to the second linear region takes place for a strain in between 0.3% and 0.5%. (The mean rupture strain of the fibres is approximately 1.2% [363]. In composites the statistical distribution of the rupture strain of the fibres in the fibre bundles has to be considered and the strain of the composite is limited to the break of the first fibres).

In composites processed by fibre coating and HIPing also two regions in the stress-strain behaviour are observed, curve a in Figure 2.44. Sometimes however curves are measured in which a point of inflection is observed between the two regions, Figure 2.44, curve b. Such behaviour is only observed in samples with a brittle matrix. Brittle fracture of the matrix in tensile test experiments is observed in composites with a high oxygen content. In the fracture surface of composites with brittle matrix only minor fibre pull-out is observed [344]. The point of inflection in the stress-strain curve should be assigned to catastrophic fracture of a larger number of fibres. The curve c in Figure 2.44 shows the stress-strain behaviour of a hot pressed composite [362]. The curves a and c are similar.

Tensile Strength: In Figure 2.45 the tensile strength of composites obtained by fibre coating and HIPing is shown as a function of the fibre volume fraction v_f separately for samples with a ductile (region A) and with a brittle matrix (region B). (Experimental details are given in [344].) For comparison some results are added from the literature for composites processed by hot pressing. As not all publications were available to discuss their results, no comparison can be made of the absolute values resulting from the two processing methods. There is no doubt, however, that for composites with a ductile matrix processed by fibre

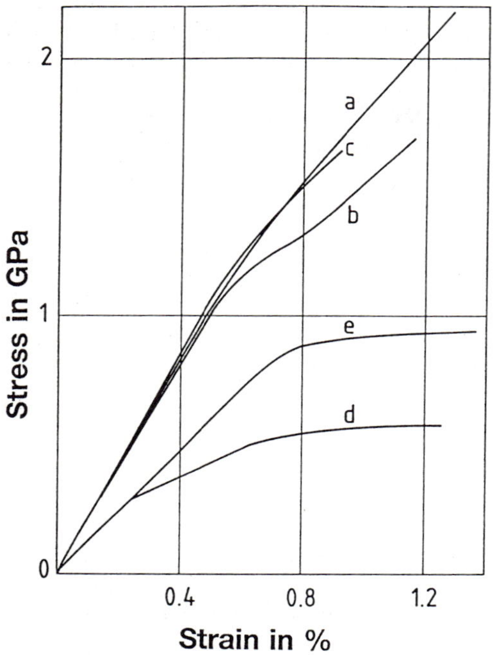

Figure 2.44 Stress-strain behaviour of a composite processed by fibre coating and HIPing with a ductile matrix (a) and a brittle matrix (b), a SiC-Ti-6Al-4V composite processed by hot pressing according to [362] (c), a SiC-Ti-6Al-4V composite tested in transverse direction (d) and for the unreinforced Ti-6Al-4V matrix with the same thermal history and sample geometry as the composites (e).

coating and HIPing, high tensile strength and Young's modulus values, according to the rule of mixture, are measured. (Young's modulus values in these composites are independent of matrix ductility, always according to the rule of mixture [344].)

The influence of matrix ductility on tensile strength was also investigated in SiC fibre reinforced Ti-15V-3Al-3Cr-3Sn alloys [335]. The ductility was varied by applying different thermal treatments to the composites. Varying the tensile strength of the matrix by thermal treatment, a correlation of the tensile strength of the composite was observed according to the rule of mixture. The reduced tensile strength of the composites with the brittle matrix in our investigations will therefore be likely explained by the reduced tensile strength of the brittle matrix.

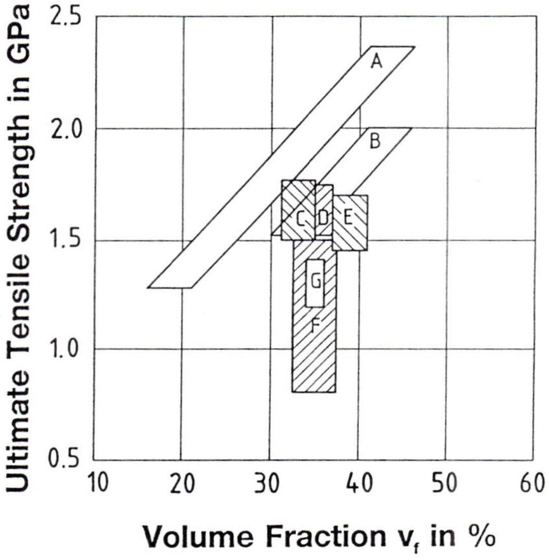

Figure 2.45 Ultimate tensile strength of SiC-SCS-6-fibre reinforced titanium based alloys, A = composites processed by fibre coating and HIPing with a ductile Ti-6Al-4V matrix, B = composites processed by fibre coating and HIPing with a brittle Ti-6Al-4V matrix (oxygen content > 3000 ppm), C = SiC-SCS-6-Ti-6Al-4V composite processed by hot pressing [350], D = SCS-6-SiC-Ti-6Al-4V processed by hot pressing [362], E = SiC-SCS-6-Ti-15V-3Al-3Cr-3Sn processed by hot pressing [362], F = SiC-SCS-6-Ti-14Al-21Nb processed by hot pressing [336], G = SiC-SCS-6-Ti-25Al-10Nb processed by hot pressing [362].

The reliability of the SiC fibre reinforced Ti-6Al-4V alloys processed by fibre coating and HIPing can be concluded from a statistical evaluation of the tensile strength and Young's modulus values obtained in the present investigations. All results were for a fibre volume fraction of 35% applying the rule of mixture. The results are summarized in Table 2.22.

Transverse Strength: For the transverse strength of SiC fibre reinforced Ti-6Al-4V alloys processed by hot pressing (Textron), a mean value of 592 MPa [350], and for SiC-Ti-15V-3Cr-3Al-3Sn composite with $v_f = 0.35$ a value of 420 MPa [364], were measured. The low transverse strength is considered to be caused by low adhesion between fibre and matrix [350]. The contribution of chemical bonding to fibre-matrix adhesion is low compared with the contribution of the mechanical "clamping" stresses [350, 365]. For the contribution of mechanical stresses to the transverse strength, a value of 310 MPa was computed [350]. Loading the composites in transverse direction results first in a reduction of the

	Ultimate tensile strength in MPa		
	Mean value	Scattering interval	Standard deviation
ductile matrix	1,820	±12.0%	205 MPa
brittle matrix	1,597	±7.5%	67 MPa
	Young's modulus in MPa		
	Mean value	scattering interval	standard deviation
ductile matrix	197,507	±7.8%	9,250 MPa
brittle matrix	198,507	±3.9%	6,580 MPa

Table 2.22 Scattering interval and standard deviation of UTS and Young's modulus of SiC-Ti-6Al-4V composites processed by fibre coating and HIPing

intrinsic mechanical stress, after which only the matrix of the unreinforced regions carries the load.

The stress-strain behaviour of a composite loaded in transverse direction shows three regions, curve d of Figure 2.44. After a linear elastic behaviour the curve has a knee and a second linear region is measured. After a second knee a region with a low slope follows. The first knee is related to a debonding of fibres, the second linear region to the elastic behaviour of the matrix [350].

Shear Strength: For determining shear strength mainly three methods are presently applied [366, 367]: the fibre fragmentation test [363, 368-370], the fibre pull-out test [371] and the indentor test [365, 372-374]. In SiC-Ti-6Al-4V composites processed by hot pressing using the fragmentation test shear strengths of 295 MPa [368], 180 MPa [363] and 120 MPa [369]have been observed. For the indentor method values of 156 MPa [369] and 276 MPa [350] are reported. Thermal treatment influences shear strength: In [371] nearly constant values are measured from 156 MPa at RT to 165 MPa at 800 °C/50h. In [350] an increase from 276 MPa to 358 MPa for 900 °C/10h was observed. A load of 2000 cycles between 27 °C and 427 °C reduces the shear strength of the composite to 208 MPa [350].

In [369] shear strength is measured by the fragmentation method as a function of the test temperature, and a linear decrease of shear strength from 200 MPa at RT to very low values at 800 °C was observed. This result is explained by a reduction of the radial compressive stress in the matrix.

The shear strength τ and the tensile strength σ of a fibre with critical length L can be computed according to

$$\tau = \sigma \cdot \frac{d}{2L} \tag{2.12}$$

where d is the fibre diameter. (For the dependence of critical length on test temperature see [369].) Measuring the shear strength τ by the indentor method and the mean value of the critical fibre length in a polish parallel to the fibre of a tensile test sample the "in situ" fibre strength can be determined [371]. For SCS-6-SiC fibres in a Ti-6Al-4V-matrix, composite strength was determined after a thermal treatment of 800 °C for 50 h and 100 h. Values of 3 GPa for 800 °C/50h and 2.5 GPa for 800 °C/100h were obtained [371]. This is in disagreement with observations, in which SCS-6-fibres coated with titanium annealed at 850 °C/8h show no strength reduction [375]. From analytical TEM results one has to expect that the SCS-6 coating is stable enough to prevent fibre strength reduction [355] during that treatment. It is not clear whether the above method for "in situ" fibre strength determination leads to false results [363] or whether the processing of the composites has reduced the fibre properties. In the second case, the reduced values for fibre strength have to be applied in the rule of mixture.

The shear strength of SCS-6-SiC-Ti-6Al-4V composites processed by fibre coating and HIPing using the indentor method was investigated. The microhardness diamond and a cylindrical stainless steel flat-bottomed punch with a diameter of 110 µm were applied, and thin slices of the composite cut perpendicular to the fibres were used. The fibre push-out test yielded the following observations [372]:

1. The results of the shear strength measurements can be distorted by fibre loosening during surface preparation. The thickness of the composite slices should therefore not be lower than 0.8 mm.
2. Shear strength depends on the fibre coating method. In composites with a brittle matrix processed using magnetron sputtering for matrix deposition (temperature of the fibre during deposition 250 °C), a shear strength of 150 MPa is measured. Composites obtained using EB-PVD (fibre temperature during deposition 800 °C) have a shear strength of 180 MPa.
3. Thermal treatment of 650 °C/170h results in an increase of shear strength to 205 MPa.
4. During indentor loading the fibre is squeezed out with the coating as well as without the coating. The shear strength in both interfaces should therefore be of the same order (compare [371]).

Shear strength in Ti-SiC composites can be influenced by chemical and mechanical contributions. The following mechanical contributions are considered to be in action [350]: a) radial pressure on the fibre, which is built up during cooling after processing the composite. It is caused by the difference in the thermal expansion coefficients of fibre and matrix, b) pressure or tensile load due to the growth of new phases with a different specific volume during consolidation, c) radial interfacial stress on the fibre during composite loading in fibre direction due to difference in the Poisson coefficients of fibre and matrix. d) During tensile loading in fibre direction a contribution from friction between the fibre and the matrix should be taken into account. Using the indentor method an amount of 48 MPa was determined for the SiC-Ti-24Al-11Nb composites [365].

Due to the small thermal expansion coefficient of the fibre in comparison to the matrix (SiC: $5 \cdot 10^{-6}$ K^{-1}, Ti-6Al-4V: $10 \cdot 10^{-6}$ K^{-1}, [369, 376], composite: $6.5 \cdot 10^{-6}$ K^{-1}, [377]) a compression load on the fibre of 310 MPa is computed. According to [350, 364] the shear strength of the Ti-SiC composites is mainly determined by mechanical contributions. The influence of the chemical composition of the matrix, of the thermal treatment and of the chemistry of the fibre surface [363] on shear strength [371] suggests however a chemical contribution to shear strength.

The shear strength in the Ti-SiC composites is lower than the shear strength of the matrix, which has a value of 450 MPa (1/2 of the matrix tensile yield strength). The fibre-matrix interface is therefore the weakest link in the composite for a shear loading. This is important for the mechanical properties, e.g. for the toughness.

Fatigue Strength: A direct comparison of the published fatigue strength results for Ti-SiC composites is difficult, because the material and testing conditions, such as fibre volume fraction, interface adhesion, loading conditions and geometry, are very different. In comparison to unreinforced material, for the Ti-SiC composites both lower [335], as well as improved [332, 378-380] fatigue strength is observed. According to [332, 381] the kind of fibres used for the composites has a decisive influence on fatigue life. In the SiC-Ti-6Al-4V-composites a lower fatigue strength was observed compared to the B_4C/B-Ti-6Al-4V composites. These results are explained by a larger reaction zone in the SiC reinforced material. During fatigue loading, cracks are initiated in the reaction zone which are responsible for the reduction of fatigue life [369].

Composites processed by fibre coating and HIPing have an improved longitudinal fatigue life in the low cycle fatigue region compared with the unreinforced material, as is shown in Figure 2.46. A sample with a ductile matrix loaded by the stress amplitude of ±950 MPa (R=-1) has a life of 26,500 cycles, a sample with a brittle matrix still 12,000 cycles, the unreinforced material however only 459 cycles. This is simply explained by the higher tensile strength of the reinforced material. For composites with a ductile matrix, the fatigue life at low stress amplitudes (below ±600 MPa) is however also higher. In the range of medium stress amplitudes, a larger scattering of the fatigue life is observed. For samples with a brittle matrix the fatigue life at low stress amplitudes is lower than for the unreinforced material.

Stress amplitude probably influences crack initiation and crack propagation [344]. At high load amplitudes crack initiation and propagation will likely take place in the matrix. The fatigue life in reinforced materials at these stresses is increased by the bridging effect of the fibres. For stress amplitudes below approximately ±900 MPa the tensile strength of the reaction zone limits fatigue life. In the composites processed by fibre coating and HIPing the reaction zone thickness is roughly 1 µm. According to [362] a reaction zone of that thickness will have a tensile strength of 900 MPa. The scattering of fatigue life values in the

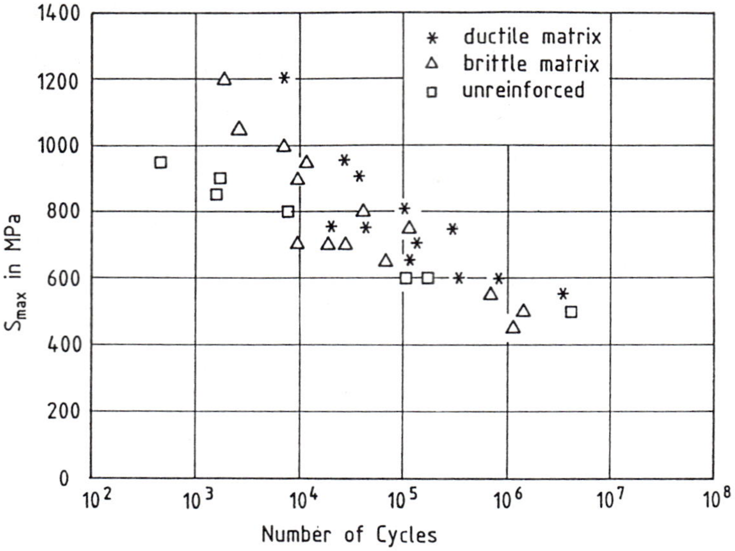

Figure 2.46 Fatigue life as a function of the stress amplitude (R=-1) of SiC-Ti-6Al-4V composites with a fibre volume fraction of 0.35 processed by fibre coating and HIPing with a ductile and a brittle matrix in comparison to unreinforced material with the same sample geometry

region between ±900 and ±600 MPa is probably due to the scatter of strength of the reaction zone. Below a stress amplitude of ±600 MPa the composites with a ductile matrix have a fatigue life comparable to those with an unreinforced ductile matrix. Composites with a brittle matrix have a lower fatigue life compared to the unreinforced ductile material. Both observations make it likely that below ±600 MPa fatigue life is mainly limited by the fatigue properties of the matrix.

The influence of the reaction zone on fatigue life can be clearly detected in the fracture surface of fatigued samples [344]. Nearly each fibre initiates several cracks, Figure 2.47a. In the polish of fatigued samples parallel to the fibres crack initiation can be identified at the reaction zone of many of the fibres. An example in shown in Figure 2.47b.

Toughness: The behaviour of notched samples under loading is influenced by the properties of the composite components, that is by the matrix, the fibres, the fibre protecting coating and the reaction zone. In the region of the notch some microscopic effects participate in the failure mechanism, such as, e.g., fibre break, matrix deformation, crack initiation in the interface, fibre bridging and fibre pull-out [350, 382].

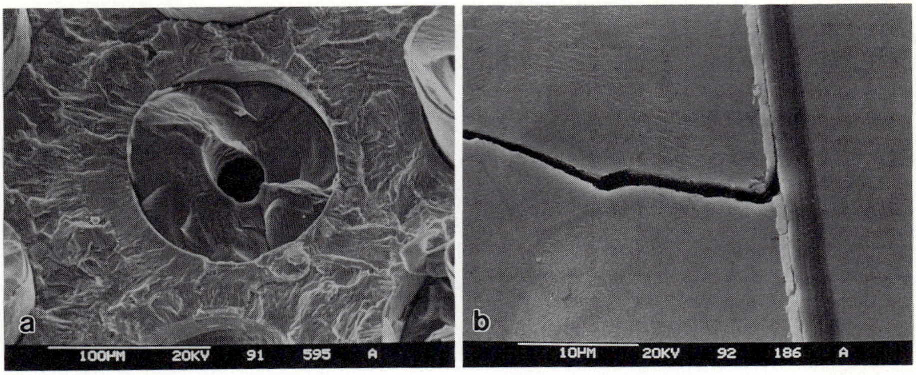

Figure 2.47 a) Fatigue fracture surface of a SiC-Ti-6Al-4V-composite with fibre volume fraction of 0.35 processed by fibre coating and HIPing loaded at ±900 MPa (R=-1).
b) Fatigue crack initiation in the reaction zone of a SiC-Ti-6Al-4V composite loaded at ±900 MPa. Metallographic polish parallel to the fibre direction.

In a three point bending test experiment with SiC-Ti-6Al-4V composites loaded by a constant increasing load, a nonlinear load-displacement diagram with a maximum is observed. The maximum is called the critical load and is related to the break of the first fibre in the composite. Crack propagation is associated with strong plastic deformation in the matrix and is followed by a crack in the matrix and breaking of the fibres in the crack region. For the crack formation energy of these composites a value of approximately 390 $J \cdot m^{-1}$, and for the fracture energy 80 $kJ \cdot m^{-2}$ have been measured [382]. Crack formation energy and fracture energy depend on the matrix and the thermal treatment of the composite, which changes the toughness of the matrix and the fibre-matrix shear strength.

For samples of the SCS-6-SiC-Ti-6Al-4V composites notched perpendicular to the fibre direction and loaded by fatigue, crack growth behaviour with increasing, as well as with decreasing crack growth rates by accompanying increasing crack length was observed [350]. According to [383, 384] a decrease of crack rate with increasing crack length predominates in these composites. The transition from subcritical crack growth to catastrophic failure is determined by the number of intact fibres bridging the crack. This number is responsible for the effective stress intensity range in the crack tip and is essential for the decrease of the crack growth rate with increasing crack length [383, 384].

High Temperature Strength: Annealing SCS-6-SiC-Ti-6Al-4V composites at 850 °C for 150 h has no influence on the longitudinal stress-to-failure measured at room temperature. In a bending test already after 20 to 50 h at 850 °C a strong reduction of the properties is however observed [376]. For SCS-6-Ti-6Al-4V composites processed by Textron with a fibre volume fraction of 0.35 and for which a RT tensile strength in the range of 1620 MPa to 2000 MPa was measured, tensile experiments at 427 °C result in a value of only 1500 MPa. Similar results are obtained in [369] in which a linear reduction of the ultimate strength from 1600 MPa at RT to 800 MPa at 800 °C was observed. The reduction of tensile strength with increasing test temperature is explained by a reduction of the tensile strength of the matrix [377] or by a reduction of shear strength in the interface [369]. Composites annealed after processing at 900 °C/10h have a tensile strength of 1300 MPa measured at 427 °C. For a cyclic thermal loading with 2000 cycles between RT and 427 °C no strength reduction was observed [350]. Thermal cycling decreases the fibre matrix reaction due to reducted adhesion in the interface [385].

A similar behaviour as for unidirectional composites was also observed for 0/90/0 laminates with the Ti15V3Cr3Al3Sn alloy as matrix and 0.375 volume fraction of SCS-6-SiC fibres [352]. After a thermal treatment of 1000 °C/1h a 30% reduction of tensile strength was observed. The fatigue life of such thermally treated samples is reduced by 50%.

The thermal treatment is considered to influence mainly the interface between fibre and matrix. The reduction in fatigue life after thermal treatment is due to increased adhesion in the interface, which results in a lowering of crack deflection, brittle crack propagation and fibre fracture [352].

The Ti-6Al-4V alloy is applicable in a composite up to temperatures of 400 °C to 500 °C. Other titanium based alloys are likely applicable up to temperatures of 800 °C [386, 387] (see section "Titanium Aluminides"). The SiC-SCS-6 fibre is also stable up to 800 °C to 1000 °C [347, 375]. The presently available fibre protecting coatings are however stable up to 600 °C [355] for long annealing times. Applications for SiC fibre reinforced titanium based alloys at temperatures up to 800 °C will therefore be possible only, when new matrix systems, new processing methods and fibre protecting coatings stable up to 800 °C are developed [354].

Outlook

The properties of SiC fibre reinforced titanium alloys were reviewed and the properties of composites processed by fibre coating and HIPing compared with those obtained by hot pressing. Careful composite consolidation during processing, as shown for composites processed by fibre coating and HIPing, results in composites with regularly distributed, well enveloped fibres with a high fibre volume fraction up to 50 %. The tensile properties of such composites fulfil the rule of mixture. The fatigue properties are as good as, or better than those of the unreinforced matrix. For higher temperatures a reduction of the mechanical properties has been observed by several authors. Up to temperatures of 600 °C a reduction of tensile strength should only be due to a decrease in the properties of the matrix.

Thermal loading increases the reaction zone width, which is connected with a reduction of fatigue life. The shear strength between fibre and matrix as measured by different authors is on the order of 200 MPa. The interface is therefore the weakest link in the composite. The weak interface can however not prevent a brittle fracture of the composite in tensile test experiments. A brittle matrix lowers tensile strength by approximately 20 %, has no influence on Young's modulus and reduces fatigue life.

Processing SiC fibre reinforced titanium based composites by fibre coating and HIPing is advantageous if foils cannot be produced for the hot pressing method and when components have to be processed by winding techniques [388-390], such as discs, castings and rings [387].

2.2.4 Interfaces in Metal Matrix Composites

H.J. Dudek

Wetting, adhesion and fibre-matrix reaction in metal matrix composites (MMC) are reviewed with the aim of compiling information suitable for optimizing interfaces for processing and applications. In nonreactive interfaces wetting and adhesion is mediated by chemical bonding at atomic dimensions without the development of a reaction zone. These interfaces are stable within a limited interval of temperature, annealing time and element concentration and it is therefore reasonable to define the interfaces in nonreactive systems as an "interface phase". Only metal matrix composites with an interface phase are suitable for long term applications. The properties of the interface phase can be influenced by alloying elements, fibre surface treatments and MMC processing parameters. In reactive MMC systems fibre protecting coatings have to be developed. Such coatings are effective if an interface phase between the coating and the fibre as well as between the matrix and coating can be realized.

Problems

In microscopic investigations of interfaces in metal matrix composites (MMC) three kinds of interfaces are recorded: i) those with poor wetting and adhesion between fibre and matrix, Figure 2.48a, ii) with a reaction zone between fibre and matrix, Figure 2.48c, and iii) composites with excellent wetting and adhesion, but without an indication of a reaction zone even at high magnifications, Figure 2.48b. Examples of composites in which insufficient wetting and adhesion can be observed are carbon fibre reinforced magnesium alloys [391] and alumina fibre reinforced pure aluminium [392]. A reaction zone will be found in composites with SiC fibres in titanium alloys [393] and in α-Al_2O_3 fibre reinforced AlCu and AlMg alloys [394]. The δ-Al_2O_3 fibre reinforced Al piston alloy of Figure 2.48b shows excellent wetting and adhesion without any reaction zone between fibre and matrix [395]. Similar interfaces were described for the δ-Al_2O_3 fibre reinforced AlMg and AlSi alloys [396].

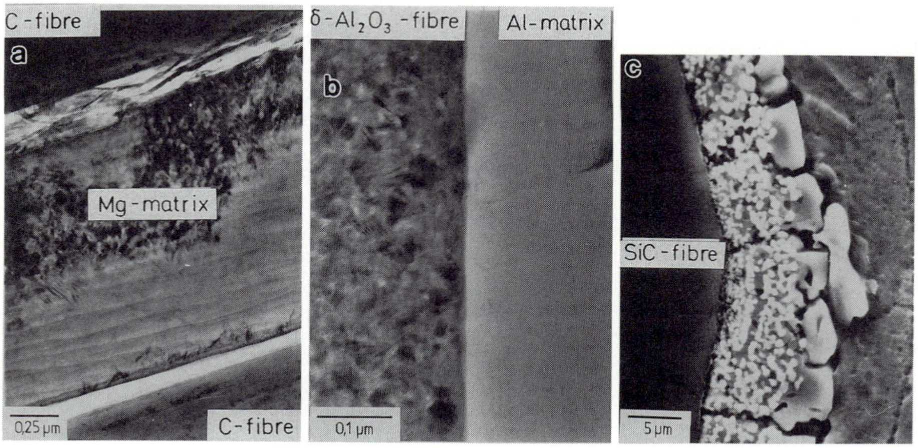

Figure 2.48 Examples for interfaces in metal matrix composites:
a) C(P100)-Mg(AZ91) composite showing insufficient wetting and adhesion [397],
b) δ-Al_2O_3-Al-12Si-Cu-Mg-Ni with excellent wetting and adhesion,
c) reaction zone in uncoated SiC-Ti-6Al-4V composites processed at 1150 °C/8h.

Poor wetting and adhesion hinder not only the transfer of load between fibre and matrix, but are often also connected with insufficient liquid metal infiltration with a break of fibres and formation of voids [395]. The fibre-matrix reaction on the other hand, with the development of a reaction zone, leads to fibre strength degradation [398, 399], which always has to be prevented in MMC development. The objective of a MMC interface design is therefore to develop interfaces with good wetting and adhesion but without the development of fibre matrix reaction zones. The subject of this presentation is to compile information suitable for finding a solution to that type of problem [3]).

Classification of MMC Systems

Fibre-Matrix Reaction: In sessile drop experiments to determine of wetting of an oxide ceramic surface by a liquid metal melt, Naidich [400] has distinguished between a nonreactive and a reactive wetting system. In the redox reaction

$$Me + MO_n \leftrightarrow MeO_{n'} + M \qquad (2.13)$$

[3]) As discussed in the section on "Titanium Matrix Composites", at the example of Ti-SiC composites the amount of adhesion strength is critical for the mechanical properties of composites. This will however not be the subject of this section.

the metal melt Me is oxidized to $MeO_{n'}$ and the metal of the oxide is reduced to metal M. If the equilibrium of reaction (2.13) is at the left side with the change of Gibb's free energy $\Delta G_R^0 > 0$ than a nonreactive wetting system exists. With the emphasis on the right side in reaction (2.13) and $\Delta G_R^0 < 0$ a reactive wetting system is present (for definition of ΔG^0 and the determination of the cohesive energy, see [401]).

The nonreactive wetting systems are characterized by a fast adjustment of equilibrium of wetting and a low negative temperature coefficient of the wetting angle [402]. Typical representatives of the nonreactive systems are Au on Sapphire [403], Au and Ge on SiC [404] and Al on Al_2O_3 [405].

In reactive wetting systems a reaction in the interface between the metal melt and the ceramic takes place, and new phases originate. Depending on the reaction speed the equilibrium of wetting with a constant wetting angle is reached in minutes or hours. For the wetting of a MgO with an aluminium melt, e.g., a constant wetting angle is not reached before the MgO substrate is completely changed into the $MgAl_2O_4$ spinel [406].

In the following discussion Naidich's the classification of wetting systems will be used for all MMC interfaces. As discussed in the introduction Al_2O_3 fibre reinforced aluminium alloys and C-fibre reinforced magnesium alloys can be considered as nonreactive MMC systems. C-Cu is also a nonreactive system [407]. The SiC-fibre reinforced titanium alloys, the Al_2O_3-fibre reinforced magnesium, the SiO_2-fibre reinforced aluminium alloys [408] and the C-fibre reinforced aluminium alloys are reactive MMC systems [409, 410].

Dissolution of the Ceramic in the Metal Melt: Beside reaction (2.13) a dissolution of the ceramic in the metal melt can take place. In the metal melt a considerable amount of the ceramic can be dissolved. In nickel e.g. 25% of WC can be dissolved at 1350 °C [411]. During the wetting experiment at the position of the metal drop a deepening in the ceramic is observed. In oxide systems dissolution of the ceramic in the metal melt during wetting experiments was observed for the Al-Al_2O_3 system [404, 412] and for the Nb-Al_2O_3 systems [413-415]. The solubility of some carbides in melts of Cu, Co, Ni and Fe was described by [411].

If fibre material is dissolved in the metal melt, the fibre surface will be damaged. In a similar way as in reactive systems, a reduction of mechanical properties of the fibres and composites will result. In addition to Naidich's subdivision it is therefore appropriate to add the dissolution of the ceramic in the metal melt to the list of conditions for a reactive MMC system. As will be discussed later the thermal stability of some MMC systems is limited by a dissolution of the fibre components in the matrix and the reaction of these components with the alloying elements.

The previous considerations concern the liquid-solid interface only. In the sessile drop experiments the gas phase is present at the drop edge. The presence of the gas phase influences the behaviour of the sessile drop experiments. In the Al-Al_2O_3 system, e.g., a reaction of the metal melt with the ceramic takes place according to equation (2.16) [404, 405, 412].

Nonreactive MMC Systems

Correlation with the Energy of Oxide Formation: In wetting experiments the work of adhesion W_a is determined by measuring the wetting angle Θ according to the Young-Dupre equation

$$W_a = \gamma_{lg} \cdot (1 + \cos\Theta) \tag{2.14}$$

in which γ_{lg} is the surface tension of the melt [4]). The work of adhesion W_a for different metals on sapphire determined in wetting experiments correlates linearly with the change of the free Gibb's energy ΔG_F^0 in the oxidation reaction [417]

$$Me + nO = MeO_n \tag{2.15}$$

The work of adhesion W_a increases with increasing $|\Delta G_F^0|$.

A similar correlation is also found for the formation of thin metal film deposition on ceramics. Thin monatomic layers show a similar behaviour during deposition as the melt in the sessile drop experiment. In the case of a strong interaction of the deposited atoms with the substrate (which is comparable with a high work of adhesion W_a) a homogeneous film of the metal develops and the thin film grows layer by layer. In the case of a larger interaction between the atoms of the deposited metal compared to the interaction of the metal with the ceramic, islands of the deposited metal grow. (In metal-metal systems intermixing between the deposit and the substrate will be an additional interaction possibility). This can be quantified by measuring the activation energy for desorption, which correlates to the change of the Gibb's free energy ΔG^0 of oxide formation for the deposited metals [418].

Influence of Alloying Elements: The alloying elements influence considerably the surface tension of metal melts [419]. The same has to be expected for the work of adhesion. Crispin, et al. [420] have investigated the influence of the alloying elements in nickel on wetting and adhesion on polycrystalline Al_2O_3 and on sapphire. With increasing content of the elements Cr, V, Ti, Al, Th and Y, the bond strength measured using the Sutton push-off test [420] runs through a maximum. The bond strength in the maximum of every alloying element correlates linearly with the change of the Gibb's free energy ΔG_L^0 of oxide formation of this alloying element such that a higher ΔG_L^0 is connected with a larger work of adhesion W_a.

[4]) For the difference in the definitions of the work of adhesion, the work of spreading and the work of immersion, see [407], for the definition of work of adhesion, [416]. In this paper the definition for the work of adhesion applied in wetting experiments is used.

(The lower indices R, F and L in the change of the Gibb's free energy refer to reactions (2.13), (2.15), and to an equation of oxidation like (2.15) for the alloying element, respectively.)

Surface Oxidation: In most sessile drop experiments the interaction of the surface of the melt with the residual oxygen in the equipment cannot completely be prevented. For melts forming an oxide layer, e.g. aluminium, the wetting experiment is modified by oxide formation on the drop of the melt and the measured wetting angle refers to the oxide rather than to the metal [421] (For the Al-Al_2O_3 system this subject is reviewed in [412]). With increasing temperature the oxide layer is dissolved in the melt or evaporated and the wetting angle changes. For pure aluminium melts on Al_2O_3 at 700 °C, Lijun, et al. [422] measure a wetting angle of 145° and at 1100 °C an angle of 80°. The alloying elements Si and Mg influence the composition of the oxide layer as well as the wetting angle. With increasing temperature the Al_2O_3 reacts with the Al melt forming a volatile Al_2O according to the equation [405, 412, 423]

$$4Al + Al_2O_3 \leftrightarrow 3Al_2O \tag{2.16}$$

This interaction takes place also with the Al_2O_3 substrate and causes an oscillation of the Al-drop when the experiments are performed under vacuum [412]. The reaction (2.16) slows down when the experiments are performed at 1 bar using a reducing gas atmosphere [424]. This discussion shows that the subdivision into reactive and nonreactive systems depends on experimental conditions.

Dissolution of Oxygen in the Metal Melt: According to Naidich [400] oxygen in solution in the metal melt has an influence on the wetting behaviour. In metals such as Cu, Ni and Ag high amounts of oxygen are dissolved in the melt. As shown by Mehrotra et.al. [425] the interface energy decreases with increasing partial pressure of oxygen in the melts of these metals. The decrease of interface energy and the increase of adhesion is explained by Naidich by the formation of oxygen-metal clusters (having an ionic character) in the melt and their segregation in the interface [400]. Supplying such metals with a high oxygen content and using processing conditions which favour oxygen segregation in the interface, the formation of an oxygen rich glassy phase will be observed [426].

In [427] the influence of the partial pressure of oxygen on the development of interfaces during deposition of Ni on an Al_2O_3 substrate is observed. For a very low partial pressure of oxygen in the deposition chamber an interface between the nickel and the substrate develops with bonding caused by intermetallic forces between Ni and Al atoms. If the partial pressure of oxygen in the deposition chamber is increased to $5 \cdot 10^{-9}$ torr, then $NiAl_2O_4$ spinels are formed in the interface, with bonding between nickel and oxygen atoms in the Al_2O_3.

Influence of the Ceramic: The kind of ceramic has an influence on the wetting behaviour of a metal melt. According to Ramquist [411, 428] the work of adhesion of different metals on carbides decreases with increasing of the stability of the carbides. The wetting angle Θ of the metals Cu, Fe, Co and Ni decreases according to Samsonov [429] on the order of the carbides ZrC, NbC, WC and Mo_2C. This order corresponds to an increase of the valence band density of the carbides. With the increase of the metallic character of carbides the wetting angle decreases and the work of adhesion increases, as was shown e.g. for the wetting of copper on TiC with different stoichiometry, corresponding to a different metallic character [430, 431].

Interface Phase

Chemical Bonding: For the adhesion of a metal with a ceramic, physical or chemical forces may be responsible. The physical bonding energy is on the order of 500 $mJ \cdot m^{-2}$, the chemical bonding energy of 1000 $mJ \cdot m^{-2}$ and larger. Comparing the work of adhesion of different metals on sapphire (Al: 955 $mJ \cdot m^{-2}$, Ni: 1114 $mJ \cdot m^{-2}$ [402], Si: 875 $mJ \cdot m^{-2}$ [430]) with the physical and chemical bonding energy height, clearly a chemical bonding for the most important metals, such as Al, Si, Ti and Ni reinforced with Al_2O_3 fibres, is in action [402].

In investigations on the interaction of monatomic evaporated metal films with a ceramic substrate by surface analytical methods a chemical interaction could be proved [427] and the development of the phases in the interface investigated [432]. This interaction is restricted in nonreactive systems to the first monolayer. The example of Ni-Al_2O_3 and the influence of the oxygen on that interaction was already discussed. Other examples are Ti-SiC [433], Ni-ZnO [434], Y-SiO_x [435] and Sc-SiO_x [436]. In the reactive system of Al-CuO even at room temperature the deposited aluminium is oxidized through several atomic layers and the copper oxide substrate is reduced [437].

XPS analysis of thin monatomic layers of Ti and Nb on Al_2O_3 show that the interaction takes place between the metal and the oxygen of the oxide, as well as between the metal and the metal of the oxide. In the Ti-Al_2O_3 system the aluminium of the alumina is reduced and a titanium aluminate Ti_3Al develops. In the Nb-Al_2O_3 system the reduced aluminium segregates in the interface [427]. In the case of a strong interaction between the evaporated thin film and the ceramic substrate, as observed in the Nb-Al_2O_3 system, the surface structure of the ceramic determines the structure of the first monolayer of the deposited metal [438].

The Surface of the Ceramic: A surface has a different crystallographic structure compared with the bulk due to primary surface relaxion [439]. According to Weyl [440] the primary surface relaxion of oxides results in an enrichment of oxygen anions and a depletion of cations. LEED investigations at α-Al_2O_3 (0001) surfaces show [441, 442] surface reconstructions with oxygen as well as aluminium atoms at the upmost layer. As shown by [443] using LEED in the upmost layer of a reconstructed (0001) face of α-Al_2O_3 the Al:O-ratio is 1:1. During interaction of a metal melt with Al_2O_3 fibres, the interaction of the metal atoms with the oxygen and the aluminium atoms has to be considered [427].

Atomic Dimensions of the Interfaces: In nonreactive systems $\Delta G_R^0 > 0$ and therefore no reaction products should develop in the interface. Using high resolution electron microscopy (HREM) reaction products can be made visible. Composites for HREM investigations were processed in different ways by hot pressing [412, 413], deposition of thin films on ceramic substrate [438], surface oxidation [444], internal oxidation [445] or co-deposition [446]. In all the nonreactive systems investigated no reaction zone was found between the metal and ceramic using HREM. Such experiments were performed in Si/sapphire [447], Al/Al_2O_3 [444], Cu/NiO, Ag/NiO, Pd/NiO [445] and Nb/sapphire [438, 448]. In the Nb/sapphire system metal and ceramic have similar lattice parameters and similar thermal expansion coefficients [449]. This system is therefore especially suitable for interface analysis. Mainly using this system the most important results obtained using HREM are [438, 445, 448, 450]:

1. Metal-ceramic wetting and adhesion take place in atomic dimensions, no reaction products are observed between metal and ceramic. In the neighbourhood of surface defects, amorphous regions are formed in the metal.
2. If the system has the freedom to choose the orientation during processing of interfaces, as is nearly always the case for thin film deposition or internal oxidation, low indicesed lattice correlations are chosen [445, 450].
3. Differences in lattice parameters are compensated by misfit dislocations (quasi periodicity [445]). Depending on processing conditions these dislocations end in the metal-ceramic interface or at some distance from that position (stand-off position). No misfit dislocations are observed in the case when amorphous regions between fibre and matrix are present.
4. In the Nb/sapphire interface no hints can be found of the surface reconstruction of sapphire. This has to be expected because during interaction with the metal reconstruction of the ceramic surface is cancelled.
5. For the Nb/sapphire interface [448] a simulation of the HREM image is only in agreement with the observed image contrast when in the interface an aluminium segregation is assumed higher by a factor of 1.5 compared with the Al concentration of an unreconstructed (0001) face of sapphire [448]. This is contrary to other observations, in which only oxygen atoms are located in the interface (in the Ag-CdO [451] and in the $Cu-Al_2O_3$ [452]).

Atomic Force Microscopy: A direct way to determine adhesion in atomic dimensions is possible with atomic force microscopy (AFM) [416, 453]. When a tip with only a few atoms in front is moved in the direction of a surface over distances comparable with the atomic dimensions, a jump in the distance, the interaction force and the binding energy is observed. Moving the tip in the reverse direction introduces plastic deformations and depending on the materials in contact some of one material is transferred to the other. The forces in both cases increase with the surface energy, and the adhesion is always greater than the attraction [454]. These experiments show that strong forces come into action as soon as two materials are brought into contact at distances of atomic order.

Computations for the Interaction of Metals and Ceramics: In computing the interaction between the metals Fe, Ni, Cu and Ag with a sapphire surface it was shown by Johnson, et al. [455] that chemical bonding between the metal d-orbital and the nonbonding 2p-orbital of the oxygen of the Al_2O_3 takes place. With the increase of the number of occupied antibonding orbitals in the interface, the adhesion energy is decreased. This explains the experimentally observed decrease of the work of adhesion for Cu and Ag on sapphire.

Nath et al. [456] determined the binding energy, charge transfer from the oxygen to the metal, and the atomic distances in the interface of the (0001) α-Al_2O_3 with the first transition metal series Sc through Cu. Binding energies between 5 eV for Sc and Ti and 1 eV for Cu were computed, and the atomic distances between the oxygen in Al_2O_3 are of the order of 0.15 to 0.2 nm, which is considerable smaller than the distances of the metals in the oxide and only a little larger then the lattice constant of the metals. The charge transferred to the metal is on the order of half an electron charge. (For a critical discussion of the computation of interfaces in metal ceramic systems, see [450].)

In the description of the interface in nonreactive systems two different regions have to be distinguished: a) the interface in the region of atomic distances, in which a transfer of load between the atoms takes place and which is determined by the chemical bonding between the atoms [416, 457] and b) the region with distances of a few lattice constants in which the adaption of both lattices takes place. This adaption can be developed in different ways, such as by an amorphous region, by a new phase [432], by lattice strains or by misfit dislocations in the metal.

The Interface Phase: The amount of work of adhesion in wetting experiments, the results of HREM interface imaging, the surface analysis of monatomic metal films on ceramic substrate, the AFM observations of the interaction forces in atomic dimensions and the quantummechanical calculations of the binding energy and distances in the interface result in the following image: The interface of non-reactive MMC systems is characterized by atomic distances and chemical bonding between the interacting atoms without the development of a reaction zone. This structure and the following discussion of the thermal stability of the interface justify considering it as a special phase in the material: the interface phase [458].

The interface phase has a central significance for the development of metal matrix composites. The existence of the interface phase ensures the possibility to introduce ceramic fibres into a metal matrix with sufficient wetting and adhesion and without fibre surface degradation. Coming back to the definition of Naidich, now the nonreactive MMC systems can be defined as systems with an interface phase.

Influence of Alloying Elements on the Interface Phase: As was discussed, the work of adhesion measured in wetting experiments correlates with the change of the Gibb's free energy for the oxide formation of the alloying elements ΔG_L^0. For the influence of the alloying elements on the wetting behaviour, Naidich [400, 459] proposed the following mechanism: The alloying elements segregate in the metal-ceramic interface as a monatomic layer. This segregation layer is responsible for the increase of wetting and adhesion observed in wetting experiments and during processing of composites. Such a segregation of magnesium was, e.g., observed in the interface of δ-Al_2O_3 fibre reinforced piston alloys [460].

From the thermodynamic point of view the amount of change of the Gibb's free energy of oxide formation of the alloying elements should be smaller than that of the main element in the alloy [459]. This may increase the probability of an interaction in the interface promoting wetting and adhesion. As will be discussed in the next section the crystallographic structure of the ceramic has an influence on that behaviour. From surface analysis it is known that the reactivity of surfaces is increased when they are doped by different metals. This can be explained by the development of less stable bonding due to the interaction of the substrate with the deposited metal [432].

The segregation of alloying elements in the fibre-matrix interface depends on a number of parameters, such as the diffusion speed and the saturation concentration of the elements in the matrix, the "chemical" environment formed by the other alloying elements, the formation of phases between the alloying elements, the density of lattice defects, the interface energy, the crystallography and orientation of the fibre and matrix. These parameters interact with each other and a general description is therefore not possible. For the following discussion of the thermal stability of interfaces it is worth noting, that the concentration needed to form a monolayer of an alloying element on the fibre surface is low (approximately 10 ppm for $v_f = 0.2$ and a fibre diameter of 3 μm) as is the time needed to segregate a monolayer of atoms on the fibre surface (on the order of a few hundred seconds for Mg in an Al-Mg-alloy at 500 °C for $v_f = 0.2$ and d= 3 μm).

Stability Limits of Nonreactive Systems

The interface phase is stable only in a limited temperature interval, annealing time and concentration of alloying elements. As was already shown in the introduction, the δ-Al_2O_3 fibre reinforced piston alloy Al-12Si-Cu-Mg-Ni is a nonreactive MMC system with good wetting and adherence in the interface with no reaction products detectable [395]. When these composites are annealed at 500 °C for 60 days, reaction products at the fibre matrix interface grow, as shown in Figure 2.49a [395]. The crystals visible at the fibre surface can be identified by analytical transmission electron microscopy as the $MgAl_2O_4$ spinels. As can be seen from Figure 2.49a the fibre surface is porous in the neighbourhood of the spinel crystals. The spinel growth is connected with a fibre surface solution and a reduction of the strength of the fibres, as can be concluded from the strength reduction of the composites. This effect characterizes reactive systems. It can be argued that a growth of the spinel can be prevented if the concentration of magnesium needed for the spinel formation is low enough. The stability of the interface phase depends therefore not only on the temperature and annealing time, but also on the concentration of the magnesium. The spinel grows preferentially at positions where the larger α-Al_2O_3 crystals reach the surface of the fibre, Figure 2.49b. In this case the spinel is forming on the fibre surface as the α-Al_2O_3-crystal is transferred into a $MgAl_2O_4$ spinel. The α-Al_2O_3 crystal structure is therefore much more sensitive to a reaction with magnesium forming a spinel than is the δ-Al_2O_3.

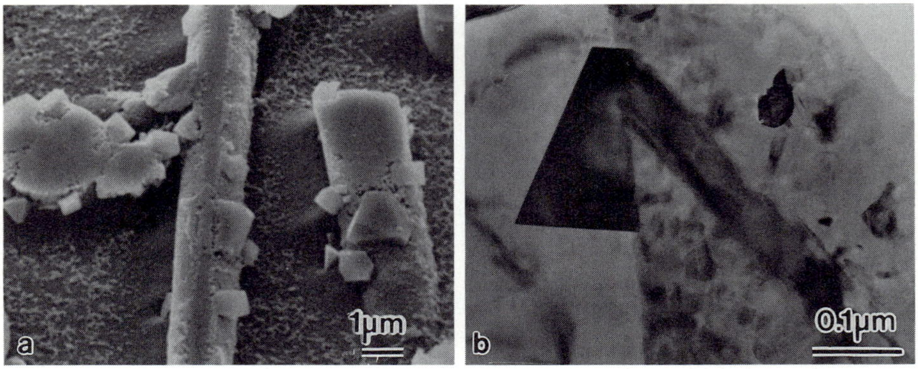

Figure 2.49 Degradation of δ-Al_2O_3 fibres in a Al-12Si-Cu-Mg-Ni alloy after a thermal treatment of 500 °C/60d, a) SEM image of a polished and etched surface, b) TEM image

As was discussed in the previous sections, the oxygen dissolved in the metal influences wetting and adhesion in nonreactive systems. For spinel growth oxygen is also needed and is taken from the metal matrix, as the growth of the spinel crystal lattice shows, and as it will be discussed in the next section for the Mg-Al_2O_3 system. The fibre degradation in Figure 2.49a can be understood as a solution of the fibre components in the melt and the diffusion of these components, especially of the oxygen to the spinel. This process can be prevented by taking a less stable oxygen donor into the metal. The role of such an oxygen donor is taken by the SiO_2 binder used to stabilize the fibre preform. Indeed, in composites with a higher binder content the crystals growing at the fibre surface after the same thermal treatment are much smaller than in the case discussed in Figure 2.49 [395].

The Reactive System Mg-Al_2O_3

In the Mg-Al_2O_3 system according to Hallstedt et al. [461] the reaction

$$Mg + 1/3Al_2O_3 \rightarrow MgO + 2/3Al \tag{2.17}$$

with the change in the Gibb's free energy of ΔG_R^0=-39 kcal/mol takes place. The reaction product MgO will react further according to

$$MgO + Al_2O_3 \rightarrow MgAl_2O_4 \tag{2.18}$$

with the change of the Gibb's free energy of ΔG^0=-47 kcal/mol. Because the change of the Gibb's free energy ΔG_R^0 is negative, Mg-Al_2O_3 is a reactive system. In calculations of the kinetics of that system, Hallstedt et al. [461] assumed that the growth of the magnesia and the spinel is controlled by the diffusion of cations, especially of aluminium through the reaction products. The calculations result in a very low diffusion coefficient for the aluminium cations under the processing and application conditions. The authors conclude therefore that a reaction in the Mg-Al_2O_3 system will be stopped after the growth of a monatomic passivation layer on the fibre surface.

For the same system Pfeifer et al [462] proposed however a reaction in which nucleation of MgO takes place only at special orientations of α-Al_2O_3. The reaction is controlled by the diffusion of the oxygen and aluminium through the metal. This is a similar reaction as was observed for the annealed δ-Al_2O_3 fibre reinforced Al piston alloy with the spinal as a reaction product, described in the last section. In the δ-Al_2O_3 fibre reinforced magnesium alloys therefore the growth of large MgO crystals can be observed [397].

This example shows that in reactive MMC systems a passivation of the fibre surface by the reaction product in general cannot be expected. However, it also shows the way out of the problem for the interface designer: the inhomogeneous nucleation of the reaction products on the fibre surface has to be prevented. This subject was investigated in surface analytical "in situ" experiments [463]. The situation of the short distances in composites can be simulated in surface analytical experiments by depositing a metal alloy film in between two ceramic films. The film thicknesses in this experiments were approximately 100 nm and the following thin film system was investigated: Al_2O_3-MgO-Mg-Al-Mg-Al_2O_3 on an Al substrate. Annealing this thin film system for different times and temperatures up to 350 °C, the Al_2O_3 in the interface between Mg and Al substrate dissolves according to the reactions (2.17) and (2.18). By contrast, the Al_2O_3 protected by MgO remains unchanged. Using AES depth profiling it can be shown that in the interface between the Al_2O_3 and the Al-Mg alloy a thin MgO layer is still present after the heat treatment. Using the considerations of Hallstedt et al. [461] this result can be interpreted in the following way: The MgO (or spinel) layer of a few nanometres is an homogeneous diffusion barrier for cation diffusion between Al_2O_3 and magnesium. No heterogeneous nucleation can therefore take place, and the Al_2O_3 layer simulating the Al_2O_3 fibre remains unaffected.

The Reactive Ti-SiC System

The Ti-SiC composites can be considered as the classical reactive MMC system. It was subject of intensive investigations for many years [393, 395, 399, 464,

465]. The results of these investigations can be summarized as follows: Already at room temperature thin titanium layers deposited on SiC react with the carbon of SiC, without however developing reaction products [433] (nonreactive interaction). Fibres annealed in the presence of titanium at temperatures of 600 °C are degradated in strength [399] and fibres tested after processing at 900 °C and removed from the composite have a tensile strength reduced to 1/10 of the original value [466], in consequence of which the composites tensile strength is lower than that of the unreinforced matrix [467]. During composite processing a reaction zone of a few micrometers is formed. Similar to the $Mg\text{-}Al_2O_3$ system in the neighbourhood of the fibre TiC is formed [468] by a fibre solution and heterogeneous nucleation of TiC, which is connected with fibre surface degradation. Some fibre protecting coatings were tested [398, 467] but in many cases the coating system was reacting with the fibre surface during fibre coating or during composite processing. From these investigations it can be concluded that fibre protecting coatings have to form a nonreactive system at the coating-fibre interface.

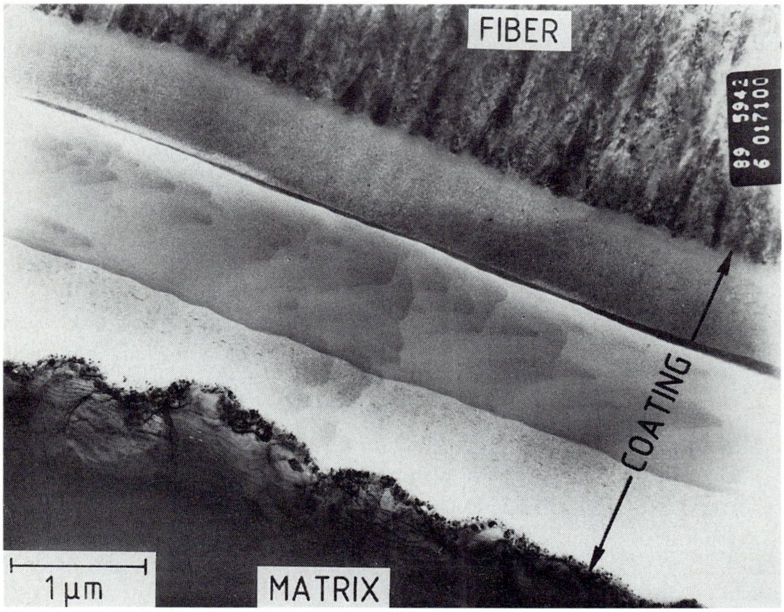

Figure 2.50 Interface in the SiC-SCS-6-fibre (Textron) reinforced Ti6Al4 alloy

The Textron SCS-6-SiC fibre is an example for the successful development of a fibre with a protecting coating for titanium alloys, Figure 2.50. The coating consists of a roughly 3 µm thick carbon layer with a few percent of small (some 10 nm large) SiC particles. The interface carbon-SiC is a nonreactive system in the range up to probably 1000 °C [469]. During processing of the Ti-SiC composite a reaction between the coating of the SCS-6-SiC fibre and the titanium alloy takes place and a reaction zone develops. The reaction zone consists of a 200 nm thick TiC layer adjacent to the unreacted coating and a region approximately a few tenths of a micrometer thick in the Ti matrix in which the carbon and silicon of the coating diffuse forming precipitates of titanium carbide and titanium silicide. As was shown by analytical transmission electron microscopy, the concentration of titanium in the carbon coating adjacent to the TiC reaction zone is below the detection limit of the EDX and EELS analysis method [470]. The TiC reaction zone and the carbon coating prevent titanium from diffusing into the direction of the fibre. This system is stable for temperatures up to 600 °C. After annealing the composite at 800 °C for 500 h the coating is partly consumed, but it is still protecting the fibre, because the titanium content in the coating remains low [470]. For temperatures below 600 °C both fibre-coating and coating-matrix interfaces can be considered nonreactive systems.

MMC Interface Design

Naidichs's subdivision into nonreactive and reactive MMC systems leads us to an understanding of fundamentals of adhesion in nonreactive systems. The existence of the interface phase in which a chemical bonding in atomic distances is realized without the development of a reaction zone allows MMC interface development. The interface phase can be influenced by alloying elements and the processing conditions. The alloying elements and the processing conditions have to be chosen according to a number of parameters, such as the change of the Gibb's free energy of oxide formation of alloying elements, the diffusion speed in the metal, the development of phases with other alloying elements, the interaction with the crystallography of the fibre surface, the interaction of oxygen with the alloying elements in the interface, and others. The stability of the interface phase is limited to an interval of temperature, annealing time and concentration of the alloying elements. As shown for the δ-Al_2O_3 fibre reinforced Al-piston alloy, the stability of the nonreactive systems is mainly limited by the fibre solution in the matrix and inhomogeneous nucleation of the reaction products in the interface. The growth of these reaction products is due to the dissolution of the fibre surface and the diffusion of the fibre components through the matrix to the place where the reaction product is growing. In some composites it will probably be possible to prevent fibre dissolution and inhomogeneous nucleation of reaction products by coating the fibre with thin layers with low diffusion coefficients for the elements taking part in the reaction. An example seems to be the reactive system Mg/Al-Al_2O_3 in which MgO or $MgAl_2O_4$ layers on the fibre surface may stabilize the interface. In the Al/Mg-Al_2O_3 system the interface can also be stabilized by supplying the composites with an oxygen donor with a lower stability compared to the fibre material. In the Al_2O_3 fibre reinforced Al alloys this task is performed by the SiO_2 binder.

In reactive systems, as in Ti-SiC, a fibre protecting coating has to be developed. The development of coatings in the Ti-SiC system visualizes the conditions which a coating has to fulfil to prevent fibre degradation: in the temperature range for the application of the composite the coating has to form nonreactive systems in the interface fibre-coating and in the interface coating-matrix.

For the development of interfaces in MMC the conditions of wetting, adhesion and reaction are important but not the only subject which has to be considered. In section "Titanium Matrix Composites" some of the mechanical points of view important for the development of fibre protecting coatings in the Ti-SiC reactive system are discussed.

3. Ceramic Materials and Ceramic Matrix Composites

3.1 Non-Oxide Materials (Silicon Nitride)

B. Saruhan

Silicon nitride is one of the best candidates among non-oxide ceramics and has a very attractive combination of properties. Such as

- high strength at room and elevated temperatures
- high impact strength
- high strength-to-weight ratio
- high specific modulus (E/ρ)
- good creep resistance
- excellent thermal shock resistance resulting from the low coefficient of thermal expansion
- good thermal stability (decomposition temperature 1900 °C)
- good chemical stability
- low coefficient of friction
- non-wettability by metals
- high dielectric strength and
- high hardness

These properties make silicon nitride applicable for the following sectors:

- cutting tools
- wear-resistant structural components (ball bearings, jets, seals, slide bearings, etc.)
- chemical and heat-resistant components
- high temperature components for nozzles and engines
- gas turbines
- turbochargers

Such properties allow the engineer to design advanced systems which can operate at high temperatures with either no, or significantly reduced, forced cooling. This results in heat engines of greater thermodynamic efficiency, and consequently, decreased fuel consumption. In order to make these properties realisable, production of a fully dense material is essential.

3.1.1 Fabrication and Microstructural Development of Non-Oxide Ceramics (Silicon Nitride)

Powder Processing

Powder processing involves four basic steps: powder production, powder preparation, fabrication and densification. The most efficient method to produce dense and flaw free ceramics is to pay great attention to each individual step. Each step has the potential for introducing a detrimental heterogeneity which will either persist during further processing or develop into new heterogeneity during densification and microstructural development.

Because strength is sensitive to stress concentrations, microstructural heterogeneities that are stress concentrators can be best observed by fracture and examinations of fracture surfaces. The processor interested in eliminating heterogeneities must first identify the dominant heterogeneity observed at fracture origins, find out how this heterogeneity is introduced during processing and then make the appropriate processing changes to eliminate the heterogeneity. This is still a trial and error method and can be rather devastating.

Many microstructural heterogeneities originate from the powder itself. Agglomerates are a major heterogeneity in powders. Agglomerates with different bulk densities can persist during powder consolidation or densification to form crack-like voids. Some powders may contain organic and inorganic inclusions in the as-received state, therefore clean rooms may remain ineffective. A good characterization and fractionation of the powders are the main steps of a perfect processing. Once a powder is fractionated and made flowable, a methodology should be available to process these powders. Among many fabrication techniques, die-pressing is commonly used but has many limitations. Traditional slip casting can be developed for simple consolidation or taken as a basis for shaping through injection molding, pressure filtration, spraying and so on. To produce a flowable slurry, one needs to understand the slurry rheology, which mainly depends on interparticle forces and particle volume fraction. Rheological changes are not without problems and can be related to strain gradients. This aspect is still being investigated in laboratories which deal with ceramic processing of all types.

Sintering of Silicon Nitride

Over the past two and half decades, there has been increasing interest in producing dense silicon nitride as a high-performance ceramic for use in high-temperature structural components, due to its intrinsic properties. "Pure" silicon nitride synthetic ceramic cannot be densified easily due to its low self-diffusivity.

Sintering of silicon nitride is carried out in a nitrogen atmosphere at temperatures up to 1800-1900 °C depending on the additive. The volume of the liquid may change with temperature in some cases, so the sintering temperature is adjusted according to composition and phase equilibria conditions. Two important processes occur during sintering: densification and α-β phase transformation. In order to achieve improved mechanical properties, it is desirable to achieve complete transformation with formation of small β grains with high aspect ratios (c/a) [471-474]. The whole process is shown schematically in Figure 3.1.

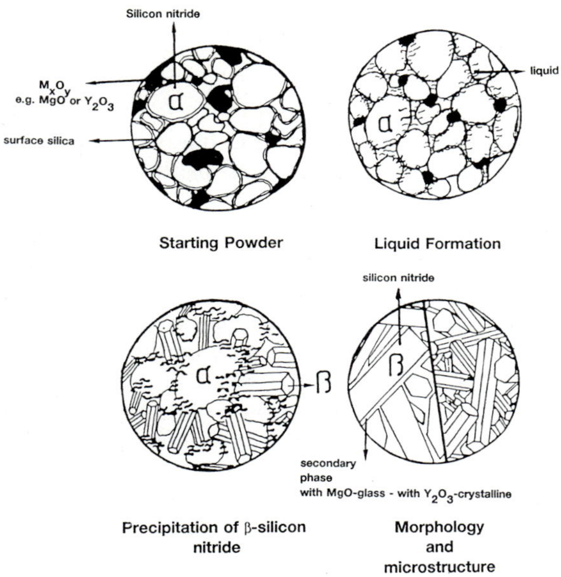

Figure 3.1 Schematic diagram of α-β transformation in silicon nitride

Sintering with Magnesia: In 1961, first attempts [475] showed that silicon nitride could be hot-pressed to near theoretical density with magnesia as an additive. The role of magnesia or in other terms, of oxide additives to silicon nitride during densification, was then investigated and it was suggested that it reacts with the surface silica which is always present around each silicon nitride powder particle [476]. During hot-pressing at 1000 °C to 1400 °C the reaction between MgO and SiO_2 gives forsterite. This combines above 1500 °C with more silica to form a silicate-type liquid, containing nitrogen, which cools down to give a glass, the devitrification products of which are silicon oxynitride (Si_2N_2O) and enstatite ($MgSiO_3$) or forsterite (Mg_2SiO_4). Then the question arose whether full-densification could be achieved via liquid phase sintering without any transformation from α to β. The following investigations were engaged in solving this question. Firstly it was shown that transformation was not necessary for attainment of full density, since this was achieved before any α-β transformation occurred [477]. In a similar way a previously fully transformed β Si_3N_4 sample containing additive at 1600 °C was successfully hot-pressed to full density [478]. Although their densification data gave a good fit to the Kingery Model [479], none of these studies involves a detailed examination of transformation kinetics.

An electron microscopy study [480] concluded that, at high temperatures, α silicon nitride dissolves into the liquid and β silicon nitride is precipitated and that densification of silicon nitride proceeds by a liquid phase sintering reaction which includes a solution-precipitation process. After the formation of liquid, dissolution of the less stable α-form in the liquid occurs allowing the nucleation of more stable β crystals on α/β interfaces. Si and N atoms previously dissolved in the liquid will diffuse to precipitate on β nuclei and allow growth of β crystals.

Iskoe and Lange [481] assumed that new β grains form as fibres during fabrication by consuming the particles and growing on existing β particles. Thus the thickness of β grains is the same as the average size of the initial particles in the silicon nitride powder and suggested that aspect ratio R (ratio of length to thickness of prismatic grains) of β grains, after transformation is completed, could be expressed as $R = 1 + \alpha/\beta$, where α and β represent the initial α and β contents of the starting powder. Due to the complexity of physical and chemical conditions in the immediate environment of the growing crystal, this expression is inadequate. Wötting and Ziegler [474], on the other hand, indicated that β grains form by precipitation of new nuclei and grow as elongated rods because of different growth rates perpendicular and parallel to the crystallographic c-axis depending on the nature of the liquid phase. They developed a method to measure the aspect ratio. This method considers some assumptions:

- All grains are prismatic and have approximately the same aspect ratio within one sample;
- 10% of the grains are cut parallel to their crystallographic c-axis and therefore show realistic aspect ratio values;
- The value d_{95} is the mean value of 10% of these highest aspect ratio values (statistical observation).

Thus, the area A and visible length l_v of each individual grain i are determined by a digitizer and the thickness d_i is calculated by $d_i = A_i/l_{vi}$. The visible aspect ratio, $a_{vi} = l_{vi}/d_i$ can be calculated and plotted against the frequency distribution of a_{vi} values. This plot gives the a-value by which the "true" length, $l_i = d_i \cdot a_{95}$ is calculated. As a result, aspect ratio, mean length and mean thickness can be obtained (Figure 3.2).

Hampshire and Jack [482], in a detailed study with 5% MgO as sintering additive to silicon nitride, showed that the densification kinetics of pressureless sintered silicon nitride can be interpreted using the Kingery Model [479], and three stages are identified as shown by the schematic log-shrinkage vs log-time plot of Figure 3.3. It is observed that with MgO, half of the total shrinkage required for full densification occurs during the first stage (particle rearrangement). During the second stage (solution-precipitation) for MgO, the slope of the curve changes to

Assumptions:

- All grains are prismatic
- The aspect ratio of all grains of one sample is ≈ constant

Analysis:

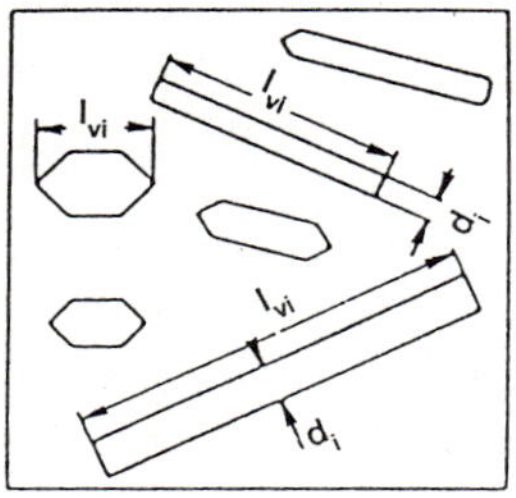

- Determination of area A and visible length l_v of each individual grain i in the microsection (digitizer).
- Calculation of thickness $d_i = A_i/l_{vi}$

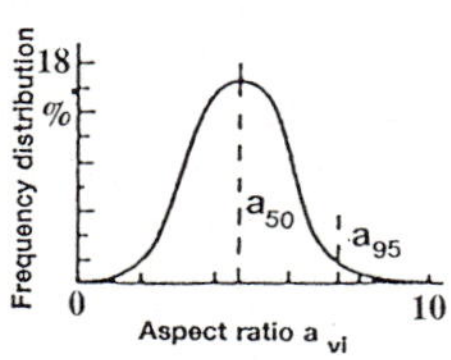

- Calculation of the visible aspect ratio $a_{vi} = l_{vi}/d_i$
- Plot of the frequency distribution of the a_{95}-values and determination of the a_{95}-value

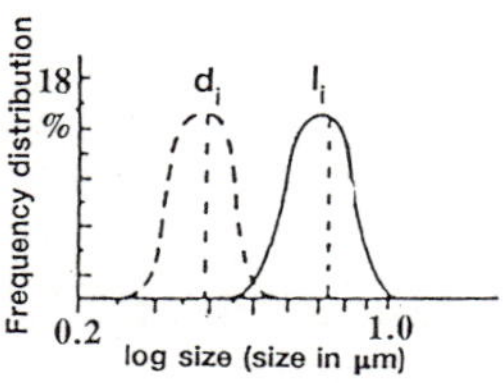

- Calculation of the 'true' length of each grain: $l_i = d_i a_{95}$
- Plot of the frequency distributions of l_i, d_i
- Determination of microstructural characteristics

Results:

- Aspect ratio a_{95}
- Mean length $\bar{l}$
- Mean thickness $\bar{d}$

Figure 3.2 Schematic presentation of the method to measure aspect ratio of β silicon nitride grains

give n=3, suggesting a reaction-controlled process. This indicates a relatively rapid transport of material through the low-viscosity liquid so that transformation is accompanied by shrinkage.

The quaternary behaviour diagram of the Mg-Si-O-N system has been presented at 1700 °C by Jack [483] and at 1600 °C by Lange [473]. Jack reported the existence of the join Mg_2SiO_4-$MgSiN_4$ while Lange assumed a binary join between

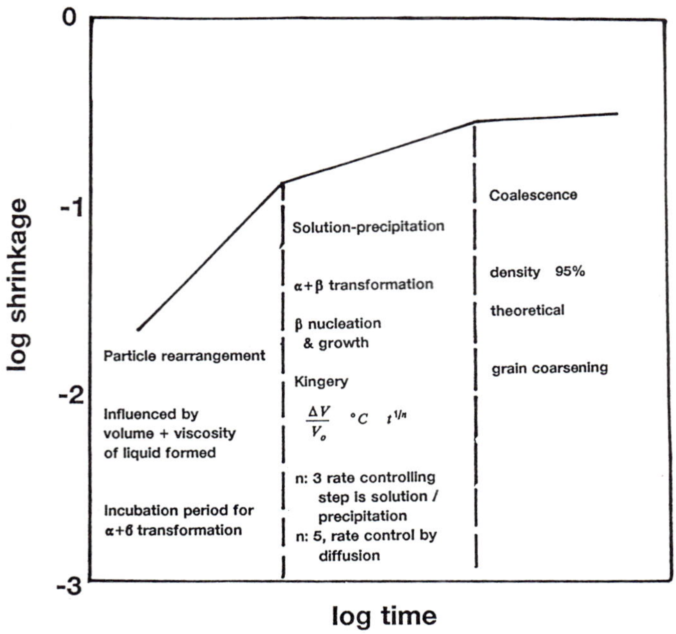

Figure 3.3 Kingery model plot for sintering of nitrogen ceramics

Si_3N_4-MgO as this represented the limit of his investigation. Tien et al. [484] in joint work with the Max-Planck-Institut für Metallforschung [485] investigated that claim and stated that a second joint between Si_3N_4-MgO can be observed depending on firing conditions. Jack and Tien et al. agree that no quaternary crystalline phase is observed in this system and therefore crystallisation of glassy phases in silicon nitride will produce only magnesium silicates, silicon oxynitride or known nitrides.

The low softening point grain boundary glass in the Mg-Si-O-N system limits the high temperature mechanical properties of silicon nitride. Their presence can effect high temperature strength, creep resistance and oxidation resistance of silicon nitride ceramic components.

Sintering with Yttria: In a search for another additive system to sinter silicon nitride which would give a more refractory grain boundary phase and in which the softening point of the intergranular phase is increased, it was found that Y_2O_3 was suitable as a densification aid [486]. The improved mechanical properties with Y_2O_3 are due to further reaction of the liquid with silicon nitride to give one or more highly refractory grain boundary phases [487]. The X-ray analyses of compositions within the system suggested that three yttrium silicon oxynitride phases exist. One of those is the binary compound $Si_3N_4 \cdot Y_2O_3$ which is later named N-melilite [488] and was also identified by and designated as phase "I" [489]. The second and third phases were called "H" phase [487], with the proposed formula of $10Y_2O_3 \cdot 9SiO_2 \cdot Si_3N_4$, and "J" phase of unspecified composition which was initially reported by Tsuge et al.[489] as $Si_3N_4 \cdot 2Y_2O_3$ but later corrected by Jack [488] to $2Y_2O_3 \cdot Si_2N_2O$.

In 1978, it was reported in a revised quaternary diagram that there were four quaternary phases at 1700 °C [490].

Later Gauckler et al. [491] explored the system at 1550 °C and stated that the apatite phase forms various solid solutions which may relate to the development of strength in silicon nitride densified with yttria.

The latest version for the Y-Si-O-N system has been developed by Thompson [492] and is given in Figure 3.4 which includes the $Si_3N_4 - Y_2O_3 - YN$ region of the diagram.

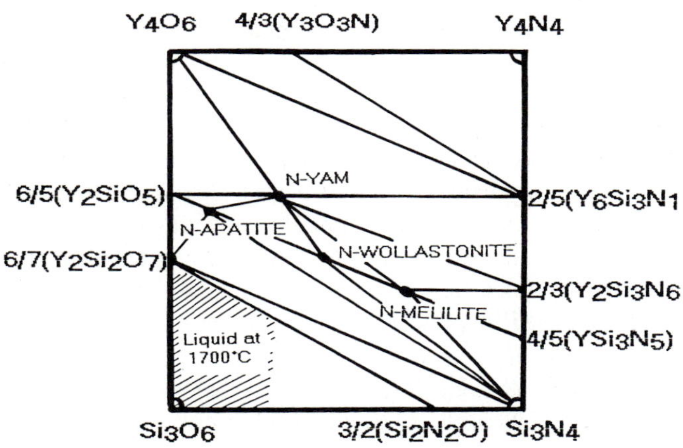

Figure 3.4 Phase relationships in the complete Y-Si-O-N system [492]

Since it was assumed that a more refractory grain-boundary phase would result when Y_2O_3 is used as an additive and concurrent phase relations have been established, it was suggested that the quaternary phases could accommodate in solid solution the impurities which would otherwise form a low-softening-point glass and thus a high strength silicon-nitride material could be realised [486].

However, some compositions of the high-strength, creep resistant silicon nitride hot-pressed with yttria fail with extensive cracking at 900-1200 °C in an oxidizing atmosphere. This is due to the oxidation of some of the yttrium silicon oxynitride grain-boundary phases to give oxides with markedly different specific volume to that of the starting material [490, 493].

The densification and transformation kinetics of silicon nitride with yttria additive is interpreted by Bowen et al.[494] using a modified form of the Coble grain boundary diffusional-creep model. The model gives a good fit for the data obtained from MgO-doped material. It is less applicable for the Y_2O_3-doped material except during the early stages of hot-pressing. The authors ascribe this to microstructural variety and the complexity of the phase reactions.

Hampshire and Jack [482], using the Kingery model, reported that with Y_2O_3, during the arrangement stage less shrinkage than that with MgO is observed. This is because Y-containing oxynitride liquids have higher viscosities than Mg-containing liquids and a smaller volume of liquid is found than in the Mg-Si-O-N system.

Sintering with Zirconia: Early studies with ZrO_2 [475, 495] had been started with the idea of incorporating tetragonal ZrO_2 into the matrix of another material and thereby producing tougher ceramics. Deeley et al. [475] were not successful in achieving high density silicon nitrides with additions of ZrO_2 as an additive, mainly because of the coarse particles used as the starting materials. Rice and McDonough [496] reported that relative densities of 97%-99% TD can be achieved by hot-pressing silicon nitride with 3% yttria stabilised ZrO_2 and zircon ($ZrSiO_4$). Their mechanical test results show that the zircon-doped silicon nitride gave higher modulus of rupture and fracture energy values at RT than ZrO_2-doped and undoped materials. No X-ray diffraction results for these materials were given. However, it is likely that, as zircon dissociates well below the hot-pressing temperatures (1770-1820 °C), a high viscosity silica-rich grain-boundary glass and ultrafine ZrO_2 remains, giving rise to increased toughness and high strength.

The first phase investigations in the Zr-Si-O-N system had shown that no reaction between Si_3N_4 and ZrO_2 exists up to 1400 °C [497]. Above 1600 °C, the reaction takes place and Si_2N_2O forms as an intermediate phase and ZrN or ZrO_2 will be the solid end product, depending on the firing conditions. It was also suggested that an addition of Al_2O_3 would overcome the reaction between Si_3N_4 and ZrO_2.

Bellosi et. al.[498] confirmed this and reported that the presence of Al_2O_3 inhibits formation of Zr-O-N phases, existence of which dramatically affect oxidation of silicon nitride. The results produced at the DLR by using very small proportions of Al_2O_3 + ZrO_2 and Y_2O_3 + ZrO_2 are presented in Figure 3.5 and clearly prove this argument.

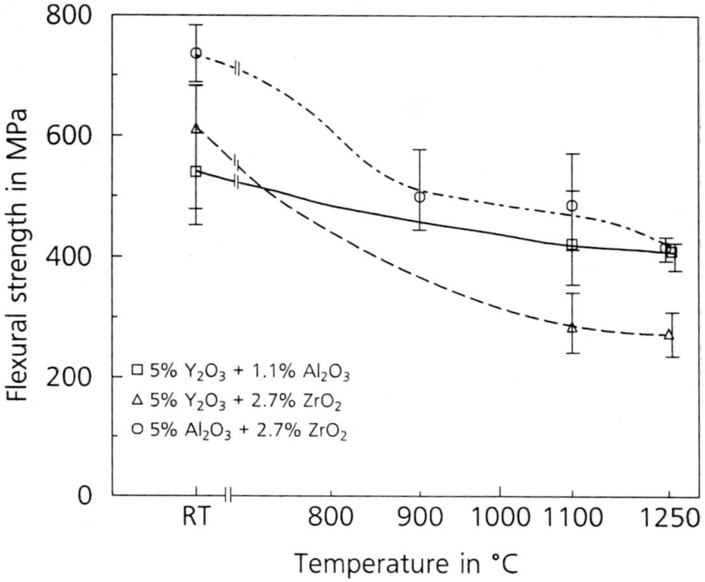

Figure 3.5 4-point-bending test results of silicon nitrides sintered with Al_2O_3/ZrO_2 and Y_2O_3/ZrO_2 in the temperature range between RT to 1250 °C [499]

The only compound reported to exist at higher temperatures (2500 °C) is $Zr_7O_{11}N_2$ [500]. McDonough et al.[501] confirmed the presence of a zirconium oxynitride phase which shows the same structure as the nitrogen stabilised cubic ZrO_2 [502]. In this system, several other phases such as Zr_3N_4, Zr_2N_2O and $Zr_7O_8N_4$ are present at low temperatures (1000-1200 °C), but decompose irreversibly to ZrO_2, ZrN and N_2 [502].

Presence of this phase as a secondary crystalline phase in silicon nitride results in better high-temperature properties [503]. Similarly, with 10% zyttrite (yttria stabilised cubic ZrO_2 with an average 12μm particle size) additions to silicon nitride, the material shows no deterioration, even at 1300 °C [504].

The formation of N-containing ZrO_2 phases during densification, however, is found disadvantageous and may be suppressed by the use of Y_2O_3 stabilised ZrO_2. At least 4.1% Y_2O_3 in ZrO_2 was necessary to prevent nitrogen uptake [505].

After Scott [506], who suggested that the presence of temperature form monoclinic ZrO_2 in the sintered silicon nitride is not necessarily beneficial, efforts concentrated on how to obtain the high temperature tetragonal form of ZrO_2. This phase acts as a metastable phase at RT and is converted to monoclinic ZrO_2 in the presence of a stress field.

TEM investigations of silicon nitride samples sintered with additions of Zr alkoxide as ZrO_2 source, with yttria stabilised ZrO_2 (+4.1%Y_2O_3) and with ZrO_2 and 1.1%Y_2O_3 show that when no yttria is added or only 1.1%Y_2O_3 is present, the significant part of ZrO_2 is in monoclinic form [507]. Only when 4.1% Y_2O_3-stabilised ZrO_2 is used as additive to silicon nitride, does the tetragonal form of ZrO_2 form as the major phase.It was also reported that ZrO_2 grains grow from the liquid during sintering, and thus the added ZrO_2 dissolves in the liquid and precipitates, indicating that it plays an important role in the liquid phase densification of silicon nitride.

Cheng and Thompson [508] studied Y_2O_3-Al_2O_3-SiO_2 liquids containing additions of ZrO_2; they observed that the solubility of ZrO_2 is around 10% at 1700 °C. This dissolved ZrO_2 precipitates on cooling from the liquid.

Formation of a Zr-Y-Si-O-N liquid is observed during sintering of silicon nitride with Y_2O_3 stabilised ZrO_2 [509]. This liquid enhances densification and also α-β transformation, mainly because Zr changes the amount and the viscosity of the liquid. The presence of this liquid phase also proves that yttrium present in ZrO_2 diffuses out and participates during the liquid-phase sintering of silicon nitride [510].

Pressureless sintering of silicon nitride with ZrO_2 or Y_2O_3+ZrO_2, by using a relatively large grain size powder (Stark LC1), results in very poor densification and formation of ZrN [511]. The similar green bodies can only be fully densified by HIPing [512]. Their resulting microstructure after HIPing shows that with Y_2O_3 stabilised ZrO_2, the tetragonal form of ZrO_2 is obtainable.

With a different combination, the mixed oxides of Y_2O_3-MgO-ZrO_2 are used to sinter silicon nitride [513]. A very sudden densification before reaching the solution-precipitation stage is observed and this leads to the conclusion that at temperatures as low as 1300 °C a low viscosity and large volume of liquid forms. The slope of the $\log(\Delta V/V_0)$ vs $\log(t)$ curve fits the Kingery Model and yields 1/3 which means that the solution-precipitation stage is diffusion controlled and the remaining densification occurs by mass transport during this second stage.

Sintering with other Oxides: Another approach to consolidate silicon nitride with reduced glassy grain boundary phase is to use $BeSiN_2$ or $BeAl_2O_4$ as sintering aids [514, 515]. In both cases Be and O are incorporated in the β silicon nitride structure as a solid solution. They demonstrated with TEM micrographs the existence of β silicon nitride with very little grain boundary glass. These small amounts were concentrated at the triple points(≈0.02 μm). Both studies report that an equiaxed β grain morphology was characteristic with this additive. Although the fracture toughness is low as a consequence, the material maintains its strength at temperatures up to 1200-1300 °C. However, Be-containing powders are toxic and therefore this type of materials did not find a commercial market.

Sintering with Mixed Oxide Additives: Tsuge et al.[516] reported that silicon nitride hot-pressed with both yttria and alumina additions densifies easier due to the formation of more liquid than with yttria alone. However, this results in more glass formation on cooling.

Similarly, Smith and Quackenbush [517] compared the mechanical strengths of sintered silicon nitrides with Y_2O_3 and Y_2O_3+Al_2O_3 and reported that the strengths at lower temperatures were higher with the additions of Al_2O_3. However, since the high volume liquid cools down to form a grain-boundary glass phase which allows shear between the silicon nitride grains above 1000 °C, the materials sintered with Y_2O_3 plus Al_2O_3 show significant strength degradation at these higher temperatures. Alumina additions may lower the glass softening points. Wötting et al. [518] reported that with certain combinations of Y_2O_3 and Al_2O_3 this secondary glass phase can be crystallized by a subsequent heat-treatment to give N-apatite and N-wollastonite. By altering the ratio between Y_2O_3 and Al_2O_3, it is possible to obtain Y-Al garnet as a crystalline secondary phase, and only a small amount of a discontinuous amorphous phase will remain at the grain boundaries [519].

Al_2O_3 concentrations in addition to Y_2O_3 lead to an increase in the amount of secondary phase and a decrease in the viscosity and the softening temperature of the liquid resulting in the precipitation of equiaxed β grains, whereas the presence of a high viscosity liquid phase slows down the material transport and results in an increase in supersaturation and eventually β grains with higher aspect ratio form [518]. In the course of producing a dense silicon nitride material by pressureless sintering, Y_2O_3 plus MgO were used as sintering aids to reaction bonded silicon nitride [520]. A 95%TD dense material with a RT-strength of 585 MPa was obtained. The presence of a secondary crystalline phase, $Mg_5Y_6Si_5O_{24}$, is reported.

Pomeroy et al. [521] studied the characteristics of the Mg-Y-Si-O-N glasses and the densification behaviour of the silicon nitride sintered with MgO plus Y_2O_3 additives. They observed that the shrinkage increases and α-β transformation decreases as the MgO content increases. The volume of the liquid formed controls the densification, whereas the viscosity affects the morphology of the β silicon nitride grains, high viscosity resulting in high aspect ratio grains. In another study [522] of the Mg-Y-Si-O-N system, it was pointed out that the surface silica around each silicon nitride particle has a significant effect on the viscosity changes in the liquid phase formed in this system. For minimum fractions of SiO_2, densification improved with increasing fractions of MgO; however, for increasing fractions of SiO_2, increased densifications were found only for low MgO/Y_2O_3 volume ratios. Similarly α-β transformation was influenced by the compositional changes of the secondary phases. The additive compositions nearest the apatite composition exhibit incomplete transformation to β Si_3N_4, suggesting formation of a crystalline phase may reduce the amount of the liquid phase available for densification.

The results presented by Saruhan et al. [523] show that combined MgO/Nd_2O_3 additions to silicon nitride are promising densification aids for the pressureless sintering, and the liquid formation region in Mg-Nd-Si-O-N Jänecke prism is extensive. Investigations on phase relationships in the MgO-Nd_2O_3-SiO_2 at 1450 °C show that a ternary crystalline phase, $MgNd_4Si_3O_{13}$, exists in this system. X-ray analyses in silicon nitride products sintered with MgO plus Nd_2O_3 demonstrate that the apatite contains a small amount of nitrogen, suggesting the presence of a solid solution range in the Mg-Nd-Si-O-N Jänecke prism [524].

Although Mg-Y-Si-O-N and Mg-Nd-Si-O-N systems are similar, the significant feature of the MgO-Nd_2O_3-SiO_2 system is that the ternary phase $Mg_5Nd_6Si_5O_{24}$ (analogue of the $Mg_5Y_6Si_5O_{24}$ phase reported by Giachello et al. [520]) cannot be prepared at 1450 °C [523].

3.1.2 Silicon Nitride Matrix Composites

Ceramic composites can be divided into two groups: Microcomposites and nanocomposites. In the microcomposites, micro-size second phases such as particle, platelet, whisker and fibre are dispersed at the grain boundaries of the matrix. The main purpose of these composites is to improve fracture toughness and strength.

The aim of the nanocomposites is to improve not only mechanical properties at room temperature but also high temperature mechanical properties such as hardness, strength, creep and fatigue fracture resistance. The additional intention with this composite is to add new functions to ceramics, such as the machinability and superplasticity common to metals.

Microcomposites

Compared with metals, ceramics have higher strength at high temperatures, and better oxidation and corrosion resistance. These properties make them the best candidates to replace Ni-base superalloys, which have almost reached their limit. Modern high performance ceramics, an important member of this group is silicon nitride, just like their traditional predecessors, however, are brittle. While this does not mean that they are weak, it indicates that the material exhibits a very small strain to failure. Metals have sufficient capacity for plastic deformation to redistribute loads, but this is not found with ceramics. As a consequence, ceramics are extremely sensitive to any surface and internal flaws which may give rise to stress concentration and stress gradients such that the local peak stresses may exceed the fracture strength of the material. There are mainly three approaches for improving the mechanical properties of silicon nitride:

- The first approach relies on reducing the processing related flaw size. This brings in the concept of fault-free fabrication.
- The second approach relies on increasing fracture strength and brings in the concept of microstructural engineering.

- The third approach aims to increase the fracture toughness of the silicon nitrides, through which its susceptibility to existing flaws can be reduced. As the former still occupies the efforts of the ceramic processors, the later two made the idea of reinforcement gain great interest in recent years, mainly because of the proposed fracture mechanisms, whose superiority has been theoretically demonstrated by many fracture mechanists [525, 526].

Silicon nitride composites are unlikely to be used in preference to simple materials unless there are positive advantages in terms of improvements to material properties and performance. The requirement to blend two distinct phases together, almost invariably means a more complex fabrication procedure together with increased cost. One of these two phases, the reinforcing component (whiskers, fibres and platelets), is produced through relatively complex processes and therefore is extremely expensive (Figure 3.6). Therefore, the economic costs of more sophisticated and controlled processing must be set against the potential advantages of the products.

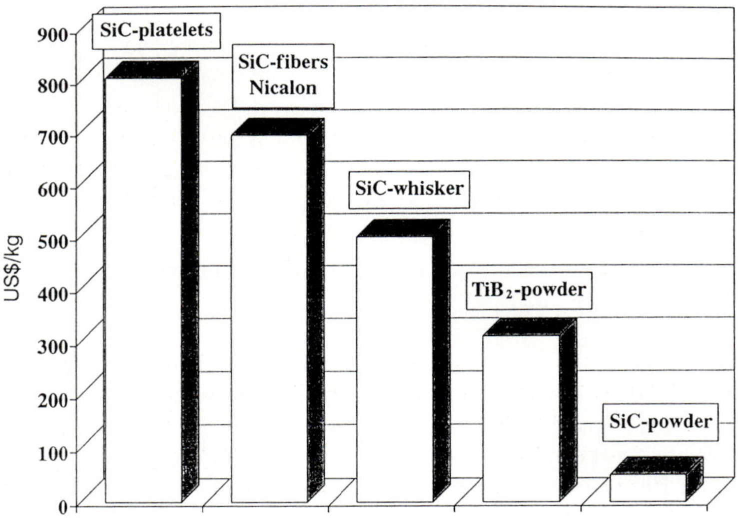

Figure 3.6 The effective costs for various reinforcing components

Although the fracture toughness or surface energy of the ceramic composites has been enhanced in recent years through various fracture mechanisms (crack deflection, branching, etc.), the material still exhibits brittle fracture. The crack deflection mechanism is based on differences in the thermal expansion coefficients and/or elastic properties of the matrix and its secondary phases. The more the dispersed particles deviate from a spherical morphology, the more pronounced the gain in fracture toughness [527] (Figure 3.7). The only reinforcing component which produces an increase in fracture toughness is the continuous fibre and fibre reinforcement offers the promise of tougher materials with a controlled fracture, some pseudo-ductility and an increased strain at fracture.

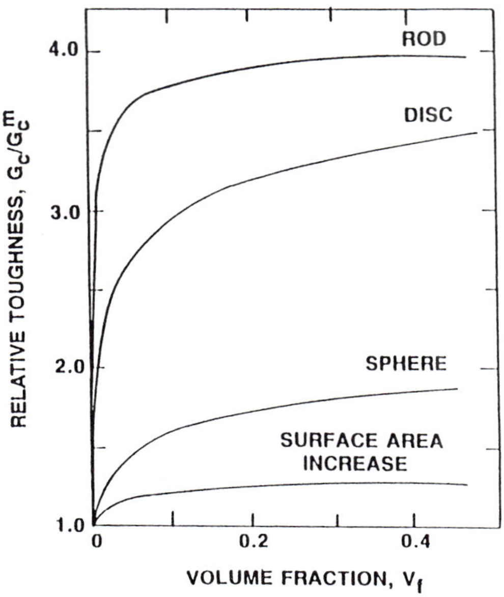

Figure 3.7 Theoretical prediction of fracture toughness gain by dispersoid particles of various morphology [527]

SiC Whiskers: Since the first research efforts in the field of whisker composites began, the commercially available SiC whiskers are increasingly visible in the market. Their production route has an important role in controlling their purity, shape and size [528]. Based on the proposed toughening mechanisms, the recommended aspect ratio for whiskers in a composite ranges from 7 [529] to 15 [528]. However, the whiskers in the as-received state consist of various sizes and shapes, having high aspect ratios (> 20 to 200) [531]. Their microstructural and analytical characteristics also differ depending on the production method [528, 530, 531].

Investigations on SiC whiskers provided by Tateho (TA), Tokai Carbon (TK) and American Matrix (AC) companies, showed that TK whiskers contain large amounts of cobalt and titanium but otherwise are very clean [531]. These whiskers consist of both smooth and contorted shaped whiskers, while the AC whiskers have surface morphologies from smooth to knobby appearance. The TA whiskers have a much lower oxygen content than either the TK or AC whiskers, but this oxide layer at TA whiskers resembles that of a Si-O-C glass. The other investigations showed that the SiC whiskers contain excess free carbon, nitrogen and calcium as impurities on their surface layer [532]. SiC whiskers are described as a random mixture of short period SiC polytypes. In Table 3.1 some properties of SiC whiskers are listed.

Whisker Grade	Impurities (%)		Length	Diameter	Aspect ratio
	LE	MET	μm		
AC	1.7	-	100	2.00	50
Nikkei	5.0	-	80	0.40	200
TK	0.9	0.15	50	0.65	77
Huber	1.1	-	140	0.90	167
TA	2.9	0.30	45	0.45	100

Table 3.1 Characterization of as-received SiC whiskers [532]
(LE: O, N, Ca ; MET: metal impurities)

The very fine morphology of whiskers is associated with a high incidence of pleural tumours following implantation [533]. Therefore, research with whiskers in many countries has been suspended.

Processing of SiC Whisker/Silicon Nitride Matrix Composites: The main concerns in producing two-component ceramic composites are to obtain a sufficient distribution of the reinforcing component in the matrix and to maintain thermal and chemical stability of the reinforcing component in the matrix, which potentially results in a relatively weak interface [534]. In order to ensure a homogeneous distribution of the whiskers in the whisker-powder mixture, it is essential to use wet processing, for instance slip casting. In slip casting, a well-stabilized and dispersed slip with a low viscosity and high solids content has to be used to reach high green densities [535]. Due to the aqueous medium, which is water in most cases, an increase of 1% in oxygen content of the silicon nitride powder has to be taken into account in the calculation of the initial composition [542]. The typical morphology of whiskers with high aspect ratios (50 to 200) causes them to tangle and form bundles, (Figure 3.8). Owing to this tendency, whiskers present large agglomerates during mixing. Milling or deaggregation of whiskers during dry processing leads to a reduction in their length [536, 537] (Figure 3.9). In relation to this morphological character of whiskers, the observed processing flaws on the fracture surfaces of the whisker composites are primarily nests or bundles of whiskers. However, additional residual contamination in the as-received whiskers are also potential failure origins in the composites [538].

During sintering, it is often observed that SiC whiskers are stable only under certain sintering conditions and/or in a certain temperature range in the Si-C-O-N system (Figure 3.10) [539]. Under nitrogen atmosphere, due to the reaction of SiC with nitrogen, degradation of whiskers may take place [528].

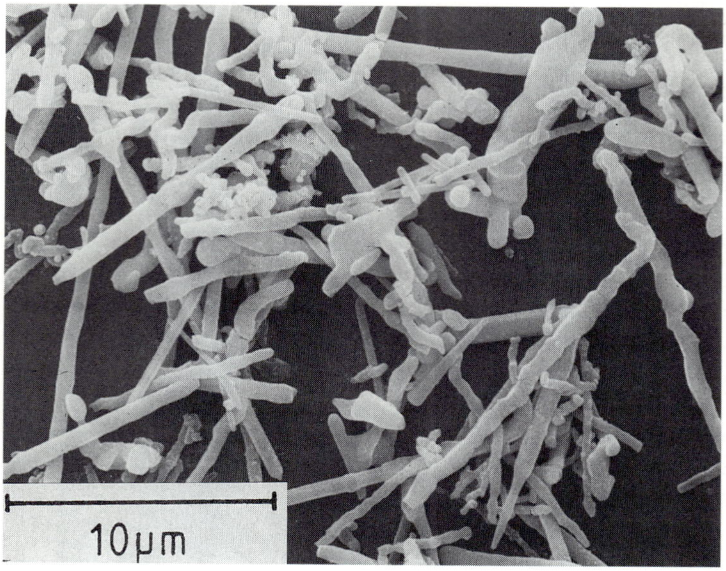

Figure 3.8 Typical morphology of SiC whiskers

The presence of the SiC whiskers retards densification kinetics. Considering these last two factors, it is critical to choose the optimum sintering time, temperature and atmosphere. Holding times over 90 minutes deteriorates the whisker, owing to the reaction with the matrix [540]. According to Bellosi and De Portu [537], the addition of whiskers affects the rate of densification substantially by inhibiting particle rearrangement during the sintering process as well as transport phenomena during the solution-diffusion-precipitation stage, since SiC is an inert phase. However, it is possible to sinter SiC whisker/silicon nitride composites to full density by using a combination of sintering techniques. One needs to adjust the sintering conditions to maintain whisker stability, for instance to optimize the sintering atmosphere. Fully densified SiC whisker/silicon nitride composites with a better whisker morphology and no apparent whisker-matrix interfacial bonding [532] can be achieved by presintering at 1500 °C in an argon atmosphere and subsequently encapsulating and HIPing at 1780 °C under 190 MPa gas pressure [541].

Another alternative technique [530] is to increase the content of the oxide additives to silicon nitride and hence to reduce the sintering temperature and ease the particle rearrangement in a larger liquid phase, which may result in formation of grain-boundary glass phases with low softening points. This affects severely the high temperature mechanical properties of the composite.

Figure 3.9 Morphology of SiC whiskers after processing in silicon nitride matrix composites

Continuous Fibres and Monofilaments: The continuous fibre-reinforced ceramic composites have stimulated many research activities because of their potential for greater structural reliability via increased toughness and reduced flaw sensitivity.

The limited availability of continuous fibres has restricted the choice of reinforcement to carbon (Toray), Si-C-O (Nicalon), Si-Ti-C-O (Tyranno), and CVD Si-C (Textron) fibres. Early work on glass matrix composites with carbon fibres provided the impetus for developing strong and tough ceramic matrix composites [543]. The carbon fibre reinforcement provided good mechanical properties at low temperature, but the fibres degraded at high temperature [544].

Nicalon fibres with better oxidation resistance than carbon have also been used for reinforcement in glass matrix composites and have been shown to have good strength and toughness at low temperatures. Nicalon and Tyranno, however, are polycarbosilane derived amorphous/microcrystalline Si-C-O and Si-Ti-C-O products, respectively, and suffer a degradation in properties as they transform to equilibrium products during exposure to high temperatures. Heat treatment tests on the as-received, polymer derived SiC-fibres in various atmospheres show that above 1200 °C their strength degrades drastically and extreme crystallization occurs. The crystallites reach >100 nm in size at 1600 °C [545, 546]. Under similar conditions, the carbon fibres show a higher strength, owing to their higher RT strength (Figure 3.11) [547].

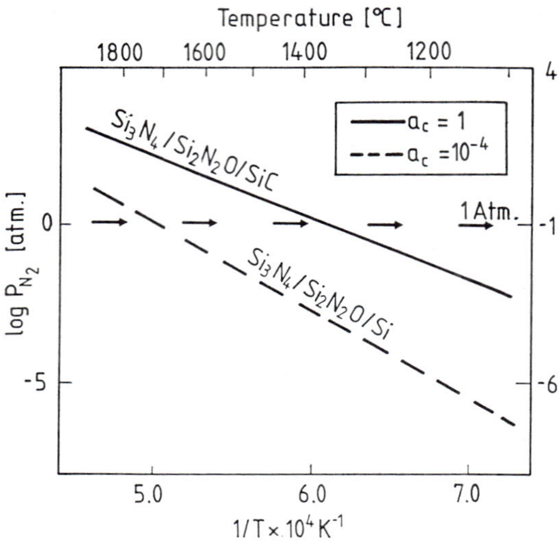

Figure 3.10 Stability region of SiC and Si_3N_4 phases in the Si-C-O-N system as a function of nitrogen pressure [539]

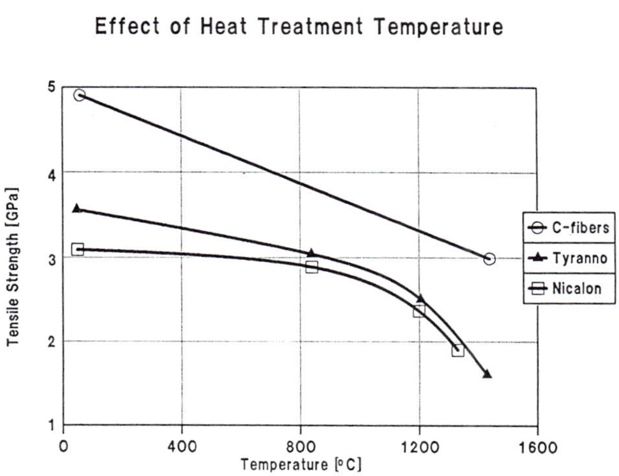

Figure 3.11 Temperature dependence of strength in polymer derived SiC and C fibres [547]

Many workers demonstrated improved strength and toughness in reaction-bonded silicon nitride [548-551] and with sintered silicon nitride [550, 552, 553] with the CVD SiC fibres provided from Textron, mainly because of their better high temperature stability. Typical commercial CVD SiC fibres on a carbon core, having a diameter of 140 μm, have a Young's modulus of 400 GPa. These fibres lie at a significantly higher strength level than the yarn fibres (3500 MPa versus 2000 MPa at 50 mm gauge length). This could be due to smaller defect sizes present in these fibres. At 1093 °C, the CVD fibre shows remarkable strength retention, while at 1204 °C significant degradation occurs. Up to 1100 °C, at least, the reduction in strength is a reversible effect and not associated with a permanent change in the fibre. The most likely explanation for the permanent degradation occurring above 1200 °C is due to crystallite coarsening rather than to the chemical reactions [545].

The strength degradation of polymer derived SiC fibres after heat treatment has been explained such that above 1200 °C, excess free carbon present in the fibre reacts with oxygen and escapes from the fibre. The evolution of CO from the fibre leaves behind a hole in the fibre and this is an on-going process. The depth of the notch is expected to be much greater in vacuum than in air or argon [554]. Heating in argon leads to exaggerated grain growth of β SiC as well as formation of some degree of chemical reactions [555]. If the fibres are exposed directly to any environment containing lower partial pressures of nitrogen, CO and SiO, these gases would evolve from the fibre. If the fibre is embedded in a solid matrix, the dissociation could occur at an appreciable rate along the fibre/matrix interface or may even induce cracks in the matrix [556].

Processing of Fibre/Silicon Nitride Matrix Composites: After the leading work of Lindley and Godfrey [557] in reinforcing silicon nitride matrix with SiC fibre, having a W core, great interest has been attracted to the area. Guo et al. [544] also reported outstanding toughening in the properties of silicon nitride reinforced with continuous carbon fibres.

A tow of carbon fibre is released through a slurry of silicon nitride containing additives and is wound onto a square board (Figure 3.12). The greatest problem in processing these green bodies was the heavy reactions taking place at the interface between silicon nitride and carbon fibre at temperatures higher than 1650 °C. The sintering temperature of silicon nitride with known oxide additives lies above 1750 °C. Secondly, the thermal mismatch between carbon fibre and silicon nitride matrix produces a high tensile strength along the length of the fibre. By using an additive combination of MgO-SiO_2-LiF, they succeeded in reducing the hot-pressing temperature as low as 1450 °C. They also suggested that the addition of ZrO_2 into carbon fibre/silicon nitride matrix composites should cause a process of energy absorption and, as a result, the tensile strength in these composites may be decreased or dispersed.

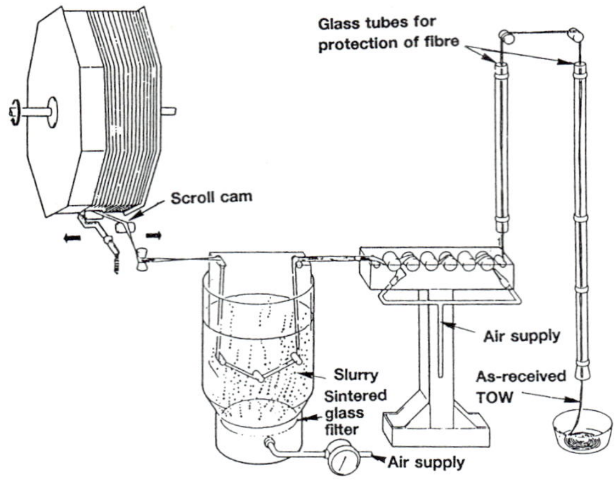

Figure 3.12 Schematic illustration of processing with continuous fibres

Processing of continuous SiC fibres with a C core fibre (monofilaments) in silicon nitride matrix is achieved by various workers through hot-pressing of multiple layers of SiC fibre and silicon nitride and oxide additive mixture powders [552, 553]. It is reported that the SiC filaments with carbon-rich layers show less or no adhesion to the matrix and exhibit the highest value for fracture toughness.

Researchers concerned about problems related to the processing of fibre ceramic composites by the powder route benefited from other alternative techniques, such as chemical vapour deposition (CVD) or chemical vapour infiltration (CVI). Deposition by CVD and CVI on fibrous preforms are shown to be an effective method in producing ceramic composites without fibre damage [552, 558]. With this technique, it has been possible to produce preforms having fibre contents up to 25 vol%. This technique requires the use of gases such as $SiCl_4$, NH_3 and H_2 (carrier gas) which are released from two separate gas inlet tubes [558]. At temperatures as low as 1100-1300 °C, a crystalline phase forms. Amorphous silicon nitride is deposited on the surface of graphite substrates at 5 torr. A reduction of the amount of reactive gas species and decrease of total pressure provides conditions favourable for the deposition of crystalline α silicon nitride [552]. Incomplete densification seems to be the main problem of this technique and efforts are required to establish optimum process parameters. The reported mechanical properties have not yet reached a satisfactory level.

Improvement of fracture toughness by incorporation of continuous fibres relies on the applicability of pull-out fracture mechanisms correlated with debonding.

The expected improvement can be by a factor of 2-10. Controlled fracture mode to an essential degree is only possible with this sort of reinforcement.

However, fabrication of silicon nitride with continuous and aligned fibres in net-shapes is rather difficult. Fibres can be employed as short and random forms as well as continuous and aligned fibres. With short fibres, having random orientation, one ends up with an isotropic material. A maximum fibre content of 20 vol% can possibly be distributed homogeneously in the slurries which contain optimum amounts of surfactants and disperser [547]. During fabrication, the fibre lengths are reduced depending on the loaded volume. The fibres tend to be aligned into the plane perpendicular to the hot-pressing direction, but within this plane, the fibres are randomly oriented. The pull-out lengths of fibres are short due to the bending stresses on the fibres caused by their random orientation.

Particles and Platelets: Whisker composites, in view of being potentially hazardous and leading to insufficient improvement in mechanical properties, have hardly found any commercial interest apart from being applied as cutting tools. This is also related to their high production costs. Therefore, alternatively, less expensive reinforcement components with uncritical geometry, such as particles, but most recently platelet-shaped crystals, have been produced.

Silicon nitride matrices reinforced with particles offer the advantages of a hard and wear-resistant ceramic material. For reinforcement of silicon nitride with particles, SiC ceramic powders are used frequently [559-562]. Particle size of the SiC powders is found to have an influence on the mechanical properties of the composites [559]. Recently TiN [544, 563], TiC [544, 561], and B_4C [544] have also been used to reinforce silicon nitride.

They may offer certain advantages over whiskers, such as simpler processing, lower production costs and better thermal stability. At present, SiC and Al_2O_3 are available in the form of monocrystalline platelets (Figure 3.13a). Production of B_4C platelets is also reported, but they are not yet commercially available. Al_2O_3 platelets are not suitable for silicon nitride matrix materials, as Al_2O_3 forms a liquid phase with the additives necessary for sintering silicon nitride.

Processing of Particle/Silicon Nitride Matrix Composites: Incorporation of SiC, TiC and TiN into silicon nitride results in a finer β grain size, hence the mechanical properties of the composites present differences to monolithic silicon nitride. The XRD analysis of the sintered products of particle/silicon nitride matrix composites shows that B_4C[544] and TiC [544, 561] react with silicon nitride to form mainly SiC in both cases and in other secondary phases, and thus decompose. The thickness of an interfacial layer of $TiC_{0.5}N_{0.5}$ which forms due to the reaction between silicon nitride and TiC increases as the sintering time and temperature increase [561]. Therefore, these two particles are not convenient reinforcing components for silicon nitride. TiN and TiB_2, however, show no visible reactions with the matrix and can be sintered easily under the conventional sintering con-

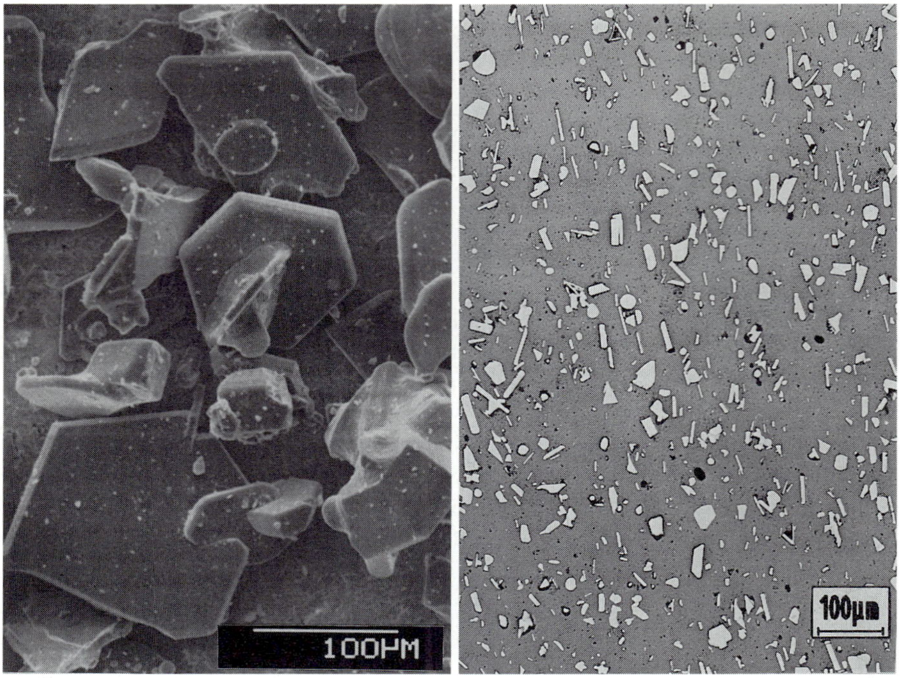

Figure 3.13 Typical morphology of a) SiC platelets (American Matrix) and b) SiC-platelet/Si_3N_4

ditions for silicon nitride and eventually lead to some improvement in RT mechanical properties (Figure 3.14). However, they oxidize catastrophically at temperatures near 500 °C, leading to degradation of the composites at high temperatures [544, 563]. XRD investigations on SiC particle/silicon nitride matrix composites show that incorporation of particles results in an incomplete α-β transformation. Depending on the characteristics of the SiC particle, under the same conditions the composition and type of secondary phase change. These changes in the matrix which have been introduced by the particles yield a mazerial completely different from the monolithic one and lead to various observations in the mechanical properties [564].

Processing of Platelet/Silicon Nitride Matrix Composites: Processing of platelet/silicon nitride matrix composites is relatively easy but requires the use of wet-processing techniques (Figure 3.13b). With platelet volume contents higher than 30 vol%, agglomerates of platelets are formed, limiting the packing density of the composite. The fraction of agglomerates depends on the diameter of the

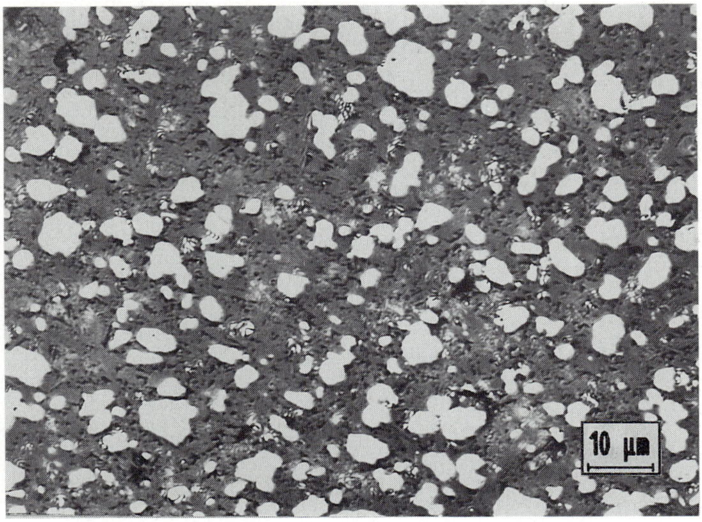

Figure 3.14 Optical micrograph of a TiB_2/silicon nitride matrix composite

embedded platelets [566]. Hot-pressing or optimized gas pressure sintering of this composite yields highly dense composites with full-phase transformation. Depending on the surface SiO_2 layer on the platelets, different secondary phase combinations are observed in the matrix [564].

SiC platelets influence the phase formations and matrix morphology and thereby induce a significant change in the matrix. Hence, it may be concluded that a matrix effect controls the high temperature deformation behaviour of the corresponding composite. This composite still shows brittle fracture, although some crack deflection and branching at the interface are observed.

Nanocomposites

The nanocomposites can be described in three groups; intragranular and intergranular composites and nano/nanocomposites. As shown in Figure 3.15 in the intra- and intergranular nanocomposites, the nano-size particles are dispersed mainly within the matrix grains or at the grain boundaries of the matrix. Niihara and his colleagues [565] observed significant improvement of mechanical properties. Ceramic nanocomposites were successfully prepared initially by chemical vapour deposition (CVD) of the TiN/silicon nitride system [567]. Although the CVD process is a preferable method to disperse the nano-size second phases into

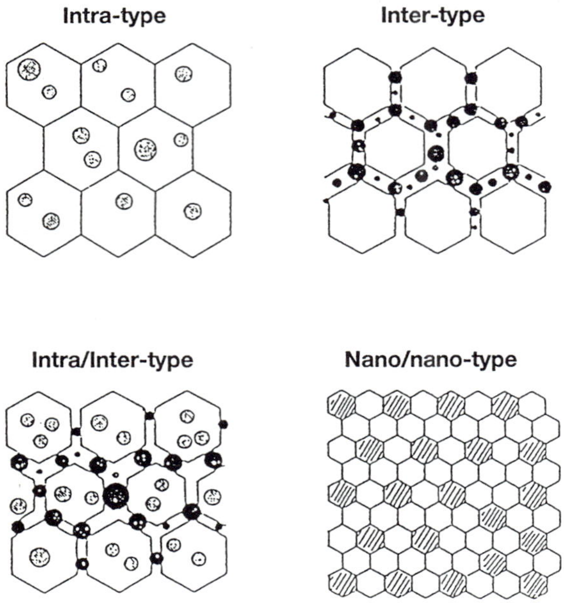

Figure 3.15 Schematic description of different types of nanocomposites

the matrix grains or at the grain boundaries, it is not applicable for fabricating large and complex-shaped component for mass production, and is also rather expensive.

Niihara [568] has succeeded in processing ceramic nanocomposites with silicon nitride matrices by usual powder metallurgical techniques. An amorphous Si-C-N precursor powder which converts to a fine mixture of SiC and Si_3N_4 particles when heated at high temperatures is also produced and employed for production of nanocomposites. In order to achieve liquid phase sintering, the precursor powder is mixed with yttria and sintered under conventional sintering conditions. As compared to monolithic silicon nitride, the nanocomposites were composed of more uniform and homogeneous elongated grains. The author reported that despite yttria additions, no impurity phases were identified at the SiC/Si_3N_4 interfaces. Further work is needed to understand and develop these composites.

3.1.3 Fracture and Fatigue of Non-Oxide Ceramics

J. Göring

In the group of nonoxide ceramic materials silicon nitride and silicon carbide are the favourite candidates for technical applications. In this section the strength and the fatigue behaviour of typical SiC and Si_3N_4 materials are discussed.

Fracture Behaviour

Fracture strength is the most usual and standardized test method for the mechanical characterization of ceramic materials [569-571]. Due to the extensive scatter of strength values, statistical methods are used to describe the strength distribution of ceramic materials. There are two main reasons to apply statistics. The first is that in many cases the scatter is too much to calculate or to design with the mean strength, the standard deviation and a security factor. Often statistical calculations are necessary to define a failure probability for a given stress or lifetime. The second reason is that the measured strength values are strongly influenced by the volume under stress. In a larger volume, the probability of a larger critical crack will increase, resulting in a lower strength value. Due to the intensive scatter of the strength values for ceramic materials the volume effect is much more intensive than in metallic materials, and not negligible. For a material with an average Weibull modulus of 7-12 the strength values measured in a three-point bend test are more than two times higher than those from a tension test. Statistical methods are necessary to describe and to calculate this volume effect,

The most common method to describe the strength distribution of ceramic materials is the two-parameter Weibull distribution [572-578]. The cumulative failure probability F for an applied stress σ_{max} is given by

$$F = 1 - \exp\left[-\frac{V_e}{V_0} \cdot \left(\frac{\sigma_{max}}{\sigma_0}\right)^m \right] \tag{3.1}$$

V_e is the volume under tensile stress, V_0 a normalizing volume, σ_0 is a scaling parameter which is dependent on the volume under stress and m is the Weibull modulus, which characterizes the scatter, independent on the volume. A supposition for the calculation of the statistical volume effect is a frequent and homogeneous distribution of failure relevant defects in the tested volume. From equation (3.1) the statistical volume effect is given by the following relation:

$$\frac{\sigma_1}{\sigma_2} = \left(\frac{V_{e2}}{V_{e1}}\right)^{1/m} \tag{3.2}$$

The volumes (V_{e1}, V_{e2}) have to be calculated with a volume integral [576, 577], including the Weibull modulus m and the stress distribution in a sample or component.

Beside the critical fracture, some oxide and nonoxide ceramic materials exhibit subcritical crack growth. A crack with a stress intensity less than the fracture toughness grows slowly until the initial crack reaches a critical length and fracture occurs.

Different experimental methods are known for determining subcritical crack growth. Some techniques deal with artificial and large cracks in the mm range [579]. Sometimes large and small cracks show different extension mechanisms influenced by the microstructure of the investigated material. In such cases it is problematic to transfer the parameters determined by large cracks to specimens or components containing small failure produced during processing (30-150 μm in general).

Dynamic fatigue tests are a method to investigate the subcritical crack growth, starting from a natural failure. If a material exhibits subcritical crack growth, the strength is related to the stress rate. During slow loading there is more time for subcritical crack extension and the strength will increase with increasing stress rates.

Basing on the well-known relation for the stress intensity factor K_I

$$K_I = \sigma \cdot \sqrt{a} \cdot Y \tag{3.3}$$

where σ is the applied stress, a the crack length and Y a geometric constant, and assuming that the crack growth velocity v is given by a power law equation [581, 580, 582]

$$v = \frac{da}{dt} = A \cdot K_I^N \tag{3.4}$$

with the crack extension parameters A and N, the following relation was found between the strength and the stress rate $\dot{\sigma}$ [583-585]:

$$\sigma_B^{N+1} = B \cdot \sigma_i^{N-2} \cdot (N+1) \cdot \dot{\sigma} \tag{3.5}$$

with

$$B = \frac{2}{A \cdot Y^2 \cdot (N-2) \cdot K_{IC}^{N-2}}$$

Including the inert bending strength σ_i and the fracture toughness K_{IC} equation (3.5) results in:

$$lg\sigma_B = \frac{1}{N+1}\, lg(B \cdot \sigma_i^{N-2}) \; + \; lg(N+1) \; + \; lg\dot{\sigma} \tag{3.6}$$

From a regression analysis of lg σ_B versus lg$\dot{\sigma}$ the crack velocity parameter N is given by the slope. The parameter A can be calculated by the intercept, the fracture toughness and the inert bending strength.

With these methods it is possible to determine the strength distribution and the static fatigue behaviour of ceramic materials. But a very important fact is that the mechanical properties of structural ceramics are strongly influenced by the microstructural characteristics of the different materials. The grain size and shape, but also the amount and composition of secondary grain boundary phase, lead to different failure mechanisms, resulting in different mechanical properties. In many cases it is necessary to consider the microstructural characteristics and the failure mechanisms to find the correct interpretation for the measured mechanical properties. In this paragraph two materials from the SiC and two from the Si_3N_4 group are selected, for discussing the failure mechanisms and demonstrating the influence of microstructure on mechanical properties.

Due to the type of ceramic, different sintering mechanisms during fabrication lead to typical microstructural features. In SiC sintering is mainly controlled by volume and grain boundary diffusion effects [586, 587], whereas a participation of liquid phase sintering is discussed. The most common sintering additives are boron and aluminium, which lead to a boron or aluminium rich silicate phase in the grain boundaries. Due to the diffusion controlled sinter mechanism in SiC materials, only a small amount of sintering additives is necessary. Thus it is possible to produce dense materials with a small portion of glassy secondary grain boundary phases, resulting in good high temperature mechanical properties, such as creep or oxidation. Figure 3.16 shows the fracture surface of a hot isostatic pressed silicon carbide (HIP-SiC) with a very low amount of aluminium as sintering additive (0.2%), produced by ESK.

The mean grain size was between 1 and 2 μm. TEM investigations of the α-SiC grains show that 6H-SiC was the dominant polytype in this SiC material. Due to the low amount of sintering additives this material exhibits a very thin but glassy grain boundary phase of only a few μm [588].

Siliconized silicon carbide is a very interesting SiC material. Starting from a green body, consisting of primary silicon carbide, carbon and a binder, any shape of a component can be formed by usual techniques like try pressing, slip casting, injection moulding or extrusion. After burning out the binder, the green body will be siliconized. At high temperatures liquid silicon reacts with the carbon to secondary SiC, without shrinkage. The residual pores are filled with silicon, leading to a 100% dense material. Because of the relatively easy and low cost fabrication, SiSiC is a favourite candidate for tubes, and burners and is suitable for heat exchangers, especially because of its high thermal conductivity.

The SiSiC displayed in Figure 3.17 is a commercially available material grade SILIT SK 501, produced by SIGRI. The microstructure shows a bimodule grain size distribution of the SiC grains. The small grains are in the range of 2-10 μm, the large ones go up to 100 μm. The density was between 3.04 and 3.08 g/cm³ and the amount of free silicon between 17 and 19 vol%. The reason for the scatter

Figure 3.16 Fracture surface of HIP-SiC showing intergranular fracture mode in the area of subcritical crack growth

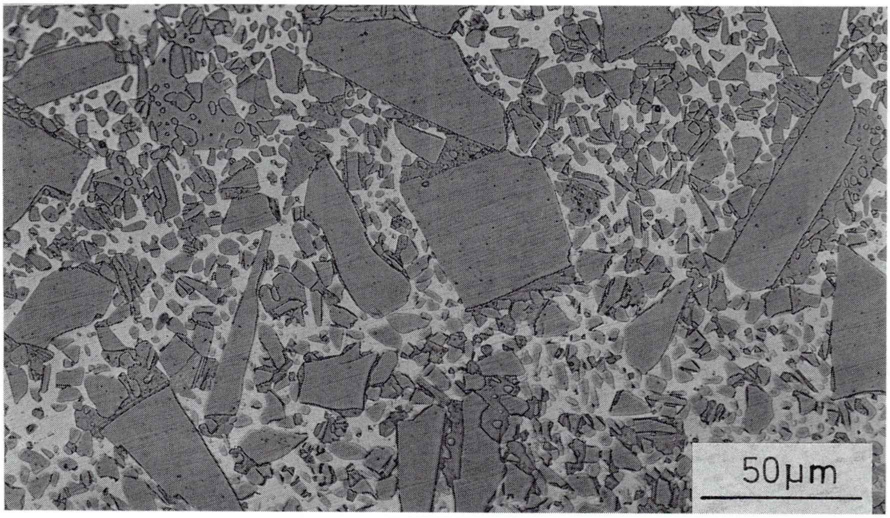

Figure 3.17 Structure of SiSiC (SILIT SK 501) showing a bimodule grain size distribution of the SiC grains (grey areas) and the metallic silicon (white areas)

of the density and silicon content was the inhomogeneous distribution of the free silicon.

The development and structure of dense Si_3N_4 materials are discussed in section "Fabrication and Microstructural Development of Non-Oxide Ceramics". Densification of Si_3N_4 occurs by liquid phase sintering. Starting from α-Si_3N_4 powder, transformation from α- to β-Si_3N_4 occurs during sintering, producing a microstructure with elongated β-Si_3N_4 grains. This special grain structure of dense Si_3N_4 materials leads to crack deflections [589], providing the high fracture toughness and strength values which are typical for dense Si_3N_4 ceramics. The high amount of additives, which is necessary for liquid phase sintering, leads to a high amount of secondary grain boundary phases, reducing the high temperature mechanical properties. Using Y_2O_3 as sintering aid it is possible to crystallize the glassy grain boundary phase in a further heat treatment. This leads to a higher viscosity of the secondary phase at elevated temperatures and to better mechanical properties at high temperatures.

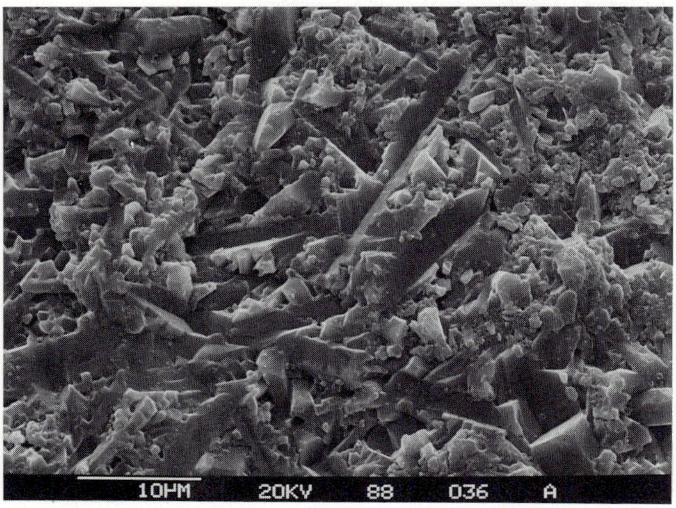

Figure 3.18 Fracture surface of HIP-RBSN with a high amount of intergranular fracture mode

Figure 3.18 shows the fracture surface of a post-hiped reaction bonded silicon nitride (HIP-RBSN) produced at DLR. The material was doped with 4.3% Y_2O_3 and the final density was 3.24 g/cm³. The grains show the typical hexagonal structure of the elongated β-Si_3N_4 grains, with a few grains reaching a length of more than 50 µm. The grain boundaries are covered with a mainly glassy yttrium rich silicate phase.

To discuss the influence of the grain boundary phase, a further material from the Si_3N_4 group was selected. Figure 3.19 displays the microstructure of a commercially available SiAlON material (SYALON 201) from Lucas Cookson Syalon [590, 591, 592].

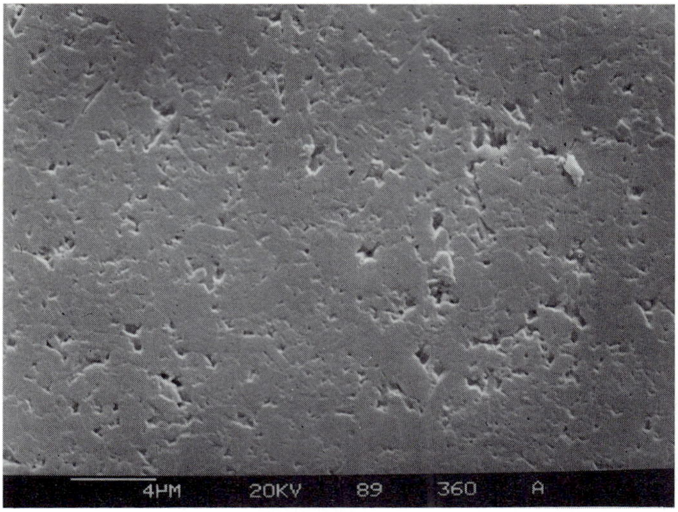

Figure 3.19 Polished and etched surface of SYALON 201

The material was doped with Y_2O_3 and by a further heat treatment the amorphous aluminium and yttrium rich glassy grain boundary phase was cristallized to yttrium aluminium garnet. The density was 3.25 g/cm³ and the SYALON 201 exhibits a fine grained microstructure with a mean grain size of 1-2 µm.

The room temperature strength values of these four materials were examined in a four point bending test. The results are summarized in Figure 3.20, and the strength values and Weibull parameter are listed in Table 3.2.

	SiSiC	HIP-SiC	HIP-RBSN	SYALON
Samples	40	30	30	30
Strength in MPa	290±38	471±69	711±72	540±114
m	9.2	8.4	12.6	5.5
σ_0 in MPa	306	500	741	585

Table 3.2 Strength values and Weibull parameter of the various SiC and Si_3N_4 materials

As expected, the Si_3N_4 materials show the highest strength values. The reasons for the high fracture toughness are investigated in [589]. When critical crack

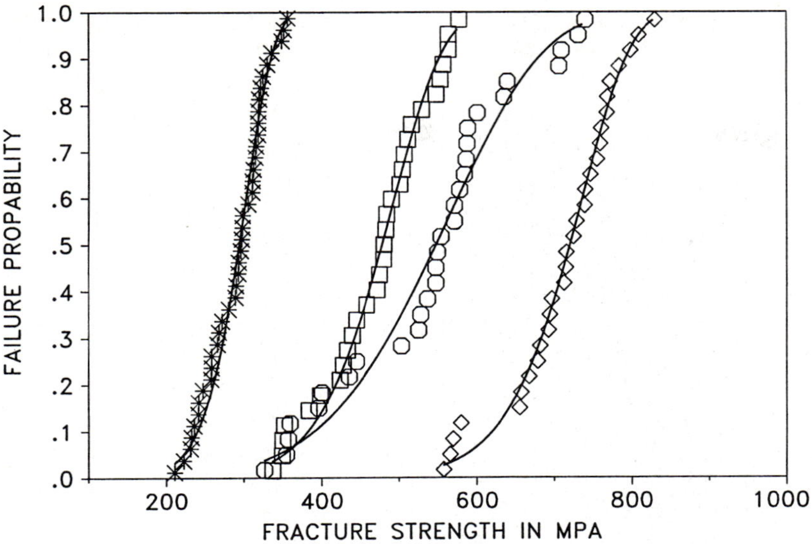

Figure 3.20 Strength distribution of SiSiC (○), HIP-SiC (x), SYALON 201 (◇) and HIP-RBSN (□) in a Weibull plot

extension starts, deflections at the crack tip occur, resulting in a mixed mode stress intensity which is less critical for a material than a pure mode I loading. The extent of crack deflections is determined by the grain shape and by the fracture energy of the grain boundary phase. Especially a higher aspect ratio (length/diameter) of the hexagonal Si_3N_4 grains leads to an increase in toughness and therefore to high strength values. Another, maybe more important microstructural influence determining strength is the fracture origin. Due to the brittleness of ceramic materials small defects lead to high stress intensities which cannot be greatly reduced by plastic deformations. So, relatively small material defects like pores, inclusions, inhomogeneities or machining damages determine the strength values of structural ceramics. In fractographic investigations the fracture origins were localized with a light microscope and characterized with a SEM. In the case of SiSiC the fracture origin could be localized at about 50 % of the broken bending bars. The cracks were starting from large SiC grains which were partially ground during surface treatment. So it is supposed that the measured strength of the SiSiC was mainly controlled by machining damages. For the other materials investigated all fracture relevant defects were material failures, and no machining damage was found determining the strength values. In some cases the correlation between the measured strength values and the fitted Weibull distribution was not sufficient, especially in the case of SYALON 201 and HIP-RBSN. The reason was a multi module defect distribution which leads to a multi module Weibul distribution. In the case of HIP-RBSN strength was determined by two different types of pores. The first four strength values were controlled by relatively large pores, created during the densification of the green body. The second type of pores was clearly smaller and the β-Si_3N_4 grains in and around the pore were very large and globular. The grain growth and the change from the hexagonal to a more globular grain shape can be referred to a local concentration of sintering additives or impurities [593]. So the step in the Weibull distribution after the first four strength values can be explained with a bimodule failure distribution.

In the case of SYALON 201 three different types of pores were identified. The strength was controlled by densification failures and by pores, referring to a local concentration of sintering additives, similar to HIP-RBSN. The third kind of pores was clearly different to the first two types but the origin could not be clarified. This multi module failure distribution leads to the insufficient correlation of the measured strength value with the fitted Weibull distribution.

Also in HIP-SiC two different kinds of critical defects were detected. About 15% of the tested bending bars failed by pores and about 85% by accumulated large grains. Due to the anisotropic thermal expansion of the SiC grains the large grains produce internal stresses [594] which are the reason for failure.

These results show that the measured strength values and Weibull parameters are characteristic values only for the tested bending bars. Calculations of statistical volume effects are not possible with the determined Weibull parameters. The SiSiC failed mainly from machining damages and the measured strength and scatter is mainly influenced by the surface treatment. In this case statistical calculations are limited to samples or components with the same surface treatment. For the other ceramics investigated, fracture was initiated by material defects and no machining damage was observed. But strength was determined by different failure types and it is supposed that the population of critical defects is not constant in different volumes under stress. This means, that the Weibull modulus is dependent on the volume under stress and cannot be used for statistical calculations. Other Si_3N_4, SiC and oxide ceramics investigated at DLR also exhibit more module failure distributions, determining the strength values. This shows that in many cases an uncritical use of measured Weibull parameters can provide wrong results in statistical calculations. The characterization of fracture origins by fractographic investigations can help find the right interpretation of measured strength values. Furthermore, fractography will help to develop an optimized material quality. This is clearly seen in the investigated HIP-RBSN. The upper strength values are controlled by pores referring to the densification of the green body. If it is practicable to eliminate these critical failures, the expected strength distribution can be calculated to 733±47 MPa with a Weibull modulus of 17.

Fatigue Behaviour

Beside critical crack extension, which is the main failure mechanism during strength testing, for some oxide and nonoxide structural ceramics failure mechanisms are known which produce continuous crack propagation at stress intensities less than the fracture toughness. The fatigue behaviour of structural ceramics is subdivided into static fatigue, if the strength is constant over time; dynamic fatigue, if the stress is introduced under a constant stress rate until failure, and cyclic fatigue if the stress is cyclic. In general it is assumed that static and dynamic fatigue is controlled by the same failure mechanisms. At room temperature the subcritical crack extension is determined by stress corrosion effects. In the presence of water a corrosion effect under support of the stress is known which leads to a continuous crack growth [585, 595, 596], especially with alumina and in some cases MgO doped Si_3N_4. The materials reported in this paragraph showed no indications of dynamic fatigue effects at room temperature. At elevated temperatures the decrease of viscosity in the secondary phase leads to diffusion effects, creating cavitations and micropores at the grain boundaries which weaken the material, and enable subcritical crack growth. At higher temperature and lower viscosity of the secondary phase grain boundary, sliding is a further effect creating damages and supporting crack growth. If the viscosity is too low and grain sliding leads to plastic deformations, the methods of linear-elastic fracture mechanics cannot be used and the damage or failure of the material can be described with the methods of creep crack growth or creep damage.

HIP-SiC and HIP-RBSN show distinct subcritical crack growth in dynamic fatigue tests at temperatures above 800-900 °C (see Figure 3.21 and Figure 3.22).

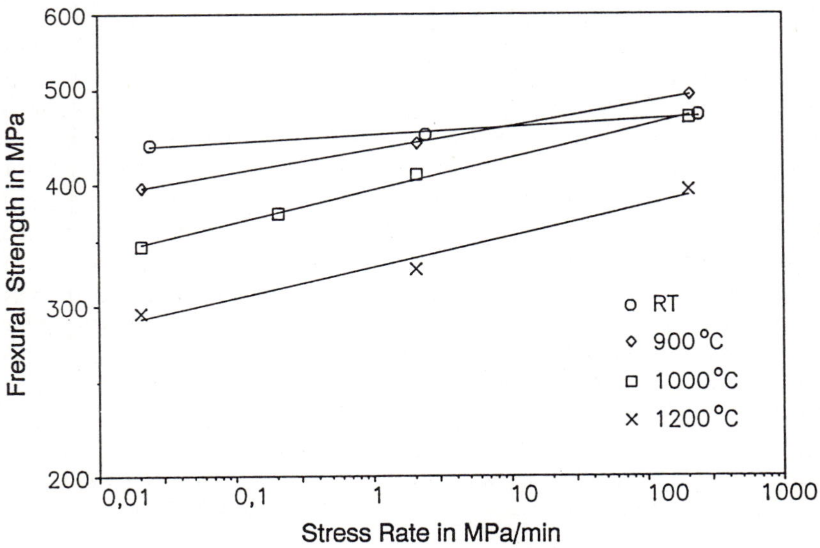

Figure 3.21 Strength versus stress rate for HIP-SiC at various temperatures

In a further HIP-SiC quality, fabricated without sintering additives, subcritical crack growth was also detected, but it was less intensive than in the case of Al-doped HIP-SiC [588]. Although both HIP-SiC qualities are produced with a minor

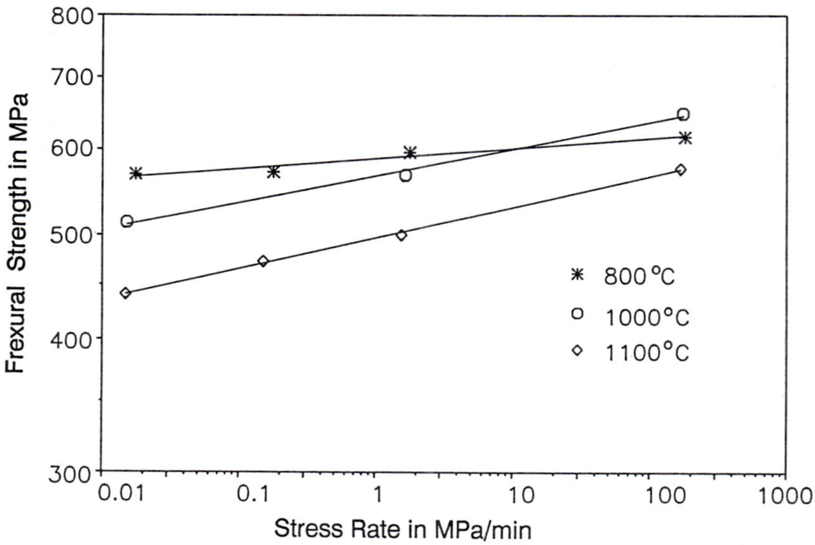

Figure 3.22 Strength versus stress rate for HIP-RBSN at various temperatures

amount or without sintering additives, TEM studies revealed a noncrystalline phase at the triple grain junctions and grain boundaries. This leads to the assumption that the amount of glassy phase has only a minor influence on subcritical crack growth. With SYALON 201 no evidence for subcritical crack extension was found during dynamic fatigue experiments up to 1200 °C. That demonstrates that the structure or composition of the grain boundary phase has a strong influence on the fatigue behaviour, much more than the amount. But not so the fracture mechanism during dynamic fatigue tests, which are different for HIP-SiC and HIP-RBSN. At the fracture surfaces of HIP-SiC in the area around the fracture origin, a 100% intergranular fracture mode was observed (see Figure 3.16). During dynamic fatigue tests a single crack grew, starting from a failure in the material. For HIP-RBSN sometimes 100% transgranular fracture was observed when the crack started from a defect. In some cases, areas with a relatively high amount of intergranular fracture were found at a distance about 50-100 μm from the fracture origin. These observations demonstrate that the reduced strength during dynamic fatigue testing is not caused by a subcritical crack extension of a single crack. Under the assumption that the amount and composition of the secondary phase is inhomogeneously distributed in the test bars, diffusion or grain sliding effects are also inhomogeneous in different areas around the crack tip. In regions which are more sensitive to diffusion or grain sliding, the grain boundaries will be weakened while the material at the crack tip itself provides no damage even under higher stress intensity. So the strength of the material is reduced by regions with weak grain boundaries showing mainly intergranular fracture mode, and the critical crack extension starts from the failure origin under transgranular fracture mode. The different failure mechanisms of HIP-SiC and

HIP-RBSN also lead to different high temperature deformation behaviour. While HIP-SiC shows strong linear-elastic deformation up to 1400 °C, HIP-RBSN exhibits distinct plasticity at temperatures above 1150 °C. Grain sliding effects are much more probable in HIP-RBSN with a high amount of viscose phase than in HIP-SiC with very thin grain boundary phases. With SYALON 201 linear-elastic deformation was detected up to 1300 °C. This shows that high temperature deformation behaviour could also be improved by a more crystallized grain boundary phase. With SiSiC no indications of subcritical damage were found at temperatures up to 800 °C. At higher temperatures, plastic deformation was observed due to the relatively high amount of metallic silicon (17-19 vol%) in this material grade. Intensive investigations of creep behaviour in SiSiC [597] show mainly primary creep up to stresses near the strength values. This deformation behaviour is caused by the stiff network of SiC grains (see Figure 3.17) where the contacts between brittle SiC grains increase with higher plastic deformations, leading to a decrease in the nonlinear-elastic part of the deformation. The creep of SiSiC is strongly controlled by the content of metallic silicon, but also by the shape and distribution of the SiC grains. Newer SiSiC material grades are being fabricated with lower silicon content to improve the high temperature properties.

These examples demonstrate that static fatigue and high temperature deformation behaviour is strongly influenced by the microstructural characteristics of the various ceramics. Sometimes it may be possible to describe different failure mechanisms with the same fracture mechanical methods, for example, the static fatigue of HIP-SiC and HIP-RBSN with the subcritical crack extension parameters A and N, determined in dynamic fatigue tests.

But static fatigue is a relatively simple lifetime study. For structural purposes cyclic loading conditions are much more important. Due to the brittleness of ceramics a first assumption was that the crack growth rate only depends on the stress intensity factor K_I [598], and the lifetime under cyclic loading can be predicted from crack extension parameters, determined in static or dynamic fatigue tests. Investigations of cyclic fatigue behaviour have not often been reported [599-605], but it is sure that some structural ceramics exhibit a real cyclic fatigue effect, beside static or dynamic crack growth. In [604] cyclic fatigue effects were detected with sintered silicon nitride (SSN) at room temperature. The creation of microcracks was observed after cyclic loading in the tested Si_3N_4 specimens, reducing the lifetime of the material. At high temperatures (1200 °C) a contrary effect is reported [602] for SSN. In this case the lifetime measured in cyclic tests was 10 times higher than predicted from static fatigue experiments. Creep effects were discussed which lead to a redistribution of the elastic stresses during cycling. Another microstructural feature is that damages at the grain boundaries like cavitation by diffusion or grain sliding are partially reversible damage processes. During compression a damage healing process can occur which is not included in the lifetime predictions based on static crack extension parameters. These different failure mechanisms at room and high temperatures are strongly influenced by the microstructural characteristics of the ceramic materials. Correlations between the microstructure, the failure mechanisms and the mechanical characterization of cyclic fatigue behaviour are necessary to understand the damage in the material and to find the correct fracture mechanical methods to describe the fatigue behaviour of ceramic materials.

The various types of fatigues lead to a limited lifetime or a reduced applied stress lower than the strength values of the material. These effects can be taken into account and do not prevent technical applications for ceramic materials which show static or cyclic fatigue effects. More critical for structural purposes is the low fracture toughness and the brittleness of ceramic materials. In recent years efforts have been launched to create tough or ductile materials basing on the well-known monolithic structural ceramics. Different concepts were developed for enhancing the reliability and damage tolerance of ceramics by incorporating various reinforcements, where a distinction is mase between strengthening and toughening. If a reinforcement exhibits high strength and a high Young's modulus (for example a fibre) with a strong interfacial bonding, the load can be transferred from the matrix to the fibre, resulting in increasing strength values but brittle failure. More important for failure tolerance are the toughening mechanisms. The aim of the intensive developments in the last few years was to create microstructures such that strength becomes insensible to the size of flaws. For toughening, a resistance curve (R-curve) characteristic is necessary where the fracture resistance increases with crack extension. A well reported method is transformation toughening in tetragonal ZrO_2. A martensitic transformation of the ZrO_2 takes place in the crack tip stress field, which is connected with an increase of the volume, leading to compression stresses in front of the crack tip. Strength values of more than 2 GPa and fracture toughness of 14 $MPa\sqrt{m}$ are reported, although the incorporation of ZrO_2 particles in a ceramic matrix can improve the strength and fracture toughness of a ceramic material.

Another method working in front of the crack tip is toughening by crack branching and deflection. By incorporating whiskers, particles, platelets or short fibres into the microstructure of the materials, crack branching and deflection will increase, resulting in higher fracture toughness. As mentioned before, the relatively high strength and fracture toughness of dense Si_3N_4 materials with high aspect ratios is caused by crack deflection and branching [589]. Because of this, dense Si_3N_4 materials with an extended aspect ratio are sometimes attached to the reinforced ceramics as monocomposites. Another type of reinforcing systems is at work in the crack wakes behind the crack tip. If reinforcing elements bridge the crack surfaces, the crack tip will be deloaded with increasing bridged crack length, leading to an R-curve behaviour. The most commonly reported reinforcing mechanism is the pull-out effect in ceramics with continuous fibres. After interfacial debonding the fibres will bridge the crack surfaces and deload the crack tip. In this stage the friction between the fibres and the matrix is negligible and the reinforcing effect is mainly elastic. After the fracture of the fibres they will be pulled out of the matrix with a bridging force in the crack wake determined by the sliding resistance of the debonded interface. The friction effects during pull-out lead to an energy loss during crack growth resulting in a strong R-curve behaviour with a tough failure characteristic.

To discuss the fatigue behaviour of ceramic composites it is necessary to divide the reinforcing mechanisms into mainly elastic or maibly frictional effects. Pure elastic reinforcing mechanisms are mostly repeatable without wear effects and it is to be expected that the material is insensitive to cyclic loading. If the reinforcing is determined by frictional effects, it is to be assumed that the interfacial roughness of the sliding surfaces will decrease during cyclic sliding. Therefore it is to be assumed that the material shows continuous damage under cyclic loading conditions. Similar pull-out mechanisms are also discussed in whisker and sometimes in particle or platelet content materials. In these cases the pull-out length is very short and the R-curve characteristic is less intensive than in the case of continuous fibres. To enhance the fracture toughness or the brittleness of a ceramic material, a stable crack growth before critical crack extension is necessary to utilize the R-curve behaviour (crack wake effects). Beside the intensity of the R-curve, stable crack extension is dependent on the initial flaw size. A small crack can extend critically without stable crack growth and without a reinforcing effect, even if the material exhibits R-curve behaviour. These discussions show that it is necessary to consider the microstructural characteristics and the failure mechanisms when developing and optimising a ceramic material with enhanced reliability and damage tolerance by introducing different concepts to reinforce nonoxide ceramics like SiC and Si_3N_4.

Parallel to the intensive activities in the field of ceramic composites the mechanical properties and fabrication techniques of monolithic ceramic materials were improved so they could be utilized in technical applications. The positive mechanical properties of monolithic ceramics are used in the fields of wear resistance (bearings, bushings, packing rings, wear plates or cutting tools), good high temperature oxidation and corrosion behaviour (burners, heat exchanger) and in the last few years for high strength and high temperature applications (rotors for gas turbines or turbochargers).

3.2 Oxide Materials (Mullite)

H. Schneider

Though mullite ($Al_{4+2x}Si_{2-2x}O_{10-x}$)[5]) occurs very rarely in natural rocks[6]), perhaps no mineral has been regarded as so basic in ceramics.

5) Mullite was named after the Isle of Mull (Scotland), where it was found for the first time.

6) Mullite is found in some rocks of the sanidinite facies [606] and in xenoliths of corundum-sillimanite rocks [607].

The immense importance of mullite in the field of traditional and advanced ceramics is documented in the enormous number of scientific and technical investigations which were carried out in recent years. Of particular value for understanding the formation and properties of mullite materials are the basic publications of Pask et al. [608, 609]. Other recent review papers on mullite ceramics were published by Somiya and Hirata [610] and by Aksay et al. [611], while Schneider [612] gave a mineralogical overview on mullite. Much recent progress in the improvement of mullite ceramics was the result of two special mullite conferences, one taking place in Tokyo (Japan) in November 1987 [613], and the other in Seattle (U.S.A.) in October 1990.

Mullite is certainly not suitable for high-strength structural applications at room temperature. This is due to the relatively low fracture toughness which limits the mechanical strength. On the other hand, mullite retains its room temperature strength up to high temperature. Moreover, mullite ceramics exhibit excellent creep resistance and outstanding thermal shock behaviour under oxidizing conditions. The favourable high-temperature properties of mullite open a wide field of applications, especially if the toughness is increased by fabricating particle or fibre reinforced mullite-matrix composites. The most important application field at the moment, however, is the use of mullite and mullite matrix composites as thermal and oxidation barriers.

Because of the low melting points of most metals and their tendency to form metal oxides at low temperatures, heat and oxidation barriers are necessary for many metal substrates exposed to temperatures $\gtrsim$1000 °C in oxidizing atmosphere. Mullite oxidation protection layers are also suitable for Si_3N_4, SiC, and $MoSi_2$ ceramics at very high temperatures (>1700 °C). These materials form SiO_2 glass layers at the surface of specimens when exposed to air at high temperatures. O_2 must diffuse through these boundary layers, which slows down further oxidation. Very low viscosity of the glass at very high temperature causes disturbance of the surface layers and therefore catastrophic oxidation of the substrate. However, the working temperatures of SiC, Si_3N_4, and $MoSi_2$ can be raised substantially by using mullite oxidation protection layers. Mullite layers on SiC substrates may also be used for the corrosion protection of SiC heat exchanger systems.

The possible application of ceramic coatings on substrate surfaces depends on several factors, among which good bonding between layer and substrate (either mechanical or chemical) is the most important one. Bonding between layer and substrate, especially at high temperature, is controlled by the fitting of mechanical and thermal properties of both materials. If both properties differ too much, sandwich-type composites with intermediate layers having properties in-between those of the substrate and of the mullite layer may be reasonable. Often, composites with a continuous range of materials properties from those of the substrate to those of the protection layer are used. Composites of this type have been designated as functional gradient materials (FGM).

Two main techniques to produce mullite protection layers have been applied in practice, or are discussed for use in the future:

- Fabrication of mullite layers by oxidation of metal or ceramic substrate surfaces (self oxidation), or by O_2 implantation
- Fabrication of mullite layers by deposition on metal or ceramic substrates

The most experience in the field of mullite coating processing is available for plasma spraying. Plasma sprayed mullite coatings on metals can successfully be applied if intermediate layers or FGMs with properties in-between those of the metal and mullite are taken into account. Possible materials for intermediate layers or FGMs are ZrO_2 or admixtures of ZrO_2 plus Al_2O_3 and mullite in changing proportions. An important application of plasma spraying is the coating of SiC heat exchangers with mullite layers in order to protect them against corrosion from Na silicate vapors at temperatures $\gtrsim$900 °C. Investigations of Butt and co-workers [614, 615] have shown that plasma sprayed mullite or mullite plus ZrO_2 coatings on SiC heat exchanger tubes extend the materials' lifetimes considerably. The good strength of the plasma coated materials was attributed to good mechanical bonding between mullite coating and SiC substrate. A scanning and transmission electron microscope study of the microstructures of mullite plasma layers has recently been performed by Braue and Schneider [616]. They found that plasma deposited mullite materials are characterized by distinct states of disequilibria on the nanometer scale, with extremely heterogeneous phase developments and pore distributions (Figure 3.23).

Besides the application of mullite or mullite matrix materials as thermal barriers and oxidation resistant coatings on substrates, they have considerable potential for:

- advanced catalyzer substrates
- creep resistant components
- advanced substrate material for multilayer packages and for SOC (silicon on ceramics) photovoltaic silicon solar cells
- optical windows to be used at high temperature, and for
- functional and structural components for turbines and engines

Considerable efforts have been made during recent years to improve the mechanical and thermal properties of mullite ceramics. Well established strategies include:

- Optimization of mullite ceramics by varying the synthesis and processing procedures, and
- optimization of mullite ceramics by fabricating mullite matrix composites and mullite coatings.

Another though less known route to improve the materials' properties is the variation of the crystal chemistry of the phase mullite. It has been shown that a change in the Al_2O_3/SiO_2 ratio, or doping the mullite structure with transition metal cations distinctly influence the materials' properties [618, 619].

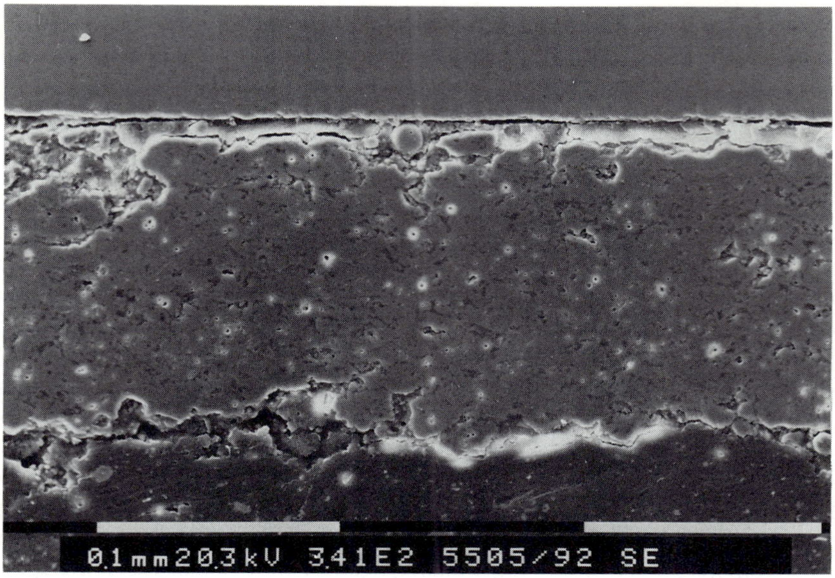

Figure 3.23 Scanning electron microscopic photograph of a plasma deposited mullite layer on a dense SiC substrate. Note the debonding between substrate and layer possibly due to the layer thickness. Chemically etched section [617].

Mechanical Properties of Monolithic Mullite Ceramics

A large variation of flexural strength values, ranging between ≈150 and ≈450 MPa[7]) is reported in literature for monolithic mullite at room temperature [611, 620-623]. It has been suggested that the mechanical strength of mullite ceramics is controlled by the mullite crystals' properties, but it also depends on their microstructure. This includes size, shape and amount of crystalline phases and amount and distribution of glass and pores. Factors influencing the microstructural development of mullite ceramics are bulk chemistry (Al_2O_3/SiO_2 ratio), presence of impurities and synthesis and processing conditions. Single phase mullite ceramics retain their mechanical strength up to high temperature (1400 °C or higher, Figure 3.24, [622-625]). Dokko et al. investigating the high strength potential of single crystal and polycrystalline single phase mullite under compressive stress-strain and creep conditions up to 1500 °C explained this favourable

[7]) Strength values ≥350 MPa should be considered with caution because they may be simulated by processing induced strain or by inadequate measuring techniques (e.g. 3-point bending instead of 4-point bending techniques)

mechanical high temperature behaviour with a nearly complete lack of dislocation glide.

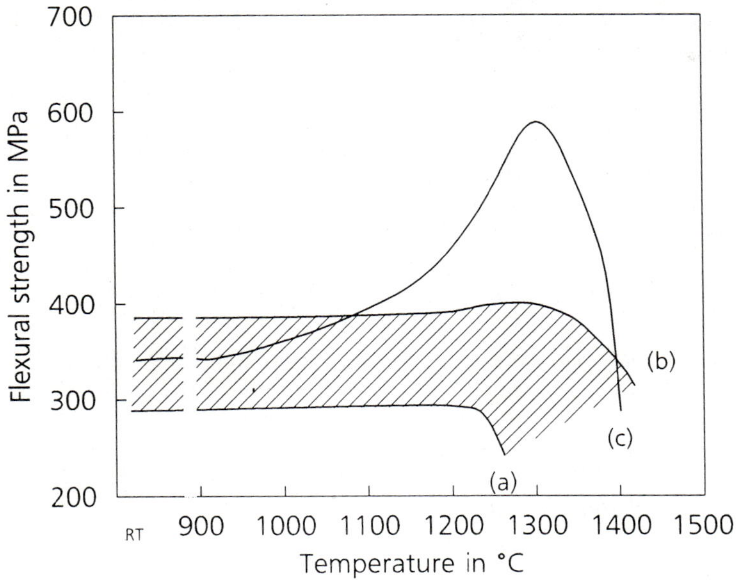

Figure 3.24 Flexural strengths of monolithic mullite ceramics as function of the temperature. (a) and (b), and the hatched area between both curves represents typical strength values of single phase mullites. The large variation is due to different microstructures (see the text). (c) Mullite containing glass phase. The "strength increase" in glass containing samples can be understood by energy consuming processes in the glass phase, which are strongly controlled by the heating velocity (references see the text).

Because mullite displays (nearly) no plastic deformation by dislocation glide, the amount, viscosity, and distribution of a possible existing glass phase play a major role in its high temperature strength: Glass phases occurring in pockets and at mullite grain triple points have a less severe influence on strength reduction than glass phase envelopment of individual mullite grains. Glass viscosity in turn depends on the amount of impurities dissolved in the glass. These impurities can decrease the softening temperature of the glass phase to ≈300 °C . The strength and fracture toughness increase frequently observed between ≈900 and 1200 °C (Figure 3.24) was explained with a blunting of the microcracks with the highly viscous glass [623].

The flexural strength of mullite ceramics slightly decreases with the bulk Al_2O_3 content of the material at room temperature. Slightly better mechanical values of Al_2O_3 poor (68% Al_2O_3) compositions are explained by the development of a skeleton of interlinked, elongated mullite crystals [621, 626, 627]. The high temperature strength of Al_2O_3 poor compositions is higher than that of Al_2O_3 rich (78% Al_2O_3) materials, due to an energy consuming plastic deformation at the crack front in the highly viscous glassy phase. The observation of a higher strength of Al_2O_3 poor mullite ceramics was not supported by measurements by Kanka and Schneider [621].

The fracture toughness of mullite is nearly constant ($\approx$ 2.2 to 2.5 $MPa\sqrt{m}$) at room temperature independent of the bulk Al_2O_3 content [611, 627]. At 1300 °C the fracture toughness of the Al_2O_3 poor (68% Al_2O_3) composition is rather high (4.7 $MPa\sqrt{m}$), but is much lower ($\approx$2.6 $MPa\sqrt{m}$) for the Al_2O_3 rich (78% Al_2O_3) composition. The composition and temperature dependence of fracture toughness values may be interpreted in a similar way as strength.

The microhardness of mullite decreases with temperature from $\approx$14 GPa at RT to $\approx$10 GPa at $\approx$1000 °C, which is a much lower temperature induced decrease than that of SiC and Al_2O_3 [628]. Again, this is explained by the absence of dislocation glide in the mullite structure. It was suggested that the complex domain structure of mullite, with its array of antiphase domain boundaries, additionally hinders the mobility of the few dislocations present in mullite.

Single phase mullite has excellent creep resistance up to high stresses and high temperatures [624, 629, 630]. As in the case of strength and microhardness, this has been explained by the lack of dislocation glide even around defects and grain boundaries. Creep velocities of mullite ceramics with coexisting glass phases are higher than those of single phase mullite, and creep curves display three temperature dependent branches instead of the linear curves of single phase mullite (Figure 3.25). Nixon et al. suggested that an important factor controlling the course of creep curves is the viscosity of the grain boundary glass phase: At relatively low temperature, creep rates seem to be mainly influenced by grain boundary glide, while at intermediate temperature there is an additional influence of cavitation. At high temperature, creep is controlled by grain boundary glide due to viscous flow, and diffusion through the glass phase.

Mechanical Properties of Mullite Matrix Composites

Because of its inherent high temperature stability, oxidation resistance and low thermal expansion, mullite is a promising baseline material for structural ceramics. Whiskers, fibres, platelets, and particles have been used to improve the mechanical properties of mullite matrices. Mullite additions to other ceramics in order to improve the thermal isolation, thermal shock resistance and high-temperature strength properties of the matrices have also been described.

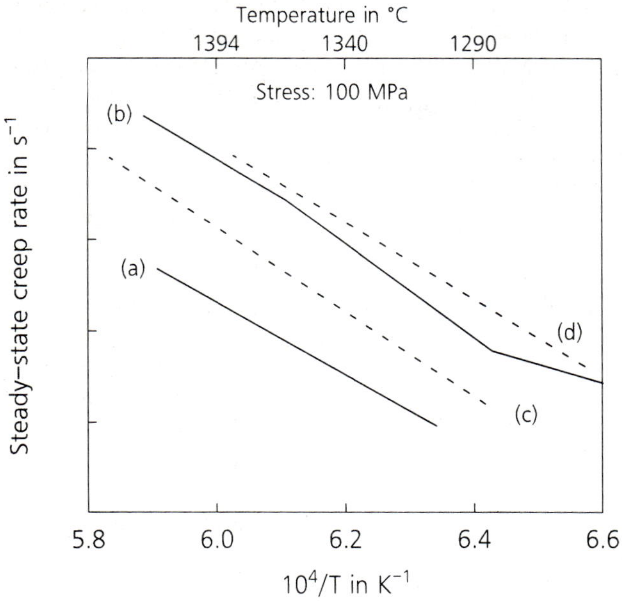

Figure 3.25 Steady-state creep rates of monolithic mullite ceramics as a function of the reciprocal temperature (a), (c) Single phase mullite [629, 630]; (b), (d) Mullite-containing glass phase [629, 630]. The three temperature-dependent branches of curve (b) correspond to the three activation energies of creep.

Whisker as well as short and continuous fibre reinforced mullite systems have been investigated. Whisker and fibre reinforcement bases on matrix microcracking, crack deflection, crack bridging, and on pull-out, with effectiveness being controlled by the geometrical properties of the fibres, existence of interfacial glass phases between fibre and matrix, and fibre distribution. Different load deflection behaviours of the composites depend on coating and surface treatment of the fibres. This is attributed to different fibre matrix interactions and interfacial stress.

SiC whisker toughening of mullite has attracted intense research effort. Crack bridging and SiC whisker pull-out have been evaluated as the dominant toughening mechanisms [632]. A linear dependency exists between SiC whisker content and fracture toughness: For a 20 vol% SiC whisker-mullite material, a fracture toughness of $\approx$ 4.5 MPa$\sqrt{m}$ versus $\approx$2 MPa$\sqrt{m}$ for the baseline material was established [631-635]. For the same material type an increase of flexural strength from $\approx$200 MPa (baseline material) up to $\approx$450 MPa was observed [631, 636], (Figure 3.26, 3.27). This improved strength is retained up to $\approx$1200 °C. The temperature dependence of the fracture toughness K_{Ic} in SiC whisker-mullite ceramic systems shows little variation up to ≈1100 °C. As a consequence of the close fit in thermal expansion of SiC whiskers and mullite, only relatively small compressive residual stresses were obtained at the SiC whisker-mullite interfaces: These are much lower than the high compressive stresses found in SiC whisker-Al_2O_3 systems, or the high tensile stresses measured in SiC whisker-Si_3N_4 composites. During high temperature exposure of SiC whisker-mullite composites oxidation of the material occurs [637, 638]. However, SiC whisker-mullite materials show a lower weight gain as compared to other whisker reinforced oxide based composites, emphasizing the potential of mullite matrix materials under oxidizing conditions.

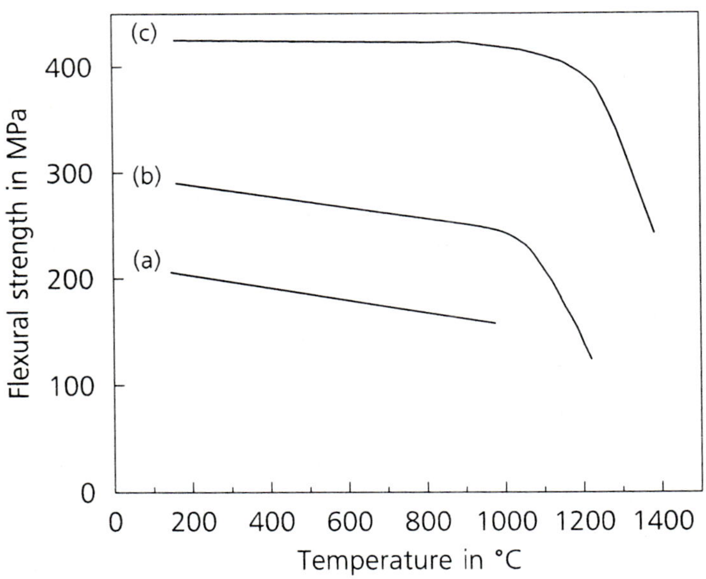

Figure 3.26 Flexural strength of mullite composites as a function of temperature. (a) 20 vol% ZrO_2 particle-mullite composites [640]; (b) 20 vol% ZrO_2 particle (+ 3% TiO_2)-mullite composite [641]; (c) 20 vol% SiC whiskers-mullite composite [631].

Hirata et al. [639] provided data on flexural strength and fracture toughness of Si_3N_4 whisker-mullite and mullite whisker-mullite systems. The addition of 10 vol% Si_3N_4 whiskers to a mullite matrix increases the density of the sintered bodies from ≈97.5% to nearly 100%. It was suggested that a SiO_2 glass film coating whisker was responsible for the densification by means of viscous flow deformation. Hirata and co-workers published flexural strength values of 630 MPa

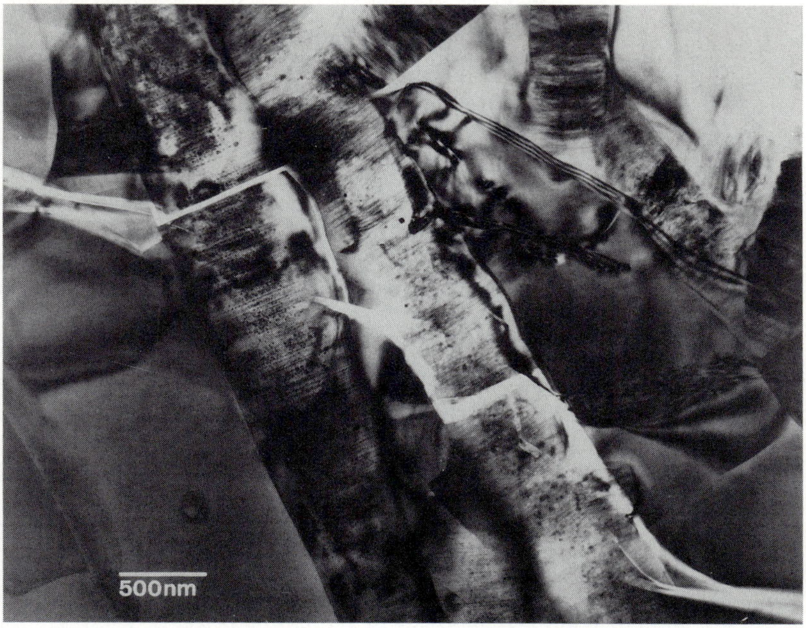

Figure 3.27 Transmission electron microscopic photograph of a 20 vol% SiC whisker-mullite composite showing crack-whisker interaction. The crack was introduced by microindentation (Micrograph courtesy Dr. T. Tiegs)

of 10 vol% Si_3N_4 whisker-mullite composites versus 540 MPa in the baseline material[8]). Simultaneously the fracture toughness K_{Ic} increases from 1.3 $MPa\sqrt{m}$ in the baseline material to 3.4 $MPa\sqrt{m}$ in the SiC whisker-mullite composite. Addition of 10 vol% mullite whiskers to a mullite matrix decreases the strength to a value of ≈450 MPa, while the K_{Ic} - value slightly increases to 2.6 $MPa\sqrt{m}$. Mullite whisker-mullite systems exhibit slightly lowered densities (96.5% versus 97.5% theoretical density). It was suggested that the absence of a glass phase coating the mullite whiskers was responsible.

Though reinforcement of mullite matrices with high strength SiC whiskers yielded improved mechanical properties [635], activities were stopped in Germany because of health hazards. Instead, platelets were used, which are thought to provide a similar reinforcement. Compared to whisker reinforcement, platelet-rein-

[8]) The flexural strength of monolithic mullite given by Hirata et al. (540 MPa) is much too high, while the K_{Ic} value is too low. Nevertheless, it is suggested that the change between monolithic and reinforced materials is correct.

forced mullite matrix composites have the advantage of higher thermal stability, easier processing and, last not least, produce no health problems [642]. Toughening increase of platelets reinforced mullite matrices is based on debonding, crack deflection and bridging and pull-out [643]. Nischik et al. found that the fracture toughness of 20 vol% SiC platelets reinforced mullite was increased up to ≈30% (3.2 versus 2.5 $MPa\sqrt{m}$) with respect to monolithic mullite. Simultaneously, flexural strength decreased (264 versus 307 MPa), which was explained with increased critical flaw sizes. However, flexural strength values increase if preoxidized SiC platelets were used (≈344 MPa: 10 vol% preoxidized SiC platelets versus ≈274 MPa, 10 vol% as received SiC platelets). Load transfer mechanisms caused by favourable particle-matrix bonding may be the reason.

Considerable efforts have been focused on ZrO_2-mullite composites using unstabilized or partially stabilized ZrO_2 as reinforcing particles. This is true for the different fabrication processes and characterization of interfacial structures via high resolution electron microscopy, but also for the measurement of mechanical properties. Most authors (e.g., [640, 641]) found that a ZrO_2 addition to mullite matrices does not increase strength, and sometimes even decreases it (typical strength values are in the range between ≈250 and 400 MPa, Figure 3.26). Other authors using Y-doped tetragonal ZrO_2 polycrystals (Y-TZP) in mullite composites noted a continuous, though nonlinear increase of the strength [645]. Rundgren et al. [646] found that the flexural strength of sol-gel prepared ZrO_2-mullite systems containing 30% ZrO_2 and 3 mole % Y_2O_3 was very high (>600 MPa) at room temperature. The sintered specimen had a very high density, probably due to liquid phase sintering, while Y_2O_3-free materials displayed a lower-rate solid-state sintering process. However, the high temperature flexural strength of composites containing Y_2O_3, dropped drastically above ≈1300 °C, whereas systems without Y_2O_3 showed a slow and continuous strength decrease. Probably the microstructure, including glass phase content, mullite and ZrO_2 crystal growth, and distribution of phases and porosity have stronger effects on the mechanical properties than the amount of ZrO_2 itself.

The room temperature fracture toughness of ZrO_2-mullite systems is considerably increased. A toughening increase up to 50% from ≈3 to 4.5 $MPa\sqrt{m}$ has been reported [644, 640, 641]. Good properties can only be achieved via a careful optimization of the starting batch, with the Al_2O_3/SiO_2 ratio of the material being a critical parameter.

Reinforcement of mullite matrices combining different particulates or combining particles with fibres or whiskers was also studied. Adding tetragonal ZrO_2 (t-ZrO_2) particles to a SiC whisker-mullite system, thus activating the ZrO_2 transformation, exhibits a more pronounced toughening effect than in the presence of just monoclinic ZrO_2 (m-ZrO_2), which basically results in microcracking only. The combination effect of ZrO_2 particles and SiC whiskers is regarded as additive. However, there is still uncertainty whether transformation toughening and whisker reinforcement are interactive or independent processes. A fracture toughness value of 6.2 MPa$\sqrt{m}$ in a 35 vol% ZrO_2 particle (+1.7 vol% Y_2O_3)- 30 vol% SiC whisker-mullite composites was reported by Ruh et al. [647]. Kamiaki et al. [648] added up to 40 vol% SiC particles to ZrO_2-mullite composites. An addition of 40 vol% SiC particles yielded almost constant strengths, regardless of the amount of ZrO_2 at room temperature. On the contrary, ZrO_2-mullite systems containing 20 vol% SiC particles displayed lower strength values with increasing ZrO_2 additions. Strength decreasing with the amount of ZrO_2 addition in low amount SiC particle samples was explained by the formation of large cracks, which in turn were caused by the occurrence of large mullite and ZrO_2 grains in such samples. Samples with high SiC particle amounts usually have more favourable microstructures with small and equiaxed mullite and ZrO_2 grains. A conclusion is that the SiC particles in addition to ZrO_2-mullite composites hinder mullite and ZrO_2 crystal growth.

The slopes of the fracture toughness curves of combined ZrO_2 plus SiC particle addition to mullite are similar to those with ZrO_2 addition alone. A fracture toughness of ≈5.7 MPa$\sqrt{m}$ in a 40 vol% SiC particle- 20 vol% ZrO_2 particle-mullite reinforced composite versus ≈3.8 MPa$\sqrt{m}$ in 20 vol. % ZrO_2 particle-mullite composite without SiC particles was obtained by Kamiaki et al. [648]. However, different toughening mechanisms were believed to be active: In the low amount SiC platelets case, pinning of the crack front by the large mullite and ZrO_2 grains and crack branching seem to increase fracture toughness, while crack deflection and branching causes improvement in the high amount SiC platelets case [648]. Okada and Otuska [649] measured the fracture toughness of a 25 vol% YTZP ZrO_2-75 vol% mullite composite reinforced by 15 vol% mullite whiskers. It was shown that the fracture toughness of the mullite-whisker-reinforced composite was ≈30% higher than that without mullite whiskers (≈5.1 versus ≈4.0 MPa$\sqrt{m}$). The increase of fracture toughness was ascribed to thermal stresses between mullite whiskers and ZrO_2 particles. The very different thermal expansion coefficients of both phases may cause intergranular fracture and crack deflection. The reinforcement effect is considerable despite the fact that most mullite whiskers are in contact with mullite grains and not with ZrO_2 particles.

Beside the composite systems discussed so far, quite unusual combinations have also been studied. Because of their excellent thermal shock resistance, mixtures of mullite and cordierite have been the subject of several investigations [650, 651]. It has been shown that the unfavourable low strength of cordierite (≤100MPa) can be improved by mullite additions. Simultaneously, the Vickers hardness of a 40% cordierite- 60% mullite system increased to ≈25% with respect to monolithic cordierite (11.4 GPa versus 8.8 GPa). The fracture toughness of the cordierite-mullite composite was similar to that of baseline mullite (≈2.5 MPa$\sqrt{m}$). Another way to improve the good thermal shock resistance of mullite ceramics is to add the Al titanate tieillite ($Al_2O_3 \cdot TiO_2$) to the mullite matrix. Tieillite is suitable for the purpose because it has a very low mean thermal expansion coefficient. However, the strong structural anisotropy of the thermal expansion of tieillite which causes its extreme low mechanical strength (≈20 MPa) restricts the application of tieillite-mullite composites.

Compared to the unusually high creep resistance of pure, glass-free mullite, the creep resistance of a 20 vol% SiC whisker reinforced material decreases (Figure 3.28). On the other hand, there is a decrease in steady-state creep by one order of magnitude with respect to monolithic mullite containing glass phase (see also Figure 3.25). Activation energies of creep are distributed within three temperature ranges, as in the case of monolithic, glass-containing ceramics. Similar creep controlling mechanisms were suggested in both cases [611, 629].

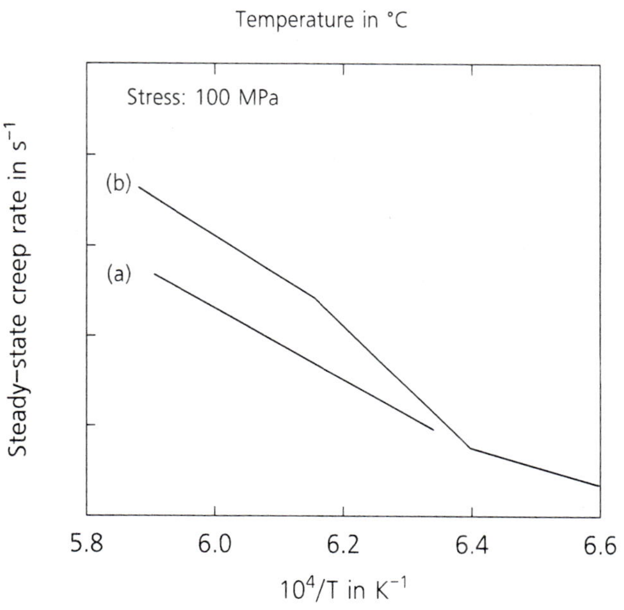

Figure 3.28 Steady-state creep rates of mullite composites as a function of the reciprocal temperature. (a) Single phase mullite (b) 20 vol% SiC whisker-mullite composite. The three temperature-dependent branches of curve (b) correspond to three activation energies of creep [629].

Creep experiments carried out on reaction sintered ZrO_2-mullite systems yielded two different temperature regions: At lower temperature relatively low creep activation energies were found, being attributed to microcrack formation. At higher temperature higher creep activation energies were reported. The latter was interpreted in terms of a solution-precipitation mechanism at the glassy grain interfaces. Different activation energy values as reported in literature, may correspond to different interfacial glass compositions and related particular microstructures [652].

Dechamps et al. [652] measured the creep behaviour of reaction sintered ZrO_2 particle-mullite systems using TiO_2 or MgO as additives or highly reactive ZrO_2-mullite powder mixtures obtained by the ultra-rapid quenching technique. The creep mechanism of all materials was controlled by the presence of intergranular glass phases. The creep data show two regions: At high stresses and low temperatures low creep activation energies were determined. This was explained by microcrack formation. The higher creep activation energies at lower stresses and higher temperatures, in turn were interpreted as due to solution-precipitation process accommodating grain boundary gliding. The better creep resistance of MgO added ZrO_2-mullite systems was explained by the occurrence of an elongated interlinked mullite grain lattice.

Thermal Properties of Mullite Materials

The thermal conductivity corresponds to the amount of heat conducted through a specimen per unit temperature gradient. Mullite has a low thermal conductivity, which is nearly independent from temperature up to ≈800 °C, and then rapidly decreases (Figure 3.29), [650, 653, 654]. The low thermal conductivity of mullite may further be decreased by doping of the phase by foreign cations, by which additional scattering centers are formed at high temperature.

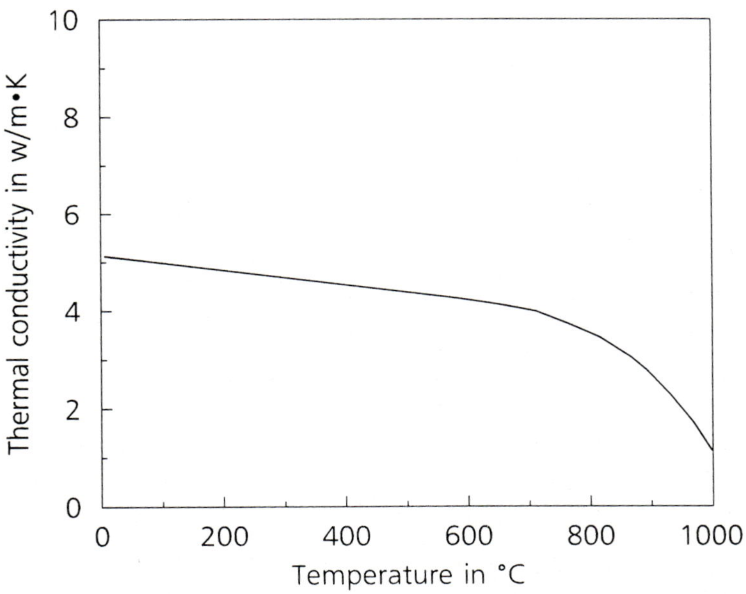

Figure 3.29 Thermal conductivity of mullite as a function of temperature [650].

Thermal expansion data are essential because they give information on the high temperature shape stability of specimens, and of temperature produced strain in a ceramic body.

Thermal expansion coefficients of fused-mullite, and of pure and Cr- and Fe-doped sinter-mullite were measured up to ≈900 °C by Schneider and Eberhard [619] using high temperature X-ray diffractometry. The mullites display very low and nonlinear a, b and volume expansions below, and low though linear a, b, c and volume expansions above ≈300 °C. Dependent on the Al_2O_3 content of the undoped mullite and on transition metal incorporation, a rather large variation of thermal expansion coefficients is observed (Table 3.3). The X-ray high temperature measurements indicate a good resistance of mullite ceramics to sudden temperature changes. Obviously this favourable expansion can further be improved if Cr-doped instead of pure sinter-mullite is used. This actually comes true, since recent mass-spectrometric investigations have shown that Cr-doped mullite neither melts nor decomposes below 1500 °C [655]. In [619] it was also shown that the thermal expansion anisotropy decreases considerably in the sequence 3/2-mullite and 2/1-mullite. This means that ceramics made from Al_2O_3 rich mullite display lower temperature-induced internal strain than stoichiometric mullite. As a conclusion of their studies Schneider and Eberhard suggested that it would be most interesting to combine the lower volume expansion of Cr-doped 3/2-type mullite with the reduced thermal expansion anisotropy of 2/1-type mullite.

Sample key	Compositional state x	Expansion coefficients 10^{-6}°C α(a)	α(b)	α(c)	α(V)	Anisotropy factor A	Ref.
			Mullite				
SM-O	0.24	3.9	7.0	5.8	158	6.2	
SM-Cr	0.21	3.1	6.2	5.6	108	6.2	[619]
SM-Fe	0.24	3.2	7.0	5.6	125	7.6	
FM-O	0.39	4.1	5.6	6.1	110	4.0	
			Sillimanite				
	0	2.3	7.6	4.8	84	10.7	[656]

Table 3.3 Temperature-induced structural expansion of mullite (300-900 °C)

SM = Sinter-mullite, FM = Fused-mullite, O = Undoped, Cr = Cr-doped, Fe = Fe-doped, α (a), α (b), α (c): Thermal expansion coefficients along a, b, and c α (V): Volume expansion alpha.(a) · α(b) · α(c). Anisotropy factor of thermal expansion: $A = [|\alpha(b)-\alpha(c)|+|\alpha(b)-\alpha(a)|+|\alpha(c)-\alpha(a)|] \times 10^{-6}$

Above ≈300 °C a direct correlation exists between the ZrO_2 content of the composites and thermal expansion values. A nearly linear relationship also exists between thermal expansion coefficients and phase contents of cordierite-mullite and tieillite-mullite composites (mullite (100%), $\alpha \approx 4.5 \cdot 10^{-6}$ °C^{-1}, cordierite, tieillite (100%) $\alpha \approx 1.3 \cdot 10^{-6}$ °C^{-1}). Composites of that type are important components for high performance catalyzer substrates for automotive and diesel engines because of their good thermal stability, good thermal shock resistance, and mechanical strength, which is improved compared with pure cordierite and tieillite [650, 651].

3.3 Imaging Microstructures of Monolithic Carbons and Carbon/Carbon Composites in the TEM

R. Pleger and W. Braue

Carbon is a unique element indeed. The electron configuration of carbon gives rise to two allotropic modifications, diamond (cubic, sp^3 hybridisation) and graphite (hexagonal, sp^2 hybridisation) which still today are the long-time favourites of solid state science. Besides these ordered structures, a large variety of carbon and graphite materials exhibits a significant amount of structural disorder which translates into highly anisotropic microstructures with interesting properties. Although this class of carbon material does not fully utilize the strong covalent C/C bond, electrical and chemical engineering cannot do without them [657].

Carbon is also a very promising structural material. Wood, nature's cellular polymeric fibre-composite, is a carbon-based material with an outstanding strength to density ratio. The roaring of a wooden Tiger-Moth biplane in the sky of a veteran air show today gives convincing evidence for the potential wood had for the aeronautic industry just a couple of decades ago [658].

It became a different composite world when high-strength/high-modulus carbon fibres were offered on an industrial scale [659]. Carbon/carbon composites are considered the ultimate all-ceramic composite at present. They combine the refractoriness of carbon with the high strength and stiffness of carbon fibres, resulting in a composite which exhibits a non- brittle failure mode and a constant strength level superior to all superalloys and monolithic ceramics at temperatures higher than 1800 °C in inert atmosphere [660].

A careful investigation of the microstructure by transmission electron microscopy (TEM) is mandatory for evaluation of the structure/property relationship as well as the failure-related mechanisms of monolithic and heterogenous carbon-based materials. This is particularly true for the complex microstructure of carbon/carbon composites which involves a large spectrum of structural carbon variants ranging from well-ordered graphite to vitreous carbon which is amorphous on a macroscopic scale [661].

This section focuses on imaging of different carbon microstructures in the transmission electron microscope [662]. It includes carbon/carbon composites and different monolithic carbon materials which match the nanoscale microstructures observed at interfacial boundaries in carbon/carbon materials. Although this approach requires a statistical evaluation of the crystallite dimensions and the different degrees of orientation by means of digital image processing, a comprehensive phenomenological discussion of high-resolution TEM images obtained from different carbon microstructures is preferred for the purpose of this section. In the following outline the structural principles and physical properties of ideal and real graphite as well as monolithic carbon materials are briefly reviewed with particular emphasis on polygranular, pyrolytic, and vitreous carbons, respectively. The corresponding microstructures are compared to matrix carbon variants revealed from a 2D C/C composite material.

Commercially available carbon materials were investigated (courtesy Schunk Inc. and Ringsdorff Inc. (FRG)). Thin (< 50 nm) transparent TEM foils are prepared by grinding, dimpling and argon ion-beam milling. A Philips EM 430 T transmission electron microscope operating at 300 kV was utilized.

Ideal Graphite

The crystallographic structure of graphite is shown in Figure 3.30. A single crystal is formed by planes of honeycomb carbon hexagons which are stacked in an ABAB sequence resulting in the space group $P6_3/mmc$. Some natural graphites exhibit an ABC ABC stacking sequence which translates into the rhombohedral space group $R\bar{3}m$ [663].

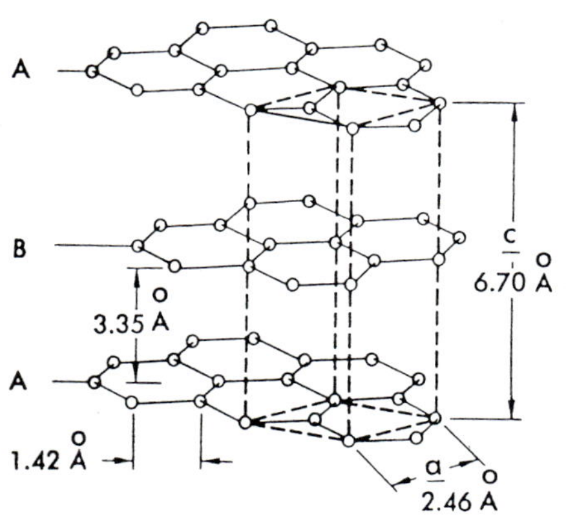

Figure 3.30 Crystal structure of graphite [665]

The in-plane C-C bond length is very short (0.142 nm) and involves strong covalent σ-bonding while rather weak π-bonding is achieved between the planes. The result is a highly anisotropic crystal (see Table 3.4) with outstanding electrical, thermal and mechanical properties along the basal planes [660, 664].

Property	‖ to basal planes	⊥ to basal planes
Bonding length [Å]	1.41	3.35
Bonding energy [kJ/mol]	430	20
Electrical resistivity [Ω · cm]	$5 \cdot 10^{-5}$	0.3
Thermal conductivity [W/cm · K]	15	0.03
Thermal expansion (at RT) [K^{-1}]	$-1 \cdot 10^{-6}$	$9 \cdot 10^{-6}$
Theoretical strength [GPa]	100	-
Theoretical Young's modulus [GPa]	1000	-
Density [g/cm^3]	2.26	-

Table 3.4 Physical properties of graphite single crystals [660, 664, 666]

Although oxidation resistance above 500 °C is poor, the general chemical corrosion stability and the thermal shock behaviour of graphite are very good. The sublimation temperature is very high (3700 °C) and the mechanical properties (σ, E) remain nearly constant up to temperatures higher than 2300 °C, emphasizing the refractoriness of graphite.

The application of graphite as a prime structural material for harsh service conditions involving both high stresses and temperatures is basically controlled by two constraints:

1. The brittle failure of monolithic graphite and the anisotropy of its properties, which can be improved by carbon materials with a highly disturbed graphitic structure and/or by continuous fibre reinforcement (see below),
2. the poor oxidation resistance at high temperatures. SiO_2- and/or SiC-based coatings are limited to service temperatures lower than 1800 °C. Long-term oxidation protection at even higher temperatures requires sophisticated multilayer coatings of refractory oxides and carbides and is still under investigation [667].

Real Graphite

The perfect graphitic structure as described in the previous section is almost never observed. Because of the low bonding energy between the basal planes, irregular, random stacking sequences between the basal layers may occur, resulting in turbostratic graphite (Figure 3.31).

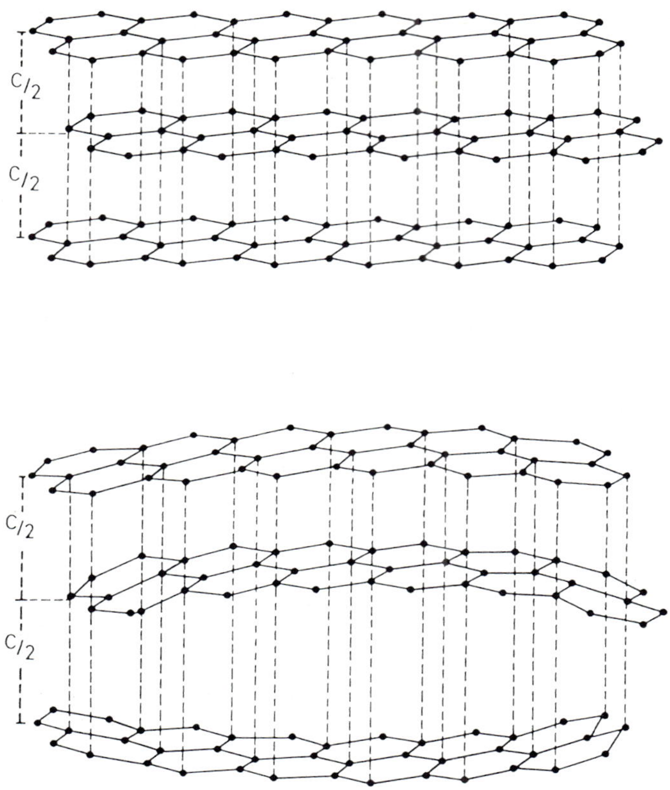

Figure 3.31 Basal a-b-planes of ideal graphite (above) and disturbed graphite (below). In the case, the lattice parameter d varies, but if the planes are also randomly rotated with respect to each other, the structure is called turbostratic [666].

Moreover, structural defects such as point defects, planar faults, dislocations and voids are introduced to the graphite lattice [668] contribute to different degrees of structural disorder and orientation as well as to the porosity observed in most carbon-based materials .

Beside the stacking sequence, the crystallite dimensions of carbon materials - viewed in-plane and perpendicular to the c-axis - display considerable variation which can be evaluated by X-ray diffraction techniques, unfortunately at very low spatial resolution [669, 670]. However, these microstructural features can be directly imaged in the transmission electron microscope which is outlined in the following for different kinds of monolithic carbon materials, such as polygranular

Property ∥ and ⊥ to basal planes	Polygranular Carbon	Pyrolytic Carbon	Vitreous Carbon	2D C/C
(∥) Electrical resistivity [Ω · cm]	10^{-3}	$4 \cdot 10^{-4}$	$3 \cdot 10^{-3}$	0.1
(⊥)	-	1	-	-
(∥) Thermal conductivity [W/cm · K]	1	1	0.2	0.6
(⊥)		0.03		0.2
(∥) Thermal expansion (at RT) [K^{-1}]	$3 \cdot 10^{-6}$	$-8 \cdot 10^{-7}$	$2 \cdot 10^{-6}$	$8 \cdot 10^{-7}$
(⊥)		$25 \cdot 10^{-6}$		$7 \cdot 10^{-6}$
(∥) Flexural strength [MPa]	15-80	100	200	200
(∥) Theoretical Young's modulus [GPa]	10	35	35	80
Density [g/cm³]	1.7	1.4-2.2	1.4	1.7-2.1

Table 3.5 Typical properties of carbon materials [666, 671, 672]

carbon, pyrolytic carbon and vitreous carbon as well as carbon composites (C/C) (see Table 3.5).

Polygranular Carbon is an important monolithic carbon material for e.g. electrical electrode material [666]. It consists of carbon black mixed with tar pitch as a binder. The green body is shaped and then pyrolyzed in an inert atmosphere. Figure 3.32a is a low-mag TEM bright-field micrograph from a thin part of this material. Because of the crystal structure (and the low degree of orientation) the basal planes a priori exhibit low contrast, whereas basal planes viewed edge-on are clearly resolved. This image *seems* to indicate that the c-axis of the graphite crystallites is preferably stacked perpendicular to the viewing direction. The corresponding diffraction pattern (see inset) however clearly displays an isotropic orientation.

Many cracks between graphite planes are visible as tiny white lines in the image. They are considered strength-limiting flaws emphasizing the poor mechanical integrity of the material. The high σ-bond strength of graphite is actually sacrificed by this microstructure.

Pyrolytic Carbon is produced by pyrolysis of hydrocarbons species and nucleation of the reactants on a heated substrate. By stringent control of the processing conditions (substrate temperature, partial pressure of the hydrocarbons in the feedstock, etc.) it is possible to achieve an orientated deposition of pyrolytic carbon with the c-axis perpendicular to the substrate . A section along the growing direction of pyrolytic carbon is shown in Figure 3.33. Cracks are visible between the basal planes with all planes extending more or less along one direction. The actual strength in this direction is about one order of magnitude higher than that of polygranular carbon (see Table 3.4). Pyrolytic carbon exhibits high anisotropy in thermal expansion, which introduces substantial residual stresses at high temperatures.

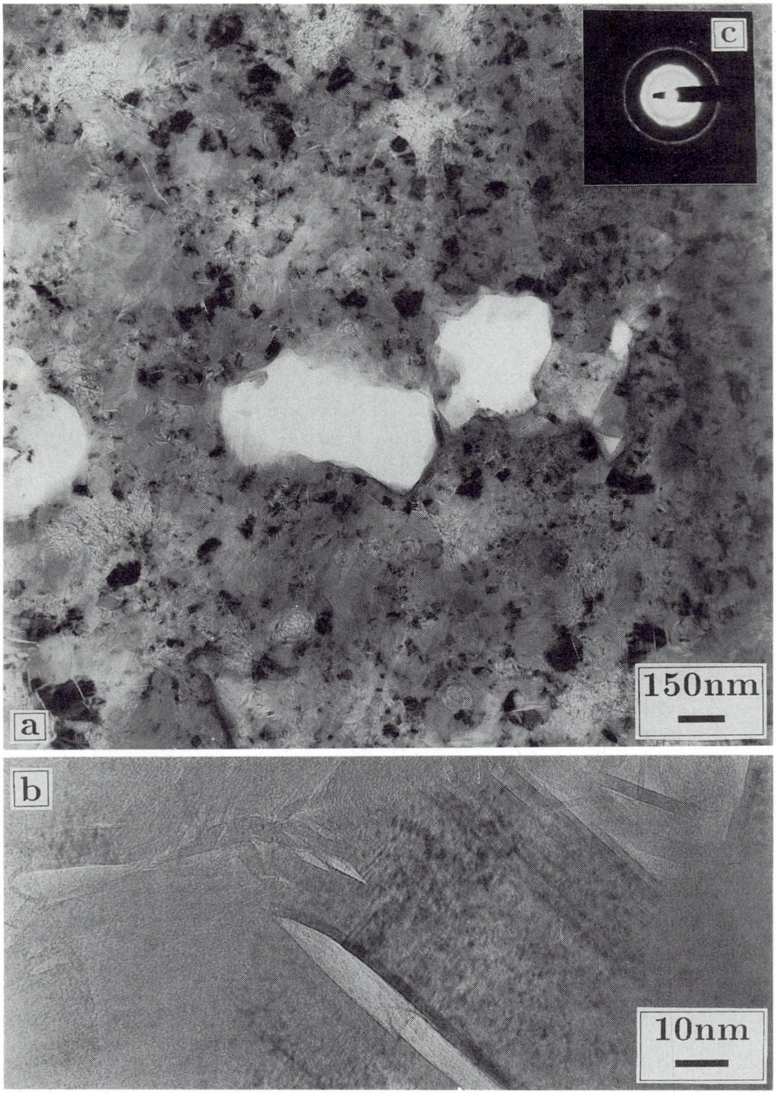

Figure 3.32 TEM micrograph of polygranular carbon based on carbon black (a). Insert (b) reveals the graphitic basal planes at high magnification. Diffraction pattern (c) demonstrates the isotropic orientation of graphite crystals on a micrometer scale.

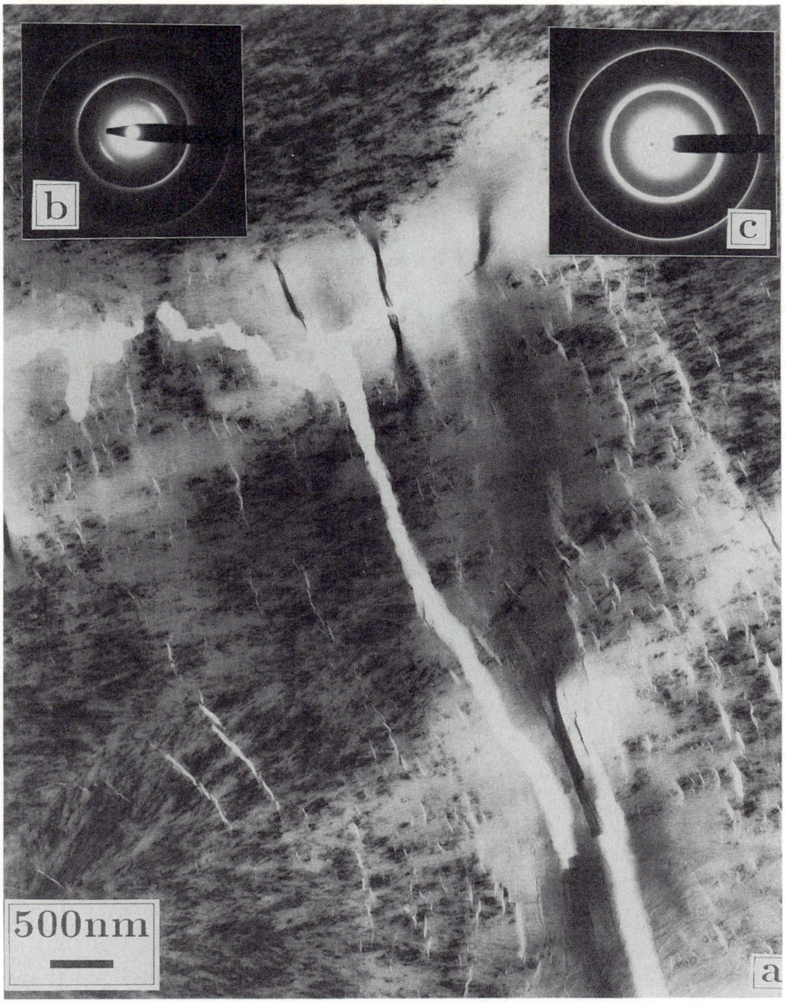

Figure 3.33 TEM micrograph (a) of pyrolytic carbon, revealing a section parallel to the growing direction. Fine cracks visible on the micrograph are separating basal planes emphasizing their preferred orientation. Graphite basal planes are following the surface orientation of the substrate (compare diffraction patterns b and c from sections parallel (such as(a)) and perpendicular to the growth direction, respectively. (002) graphitic reflections are only revealed from sections parallel to growth axis).

Vitreous Carbon: All carbons mentioned above are graphitizable (so-called soft-carbons). This indicates that they can be transformed to graphite by a thermally activated solid state transformation which requires temperatures of about 3000 °C. This is not possible with non-graphitizable carbon (or so-called hard-carbons) such as vitreous carbon under normal conditions [673].

Pyrolysis of three-dimensionally cross-linked aromatic polymers such as phenolic resins yields vitreous carbon which is characterized by glasslike fracture behaviour. An early structural model of R. Franklin (1951) [674] assumes small, crosslinked graphite crystals (Figure 3.34).

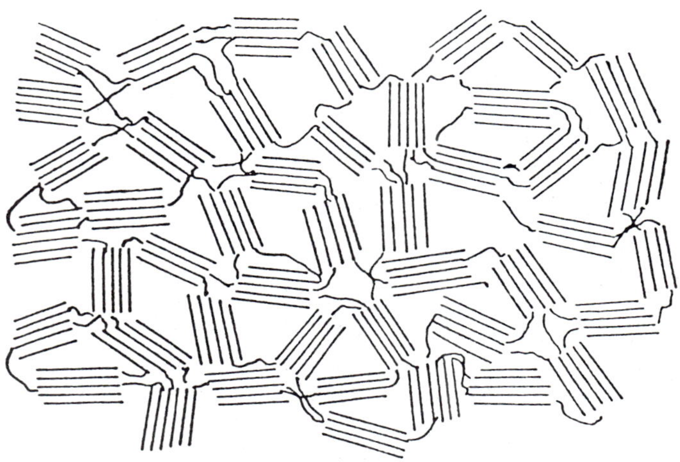

Figure 3.34 One of the first structural model of vitreous carbon ever published [674]. Turbostratic graphite crystallites with bent basal layers however were not known at this time.

The real lattice structure is revealed in Figure 3.35. The crystallites have a thickness of approximately 10 graphitic basal layers.

3D crosslinking of this configuration results in a carbon material which is isotropic on a scale of the order of 100 nm. The ribbon-like appearance of the crystallites is controlled by the imaging conditions in the microscope. The third dimension of the 3D structure is visible only as a projection on the imaging plane. An earlier structural model of vitreous carbon, neglecting these restrictions, is shown in Figure 3.36. An improved structural model of vitreous carbon accounting for the observed micropores is presented in Figure 3.37.

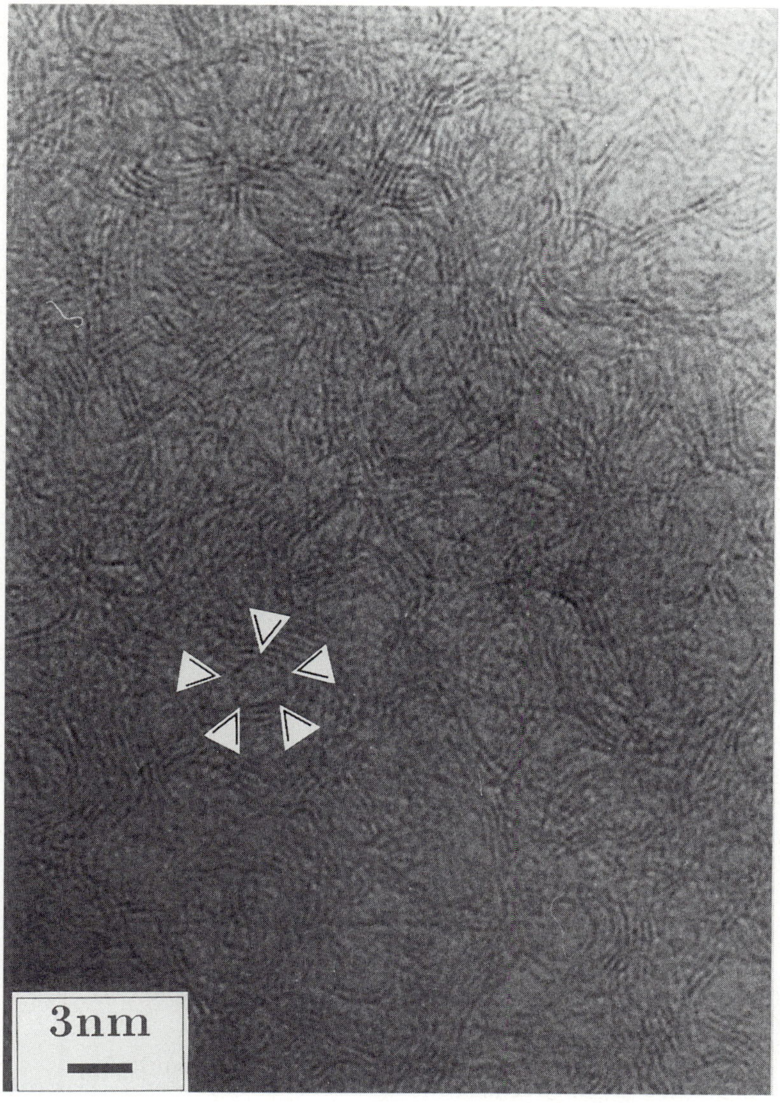

Figure 3.35 TEM micrograph of vitreous carbon. The cell walls forming the micropores (indicated by arrows) have a ribbon-like appearance depending on the imaging conditions.

As a consequence of its structure, vitreous carbon is characterized by low density and isotropic properties. Thermal and electrical conductivity are rather low,

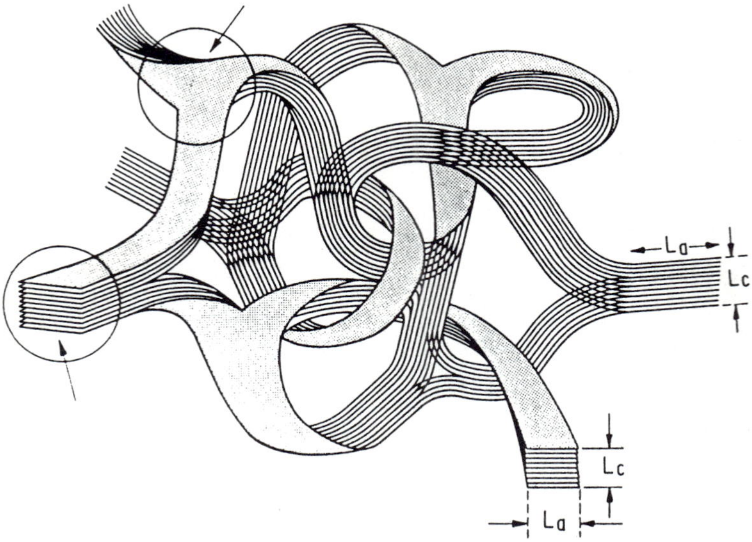

Figure 3.36 Schematic microstructure of vitreous carbon emphasising the ribbon-like appearance of crystallites [675]. This model neither explains the very low gas-permeability of vitreous carbon nor the existence of micropores (as observed by low angle diffraction [676] and TEM observations).

strength is higher by a factor of 2 compared to pyrolytic carbon along the basal planes (see Table 3.5).

The thickness of vitreous carbon materials is limited to approximately 5 mm because of the rather slow diffusion of gaseous species during processing of the material.

Carbon Fibres: High performance carbon fibres [659, 668] also belong to hard carbons, that is, they are non-graphitizable under normal conditions. Their structural configuration is similar to that of vitreous carbon, but isotropic only in two dimensions [677]. A schematic structural model is shown in Figure 3.38.

Revealing the detailed microstructure of carbon fibres has been a classical application field for high-resolution TEM work. Because of excellent reviews in the pertinent literature [661, 677, 668], this subject is not discussed in more detail in this section.

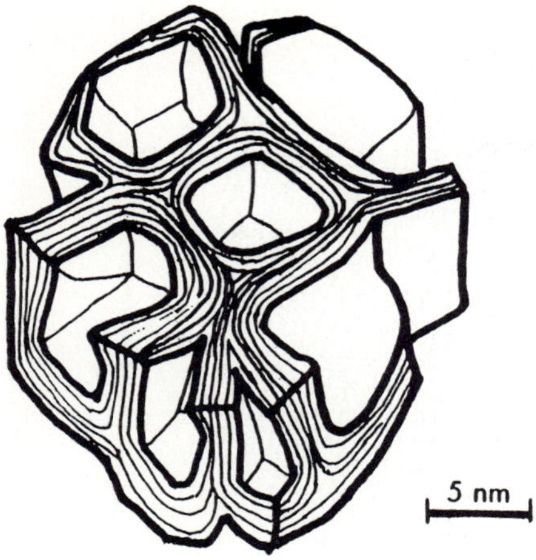

Figure 3.37 Schematic microstructure of vitreous carbon. The distance of the basal-planes is lower than 0.4 nm. About 10 basal layers are forming the cell walls of the micropores (diameter approx. 3 nm) [678].

Carbon fibres are produced by pyrolysis of either polyacrylnitril- (PAN) or pitch-based precursors resulting in carbon fibres with diameters of about 10 μm and unlimited length. The mechanical properties are outstanding. About 50% of the theoretical stiffness is reached.

Depending on the processing temperature carbon fibres with high strength (typical values: σ= 4 GPa, E = 200 GPa) or high modulus (E=450 GPa, σ=2.5 GPa) are obtained.

Carbon Fibre Reinforced Carbon

Although carbon/carbon materials are all-ceramic composites combining a brittle fibre with a brittle matrix, the failure mode is nonbrittle revealing an enormous increase of the work of fracture during loading. The dominant toughening mechanisms are load transfer, pull-out and crack-deflection [679, 680] which contribute to the high fracture toughness of this unique composite material. The strength is controlled by the very low strain to failure (<0.6%) of the matrix so the modulus of the fibre should be as high as possible [681]. For a 1D carbon/carbon material the maximum theoretical fibre content (hexagonal closed packed) is 90.6% resulting in the highest strength and Young's modulus, but of course only in one direction. 2D and 3D materials have a lower fibre content (79% and 68%, respectively). Only a fraction of the overall fibre content can be utilized along the

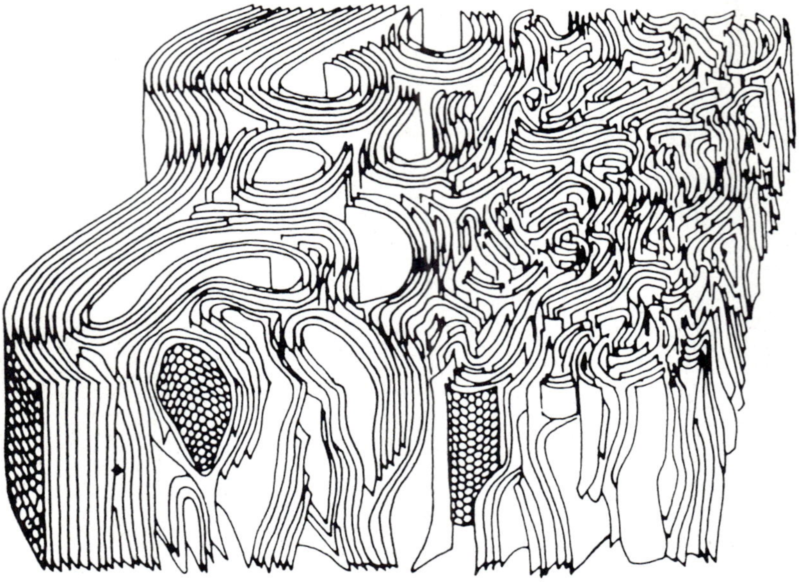

Figure 3.38 Schematic microstructure of a carbon fibre (fibre axis running from top to bottom) [682].

actual stress axis. Typical values for the strength of a 2D-C/C composite are 200 MPa up to 2000 °C [664].

Transmission electron microscopy is an ideal tool to characterize the complex microstructure of a 2D C/C composite on a nanometer scale. It is important to remember that the carbon matrix in C/C is actually a dynamic feature which is generated through pyrolysis of resin- or pitch-based precursors during processing of the composite. As an example, the microstructure of a commercial 2D C/C material employing PAN-based carbon fibres and a phenolic-resin based carbon matrix is discussed in the following. The emphasis is on the interaction of the intrabundle matrix with the carbon fibre and the resulting interfacial structures. Carbon fibre tows utilized for 2D and 3D fibre architecture contain single carbon filaments on the order of 3k to 10k. Figure 3.39 is a low magnification cross-section micrograph perpendicular to the fibre axis giving a first impression of the complexity of carbon/carbon material. The black circular structures are fibres having a diameter of about 7 μm. The main part of the matrix is of grey coloured contrast and can be identified as vitreous carbon. The light ribbon-like structure can be interpreted as vitreous carbon with a foam-like cellular structure because

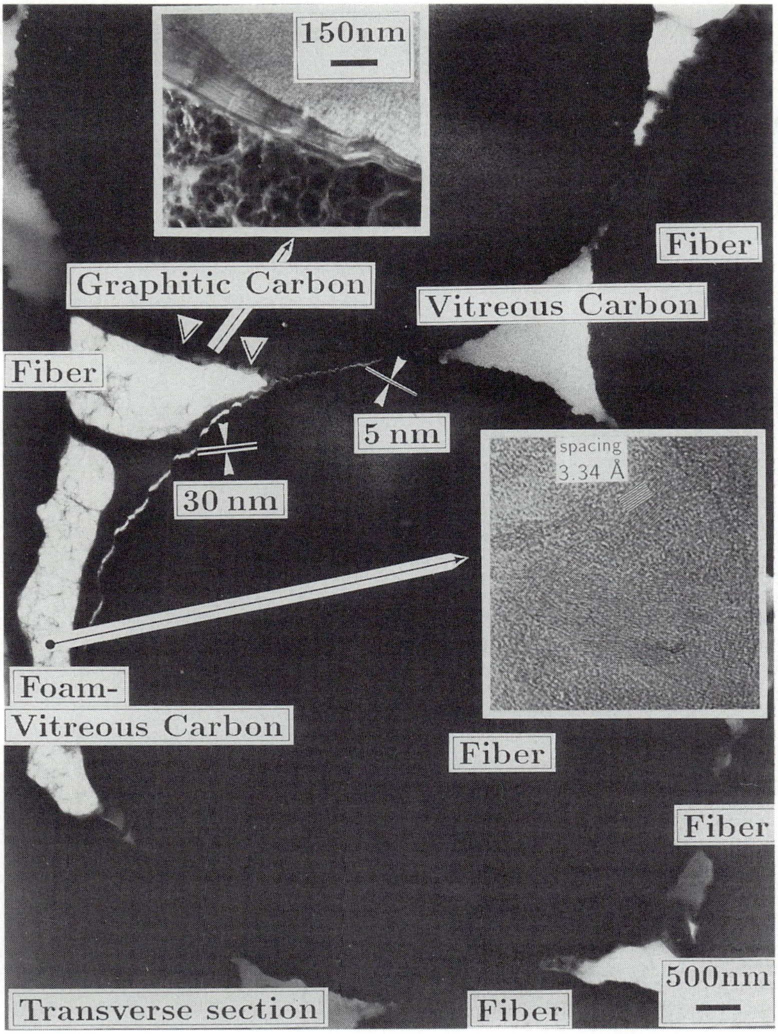

Figure 3.39 Low magnification microstructure from 2D C/C intrabundle matrix region (transverse section). Black circular regions are fibres with diameters of about 7 μm. Different structures of matrix carbon are indicated between individual fibre filaments.

of residual porosity. After processing a graphitic carbon structure displaying a high degree of orientation is found in intimate contact with the fibres. A high magnification fibre cross section micrograph is shown in Figure 3.40. Single a-b basal

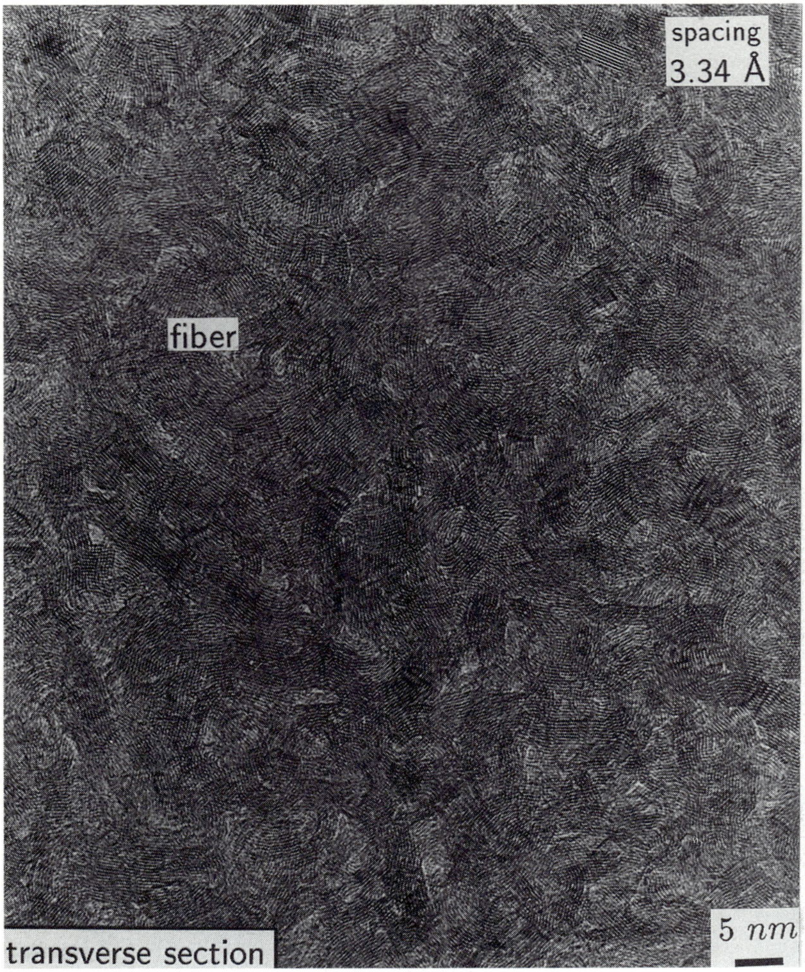

Figure 3.40 Cross section TEM micrograph of a HT carbon fibre from 2D C/C material. After processing, its microstructure is very similar to that of a typical HM fibre.

planes of graphite are resolved as black and white lines. Typically, 8-10 basal planes are forming bent crystallites. A comparison with the structures of as-received HT and HM carbon fibres clearly indicates that the structure of the fibres has changed during processing of the composite from a HT to a HM type. Thus, not only the carbon matrix, but also this transformation effect of the fibre microstructure contributes to the reported increase in stiffness of carbon/carbon composites after pyrolysis. A typical example for fibre/matrix interaction is displayed in Figure 3.41. Neither pores nor cracks are visible at this particular interfacial boundary, indicating that there is not only a mechanical interlocking of the fibre and the matrix, but that chemical bonding is actually involved between the constituents.

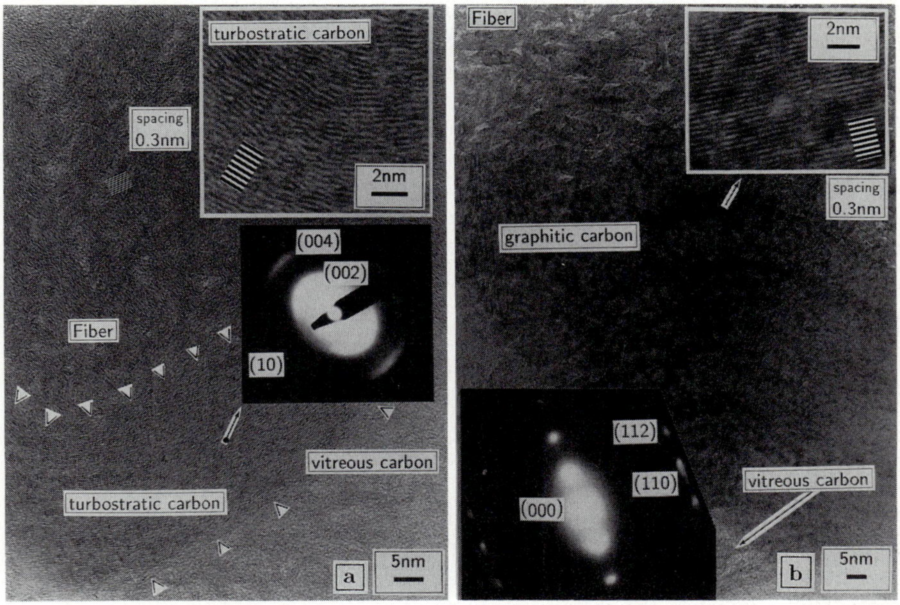

Figure 3.41 Cross-section TEM micrograph of the fibre matrix interface revealing the complexity of carbon.
The carbon fibre (top) is surrounded by matrix carbon: turbostratic carbon (a) and graphitic carbon (b), respectively and finally vitreous carbon (bottom). Both images were taken from different locations of the same fibre periphery. The degree of orientation of the preoriented matrix carbon is significantly increased from a) to b) (see inserts showing the lattice structure and electron diffraction patterns).

Displaced from the fibre periphery by approximately 30 nm, the carbon matrix has a structure very similar to vitreous carbon. Closer to the fibre a higher degree of orientation is observed. Actually more graphite crystallites are extending parallel to the fibre surface than perpendicular to it. Along the fibre/matrix interface the degree of orientation can increase to values high enough to transform the non-graphitizable matrix carbon to turbostratic carbon. In the latter variant the degree of orientation increases continuously, as shown in Figure 3.41.

This effect is known as stress-induced graphitization [683, 684]: during pyrolysis of the matrix resin the shrinkage rate reaches a maximum in the temperature range of 500 °C to 800 °C. Young's modulus also increases. A strong fibre/matrix bond at the interfacial boundary actually inhibits effective shrinkage. Because of the large difference in the thermal expansion coefficient of both fibre and matrix, tensile stresses of the order of 0.3 GPa generated along the interface are sufficient to preorient the matrix carbon. At temperatures higher than approximately 1500 °C, the carbon matrix is transformed to turbostratic and graphitic carbon variants respectively.

These selected examples indicate that the microstructure of 2D C/C composites is by no means simple. At least five structural variants of fibre and matrix carbons are revealed in the TEM. Fibre carbon (a 2D structural variety of vitreous carbon), vitreous matrix carbon and the cellular type of vitreous carbon can be attached to the group of hard-carbons. Turbostratic carbon and its continuous structural transition to graphitic carbon are considered as soft-carbons.

These microstructures connected to the intrabundle matrix/fibre interface in C/C composites are very similar to those revealed in monolithic carbon materials. For the latter class of carbon materials, the thermal and mechanical properties are well characterized. Thus the comparison with monolithic carbons offers the unique opportunity to address physical properties to local microstructures in C/C which otherwise would be virtually impossible to pursue. Following this approach a better understanding of the role of carbon matrix microstructures is achieved. This can be utilized to identify failure-related interfacial features in carbon/carbon composites, but it becomes also very important for the evaluation of C/C protection layers, because the different structural variants of matrix carbon involve different oxidation and transformation rates.

Although adressed only in qualitative terms for the sake of simplicity, this section has provided evidence that the structural variants revealed in the multiphase microstructure of matrix carbon in C/C composites are indeed equivalent to those from rather straightforward monolithic carbon materials. Following this approach, the microstructural development of C/C can be pursued in the TEM as the structural organization of the matrix carbons become more and more complex during processing of the material. This example also emphasizes that we certainly can learn a lot about complex composite microstructures by carefully revealing the basic structural patterns of the monolithic constituents.

4. Polymer Matrix Composites

P.W.M. Peters and K. Schulte

Polymer matrix composites (PMCs) are in comparison with metal and ceramic matrix composites the most widely used composites. This is caused by their good properties and relative easy processing. Depending upon the type of fibre used, the costs are low (glass fibre reinforced plastics, GFRPs) to moderate (carbon fibre reinforced plastics, CFRPs).

In the last decade polymer matrix composites were successfully introduced in secondary and primary parts of general aviation aircraft. It can be expected that the next generation of aircraft will increasingly use composite materials. Military aircraft, especially the new fighter generation, will mainly be built of composites. The advantages of composites in the design of new and improved structures allow them to compete favourably with conventional alloys and materials. This is corroborated by the fact that helicopter design is totally changing at present, with polymer-matrix composites not only being used for rotor blades, but also for the entire rotor head and fuselage.

This continuous success in aircraft structures is being accompanied by an introduction of these materials into spacecraft. Satellite and rocket design need this stiff and lightweight material in order to save weight and make space transportation economical. This process of applying composites in aerospace has been accompanied by large efforts in materials science. In order to further promote polymer composites, continuous research is necessary. However, what are the new scientific goals to be achieved?

In the past epoxies and other thermosets were the major matrix systems used. There are strong indications that these matrix materials will continue, as they allow easy handling and appropriate fabrication techniques are well developed. In order to obtain tough composites, a development towards higher matrix ductility combined with increasing service temperature and less moisture absorption has been successfully concluded. In the future, with increasing use of polymer matrix composites in general aviation aircraft, other matrix properties have to be further improved. These include electrical conductivity, as the lesson from applications up to now is that hazardous lightning damage can occur in composite parts, which necessitates this development.

Apart from structural fuselage parts, composites for the interior have not only to be light, but also extremely fire resistant. An analysis of aircraft accidents has shown that a number of passengers survived the crash, who later died during the aircraft fire. This means that for the interior parts composites have to be more versatile. In this context, light phenolic matrix systems will find increasing attention because of their superior fire resistance.

Thermoplastic matrix composites, which were the runner-up in the last decade, have not found the market share that was optimistically expected. Although these polymers possess excellent properties, production and manufacturing costs hindered their breakthrough in aircraft application.

With polymer matrix composites replacing metal parts in structural design, weight savings of about 20% can be achieved, when balanced symmetrical laminates are used. Further weight savings are possible especially on curved structures, if asymmetric laminates can be taken. However, prior fundamental research is needed.

Polymer matrix composites are not only versatile materials for structural application because of light weight, high strength and high stiffness, they are also chosen because of physical properties such as nearly zero thermal expansion. One important argument to further develop these composites is that they can potentially be used as a smart structural material, which means that a material does not only carry load, but can be used to simultaneously monitor the actual load or damage state.

In this section static and fatigue properties of fibre-reinforced polymers are treated. Before this, some aspects of fibre reinforcement are discussed.

The degree of reinforcement (stiffness and strength increase) due to fibre- or more general particle-reinforcement depends not only on the properties of the reinforcing material but also on its size and shape. This is demonstrated in Figure 4.1[685], which shows an increase of the modulus if stiff particles are embedded in a less stiff matrix. The modulus increase depends on the dimensions of the particles, especially, as indicated in Figure 4.1, on the aspect ratio l/d and on the orientation of these particles. Maximum stiffness increase occurs if l/d is large (to infinite), with the dimension l in loading direction. This is a unidirectional reinforced material with the fibres (i.e. particles with a more or less circular cross-section) parallelly oriented in the loading direction (case 7). As visible from Figure 4.1, the effect of reinforcement is only small if the fibres are oriented perpendicular to the loading direction (case 2).

4.1 Quasi-Static Strength of Polymer Matrix Composites

In this section the strength and stiffness of unidirectional fibre reinforced polymers making use of continuous fibres will be treated ($l/d = \infty$), with the fibres oriented in the loading direction as well as perpendicular to the loading direction.

Modulus and Strength of Polymer Matrix Composites

Modulus of Unidirectionally Reinforced Materials in Fibre Direction: Optimum advantage of the strength and stiffness of fibres is made when the fibres are con-

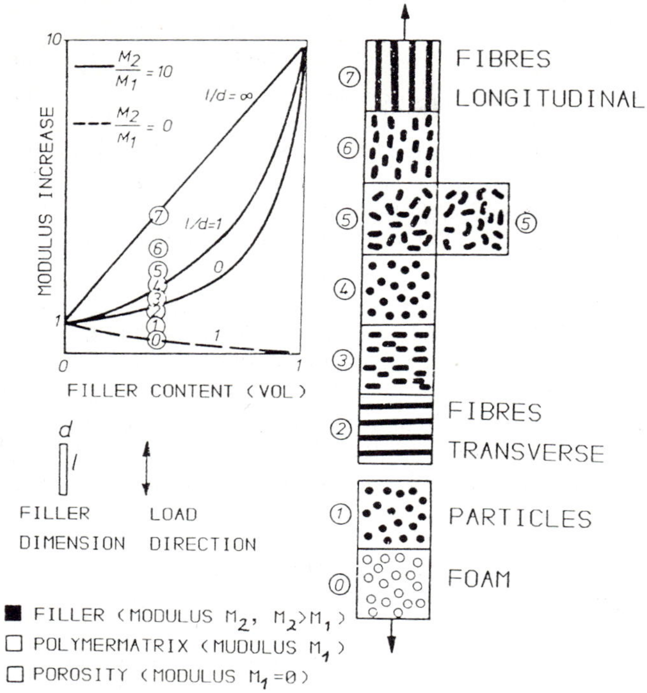

Figure 4.1 Modulus of the composite as influenced by fibrelike inclusions as a function of aspect ratio (=l/d, with l in the direction of the applied load), fibre volume and fibre modulus [685].
Reproduced with permission of Carl Hanser Verlag

tinuous and oriented in the loading direction. In this case the modulus of the composite in fibre direction, E_{11}, can simply be determined with the aid of the rule of mixtures by considering the composite to consist of parallelly oriented components leading to

$$E_{11} = E_m V_m + E_f V_f \tag{4.1}$$

where E_m, E_f are the moduli of the matrix and fibres and V_m, V_f the relative volumes of matrix and fibres. If the Poisson ratios of the matrix and fibres differ (ν_f,ν_m), there is an additional contribution to the modulus of the composite. This contribution is, however, in most cases negligible. The influence of local variations of fibre volume, nonalignment of the fibres, etc. leads to variations in the value of the modulus of the composite which are more important than, e.g., those due to the difference in Poisson ratio. In some cases the modulus of the matrix is so small in comparison with those of the fibres that the contribution of the matrix to the modulus of the composite is negligible. This is, e.g., the case for CFRP with E_f=240-400 GPa and E_m=3-5 GPa.

Strength of Unidirectionally Reinforced Polymers in Fibre Direction: The strength of unidirectional fibre reinforced polymers in fibre direction is in comparison to the modulus even more dominated by the fibres. Nevertheless, the matrix plays an important role in relation to the strength of the composite. The contribution of the matrix to the strength of the composite is, however, not directly based on its strength but on its ability to reintroduce stresses into broken fibres. This is explained in Figure 4.2, which shows a broken fibre in a composite surrounded by intact fibres. The stress in the broken fibre at the fracture site is zero. At some distance from the crack the stress is completely reintroduced by shear stresses in the matrix. These shear stresses are maximum near the fracture site and gradually reduce to zero at some distance away from the fracture site. The neighbours of the broken fibre experience a stress increase, which also reduces to zero at some distance from the fracture site. Two different approaches have been developed to account for the loss of the broken fibres. It can be assumed that the load of the broken fibre is

- equally shared by the surviving fibres (global load sharing or theory of uncoupled (loose) fibres) [686]

or that the load is

- equally shared by the neighbouring fibres (local load sharing) [687].

The stress increase for the fibres involved is larger in the case of local load sharing, so that this approach generally leads to a lower strength than for global load sharing [688]. Fracture in the unidirectional composite now can be of the noncumulative type or the cumulative type. Fracture is noncumulative if due to the stress increase next to a broken fibre neighbouring fibres fail. In this case a single crack leads to failure. This type of failure tends to occur,

- if the scatter in fibre strength is small
- if the matrix is stiff and strong.

The cumulative type of failure (statistically distributed fibre ruptures) tends to occur,

- if the scatter in fibre strength is large
- if the matrix is compliant and weak.

In all these cases the fibre matrix bond strength is considered to be ideal. Fibre reinforced polymers generally show the cumulative type of failure. This cumulative type of failure can be predicted rather well with the theory of uncoupled fibres (global load sharing) [686].

One important parameter to be determined for this analysis is the ineffective length, i.e. the length over which the stress in a broken fibre is lower than the

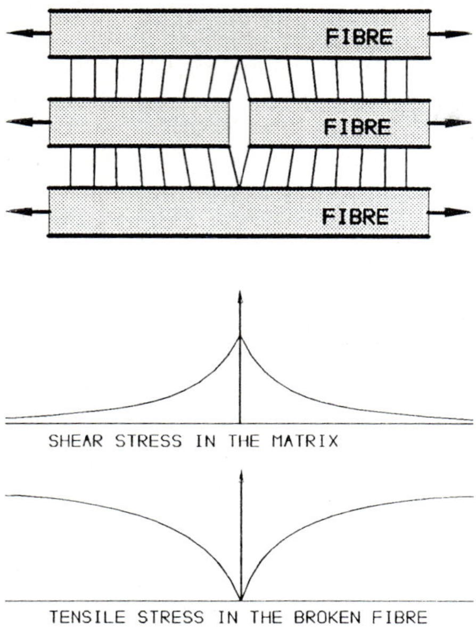

Figure 4.2 Stress distribution in a broken fibre and the surrounding matrix.

undisturbed stress (see Figure 4.2). The stress field around a broken fibre can be analysed making use of a shear lag analysis. In this analysis the matrix is considered to experience only shear stresses, whereas the fibres bear only tensile stresses. According to this approach the ineffective length δ is given by [686]:

$$\frac{\delta}{d_f} = \frac{1}{2} \cdot \sqrt{\frac{1 - V_f^{1/2}}{V_f^{1/2}} \cdot \frac{E_f}{G_m}} \cdot \ln\left(\frac{1}{1-\varphi}\right) \tag{4.2}$$

where d_f is the fibre diameter, V_f the fibre volume, and E_f and G_m the fibre modulus and shear modulus of the matrix, respectively. The stress in the broken fibres asymptotically reaches the undisturbed stress at infinity as indicated in Figure 4.2. For practical reasons Rosen [686] thus considered the ineffective length to be that length over which 90% of the undisturbed stress is reached (φ=0.9 in Eq. (4.2)). The ineffective length thus is directly related to $\sqrt{E_f/G_m}$, indicating that the larger the modulus of the matrix, the smaller the ineffective length.

If we consider the extreme case of a loose bundle of fibres (matrix modulus is 0), then every fibre in the bundle will break only once before the bundle totally fails. The bundle strength can be determined based on the strength distribution of the single fibre. A two-parameter Weibull distribution was shown adequately to describe the fibre strength distribution [689, 690]. According to this distribution the probability of failure of a fibre with a length L as a function of the applied stress σ is expressed by

$$F(\sigma) = 1 - \exp[-L\alpha\sigma^{m_f}] \tag{4.3}$$

where α and m_f are the two Weibull parameters being the scale parameter and the shape parameter, respectively. The shape parameter m_f is a measure for the scatter in fibre strength, with m_f is infinite for no scatter and m_f is small for a large scatter. The expected strength of a bundle F_{max} with a large number of fibres with fibre strength distribution according to Eq. (4.3) is given by [686]:

$$F_{max} = N \cdot A_f (\alpha L m_f e)^{-1/m_f} \tag{4.4}$$

where e is the base of natural logarithms, N the number of fibres and A_f the cross sectional area of the fibres. Effectively a reverse procedure can be followed; the fibre strength distribution can be deduced from the load-strain curves registered during loose bundle testing [691, 692]. It is clear from Eqs. (4.3) and (4.4) that the strength of the fibres and fibre bundles increases if the length L decreases. The reinforcing effect of the matrix consists in the reintroduction of the stresses in a broken fibre. Outside the ineffective length as illustrated in Figure 4.2 the stress in the fibres is practically undisturbed, so that the composite can be suggested to consist of a chain of links with a length of the ineffective length δ. The strength distribution of these links is decisive for the strength of the composite. A short length δ leads to a larger bundle strength, as is visible from Eq. (4.4). According to Eq. (4.2) δ is small if the ratio E_f/G_m is small. For polymer based composites the ineffective length measures 1-10 fibre diameters in the elastic case [686]. In the nonelastic case (plastic deformation of the polymer or fibre/matrix debonding) the ineffective length measures 10 to 100 times the fibre diameter. Thus the reinforcing effect of the matrix strongly depends upon the fibre matrix bond strenth. The reinforcing influence of a polymer matrix was found to be a factor of 4.66 as a maximum, comparing the strength of polymer impregnated carbon fibre bundles with those of loose carbon fibre bundles [693]. Another aspect of composite strength is its scatter. Relative to the scatter in fibre strength, the scatter in composite strength is much less. This is visible from the scale parameter which for a composite generally is 2-4 times larger than for the fibre [687]. According to the "local load sharing" approach this can be explained by the formation of groups of broken fibres at ultimate load. It is suggested [688] that at ultimate load i-plets (groups with a number of i broken fibres) are formed with i given by the relation

$$m_c = i \cdot m_f \tag{4.5}$$

where m_c and m_f are shape parameters for composite strength and fibre strength, respectively.

Modulus of Unidirectionally Reinforced Material Transverse to the Fibre Direction: The modulus transverse to the fibre direction is in contradiction to the modulus in fibre direction less influenced by stiff fibres. The reason is the continuous stress transfer from fibre to matrix v.v. for this fibre arrangement. The fibre and matrix are here linked in series and not parallel. The modulus in this case can be described by [694]:

$$E_{22} = \frac{E_m}{1 - \sqrt{V_f \cdot (1 - E_m/E_{f22})}} \tag{4.6}$$

with E_{f22} and E_m the modulus of the fibre transverse to the longitudinal direction and the modulus of the matrix, respectively, and V_f the fibre volume. Hashin [695] determined an upper and a lower boundary for the transverse modulus making use of the composite cylinder assemblage (CCA) model. The upper boundary was in closer agreement with the experimental data of glass/epoxy and boron/epoxy. The validity of the existent models for transverse modulus is difficult to verify by experiments. Experimental results are influenced by the actual (statistical) distribution of fibres and in many cases by the existence of an interphase. The interphase is defined as the layer (of a certain unknown thickness) surrounding the fibres in which the properties of the matrix change from a certain value at the fibre surface to the properties of the matrix in the bulk (see also section "Interfaces in Metal Matrix Composites"). Further, the transverse stiffness of the anisotropic fibres (carbon, aramid fibres) is not exactly known, adding one more difficulty to the verification of existing models with experimental data. Carbon and aramid fibres are anisotropic with a rather small transverse modulus of 15 GPa and 5.4 GPa, respectively. This gives rise to a rather small modulus in the composite transverse to the fibre (CFRP: E_{22}=8-10 GPa, AFRP: E_{22}=3-5 GPa). GFRP generally has a higher transverse modulus E_{22}=14-18 GPa due to the high stiffness of the isotropic glass fibre with E_f=73 GPa. The composite transverse moduli presented are valid for a mean fibre volume of V_f=60%.

Strength Transverse to the Fibre Direction: The strength of a unidirectional composite perpendicular to the loading direction is governed by the strain magnification, which occurs locally in the matrix. This strain magnification is the result of the (usually) higher transverse modulus of the fibre in comparison with that of the matrix. The magnitude of this strain magnification can be determined with simple models developed by Kies [696] and Chamis [697]. Based on a square array of fibres (see inset Figure 4.3) the strain magnification factor (SMF) can be determined as a function of the respective moduli and the fibre volume:

$$SMF = \frac{1}{1 - (1 - \frac{E_m}{E_{f22}})\sqrt{\frac{4V_f}{\pi}}} \tag{4.7}$$

This factor SMF, calculated for different fibre reinforced polymers according to Eq. (4.7), is presented in Figure 4.3. It shows that GFRP has the largest SMF

(SMF = 6.0 at $V_f = 0.6$), CFRP a moderate one (SMF = 3.0 at $V_f = 0.6$), whereas AFRP shows a small SMF (SMF = 1.3 at $V_f = 0.6$). This is the result of the transverse moduli for the different fibres, E_{f22} = 73 GPa, 15 GPa and 5.4 GPa for glass, carbon and aramid fibres, respectively.

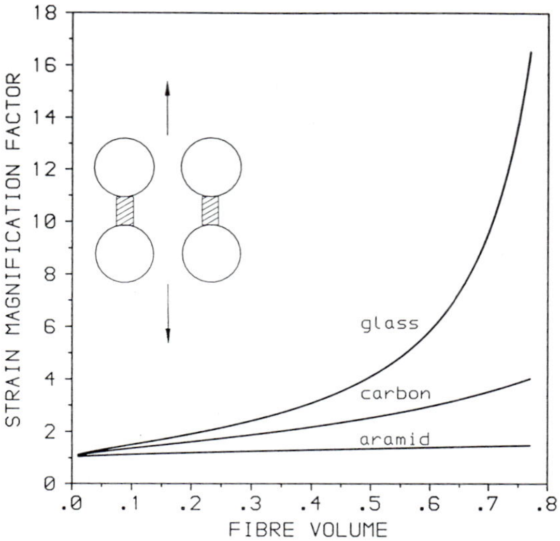

Figure 4.3 Strain magnification factor locally in the matrix (hatched area in inset) for transversely loaded composites as a function of fibre volume content.

The transverse fracture strain of the unidirectional laminate now can be determined based on:

- a strength of materials method or
- a suitable failure criterion.

Knowing the fracture strain of the matrix ε_m, the strength of materials method leads to the composite transverse fracture strain $\varepsilon_{22,\,max}$ being

$$\varepsilon_{22,\,max} = \varepsilon_m / SMF \tag{4.8}$$

For an evaluation of the transverse properties in terms of stress, reference is made to the recent work of Chamis [694].

The transverse fracture strain (or strength) based only on the aspect of maximum matrix fracture strain (or strength) neglects many other important variables. The most important variables in terms of transverse fracture strain are:

- the properties of the fibre (mainly in transverse direction)

- the properties of the matrix (modulus, fracture strain)
- the fibre volume and distribution
- the defect distribution as a result of prepregging and laminate production
- the strength of the interphase
- the constraining effect of neighbouring layers (in the case of fracture strain of a transverse ply in a multi-directional laminate).

The first three variables influence the strain magnification factor, as discussed above. Little is however known about the influence of the statistical distribution of fibres. Local variations in fibre distance influence the strain magnification on the micro-level, which is not assessed yet. Merely the influence of irregular fibre distribution on thermal strains was investigated for fibre reinforced ceramics [698].

In the following sections the influence of the other parameters mentioned is investigated with the aid of a special experimental technique, based on multiple fracture in the transverse ply of cross-ply 0/90/0 laminates.

Multiple Cracking in 0/90/0 Specimens

If a transverse ply is embedded between stronger surface plies (e.g. 0° plies), the transverse ply of many fibre reinforced materials (CFRP, GFRP, and AFRP) can be broken into many pieces (up to a few tenths of a millimeter in length) before the specimen finally fails. This phenomenon of multiple cracking in a transverse ply gives the researcher an exceptional possibility to obtain as many sets of data as cracks occurring in a *single* experiment. A technique was developed to determine a statistical distribution of transverse fracture strain based on the fracture data of a *single* experiment [699, 700]. Making use of these fracture strain distributions, all aspects influencing the transverse strength can be investigated. For this approach it is necessary:

- to measure the strain or stress at which every single crack occurs, and
- to divide the specimen in a number of elements all of which can fail.

The strain at which every single crack occurs is measured with the aid of a piezoelectric transducer in connection with a transient recorder. If the crack occurs suddenly, the piezoelectric load cell registers a small load drop as a result of a (sudden) small extension of the specimen (during a displacement controlled test). Thus, if the matrix system is not too ductile and the transverse ply not too thin, with the aid of the registered dynamic load change the occurrence of the different cracks as a function of applied strain can be determined.

In order to be able to apply statistics, it is necessary to determine how many cracks can occur. Generally, the number of cracks which occur is limited. This is caused by the fact that the stress in a transverse ply close to an existing crack is small. A certain distance from the transverse crack the stress gradually increases from zero to the undisturbed stress similar to what was shown in Figure 4.2 for a broken fibre. In the applied model it is assumed that no further transverse cracks occur in the range where the stress is lower than 90% (chosen somewhat arbitrarily) of the undisturbed stress. This length is determined with the aid of a shear lag analysis [701, 702], the element of which is given in Figure 4.4. According to this analysis the stress in the transverse ply (ply 2) as a function of the distance x to the crack is given by:

$$\sigma_{2,x} = \sigma_{2,\infty} \cdot (1 - \exp\,[-\gamma \cdot x]) \tag{4.9}$$

where $\sigma_{2,\infty}$ is the undisturbed stress in the transverse ply and γ is:

$$\gamma = \sqrt{\frac{G_m}{b} \cdot \left(\frac{1}{E_1 a_1} + \frac{2}{E_2 a_2}\right)} \tag{4.10}$$

where E_1, E_2, and G_m represent the modulus of the 0° and 90° plies and the shear modulus of the matrix, and b the thickness of the resin rich layer (= shear transfer layer, normally taken to be twice the fibre diameter) respectively. For a discussion on the shear lag parameter G_m/b, see [703]. Note the layerwise notation being different from the notation mentioned before with the modulus of the surface plies (ply 1,3) $E_1 = E_3 = E_{11}$ and of ply 2 $E_2 = E_{22}$ (the transverse modulus).

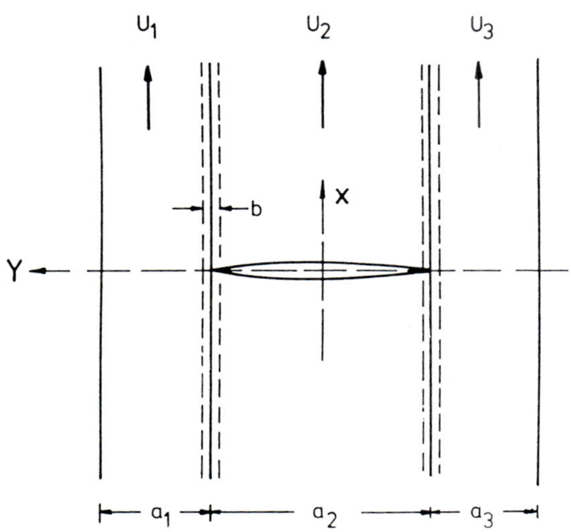

Figure 4.4 Element of the applied shear lag model with the layerwise displacement U_1, U_2, and U_3 ($U_1 = U_3$) to describe stress distribution in a 0/90/0 laminate around a transverse crack.

The element (half-)length now follows from Eq. (4.9) substituting $\sigma_{2,x} = 0.9\sigma_{2,\infty}$. Thus the number of elements in the gauge length is now known.

The experimental data (available after the experiment as a list of sequentially occurred cracks with the respective strains at which they occur) is described with the aid of a two-parameter Weibull distribution given by Eq. (4.3), but now in terms of strain:

$$F = 1 - \exp[-(\varepsilon_{f,l_0}/\hat{\varepsilon}_{f,l_0})^a] \tag{4.11}$$

where F is the probability of failure at a strain of ε_{f,l_0}. The two Weibull parameters are the Weibull shape parameter "a" and the characteristic fracture strain $\hat{\varepsilon}_{f,l_0}$ (at F = 0.632). The strain ε_{f,l_0} consists of a thermal and a mechanical component. The thermal component is determined experimentally or theoretically based on known thermal expansion coefficients of the unidirectional material in and perpendicular to the fibre directions.

An example of the evaluation of crack data from a single test performed at -40 °C according to this Weibull analysis is given in Figure 4.5a [704]. The experimental data are well described by the given straight line, which is determined with the aid of the experimental data making use of linear regression and least square analysis. Transverse cracking in different CFRP materials under different conditions was investigated, leading to two-parameter Weibull distributions. These investigations led to the model presented in the next section which describes transverse cracking in CFRP.

In some cases the crack data do not follow the two-parameter Weibull distributions completely. One reason for this can be that a condition of saturation of cracking can be reached. This is caused by the fact that the shear stress in the shear lag zone (which introduces the tensile stress in the interrupted transverse ply) reaches the maximum shear stress (matrix shear yield stress or interlaminar shear strength). According to the shear lag analysis the maximum shear stress in the shear lag zone at a transverse crack tip is given by [703]:

$$\tau_{max} = \frac{G_m}{b} \cdot \left(\frac{\sigma_{2,\infty}}{\gamma E_2}\right) \cdot \left(1 + E_2 \frac{a_2}{2E_1 a_1}\right) \tag{4.12}$$

where $\sigma_{2,\infty}$ is the stress in the transverse ply (thermal + mechanical stress) and γ as given by Eq. (4.10) on page 228. The shear stress is large if the transverse ply is thick (a_2 large) and stiff (E_2 large). GFRP generally tends to crack saturation more than does CFRP because of the higher transverse stiffness (E_2=~15 GPa in comparison with E_2=8-10 GPa for CFRP). AFRP also shows saturation of transverse cracking [705]. In this case the transverse modulus is not the dominant factor (E_2=5.2 GPa) but the low interlaminar shear strength of AFRP (as a result of the poor bond between the aramid fibre and the epoxy).

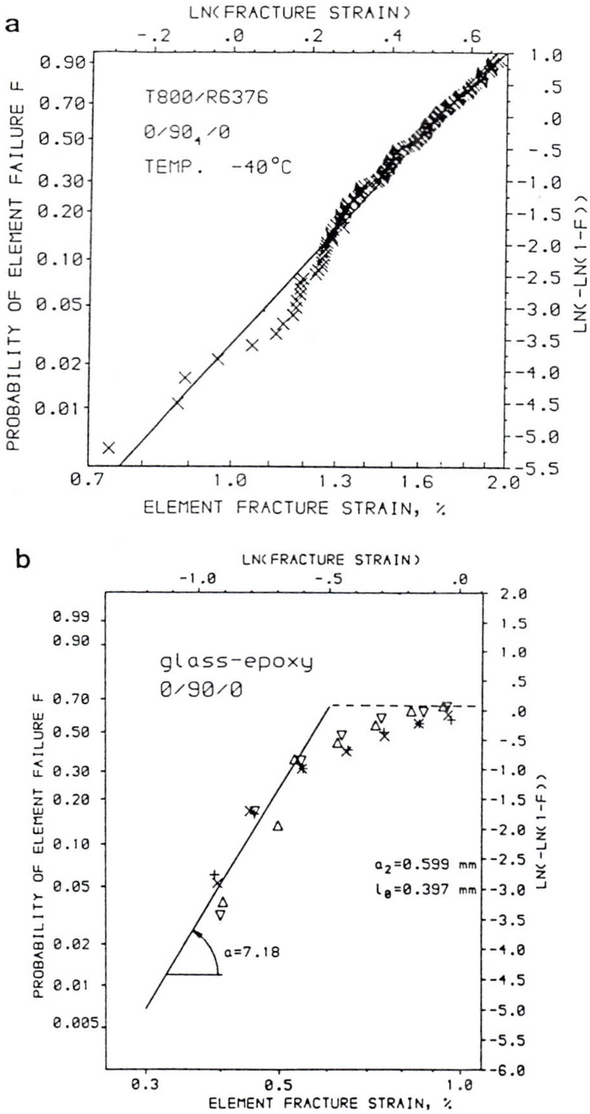

Figure 4.5 a) Example of a Weibull distribution for the transverse fracture strain of the transverse ply of a single $0/90_4/0$ specimen.
b) Transverse fracture data of (4) specimens from a GFRP $0/90_2/0$ laminate showing saturation of cracking at larger strains [706].

An example of crack saturation is given in Figure 4.5b [706] for specimens of a GFRP laminate. In this case the piezoelectric device was not used. The number of cracks in the gauge length of the specimen was determined by visual inspection of a polished specimen edge by loading/unloading at increasing applied strains. This method is applied if the piezoelectric device does not give reliable results, e.g. if the transverse ply is too thin and/or the matrix system is too ductile (thermoplastic matrix).

Model for Transverse Cracking in CFRP

The Weibull distribution for transverse cracking in, e.g., Figure 4.5 is the material response to a certain distribution of defects. These defects occur as a result of non-ideal production of the composite starting from the fibres and the matrix. In the ideal case (no defects) the Weibull distribution would be a perpendicular line ($a = \infty$) with one discrete failure strain (Figure 4.6) [707]. Defects lead to a shift and rotation ($a<\infty$) of the Weibull distribution to lower fracture strains. The actual shift and rotation depend on the distribution of defects in size and number. The material has two possibilities to reduce the effect of the defects, i.e. to shift back the curve in the direction of the ideal position. These are (i) the ductility of the matrix and (ii) the constraining effect of neighbouring layers. Both reduce the effectiveness of the defects in that the stress concentrations near the defects are diminished [708]. More about this effect is discussed in the next section.

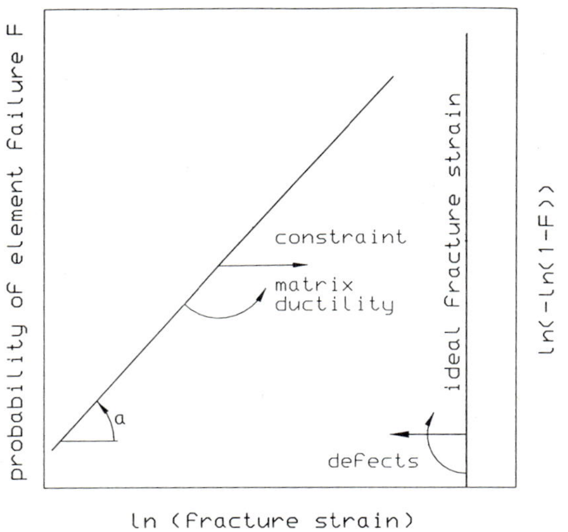

Figure 4.6 Model for transverse fracture in CFRP describing the influence of defects, constraint and matrix ductility [707].

Figure 4.6 shows schematically that the actual Weibull distribution intersects or better ends at the ideal fracture strain. For the point of intersection the material reaches the maximum (= ideal) fracture strain and thus is defect-free.

As a final consequence of the model, the maximum possible transverse fracture strain can thus be determined [709]. This fracture strain represents the weakest of either the maximum matrix fracture strain (multiplied by the strain magnification factor) or the strain at interphase failure. Which one of them is the cause of cracking should be carefully examined. This cannot always be done; therefore, in the following the maximum transverse fracture strain, for ease of terminology, simply is defined as the strain at interphase failure. The ideal fracture strain, however, will never be reached, so that for practical reasons a maximum strain at interphase failure will be determined with the aid of a measured Weibull distribution. The method is based on the assumption that the gauge length of the specimen (of 100 mm length) contains one element which is practically defect-free. Thus the fracture strain of this element represents the strain at interphase failure. Normally not all elements break, so that the fracture strain of the strongest element has to be determined by extrapolation (to F=200/201 in case the gauge length contains 200 elements).

This procedure can only be applied if the material is macroscopically and microscopically defect-free. Only in this case can it be accepted that the last breaking element is practically defect-free. In the following sections three examples of the applicability of this model and resulting strain at interphase failure are given:

- effect of constraint for $0_2/90_4/0_2$, $0_2/90_8/0_2$ and $0_2/90_{12}/0_2$ laminates
- effect of test temperature on transverse cracking in the range of -100 °C to +100 °C;
- effect of fibre surface treatment on transverse cracking.

Effect of Constraint: One important phenomenon related to transverse cracking is the effect of constraint. As shown by Flaggs and Kural [710], the fracture strain at the onset of transverse cracking increases if the transverse ply is embedded between 0°-plies. The effect is small if the transverse ply is thick, but is larger if the transverse ply is thin.

This increase in transverse strength can be predicted with the aid of an energy approach. For this energy approach the energy levels just before and after the occurrence of a transverse crack are calculated [711, 712]. Assuming that the difference in these energy levels is the energy necessary to produce the two crack surfaces seems to give a reasonable explanation of the effect of constraint. A part of the energy is, however, dissipated in the specimen (the energy balance should be considered at the crack or defect reaching a critical crack size where crack propagation becomes instable), so that constraint is not completely described this way.

More knowledge concerning transverse cracking can be achieved if cracking over the total strain range is evaluated. This is, e.g., possible with the experimental technique described above. Weibull fracture strain distributions (mean values of four specimens) produced with this technique for the transverse plies in the laminates $0_2/90_4/0_2$, $0_2/90_6/0_2$ and $0_2/90_{12}/0_2$ from [699] are presented in Figure 4.7. In addition to the expected fracture strain increase at decreasing transverse ply thickness a decrease in the scatter of fracture strain (a decreasing shape parameter "a") is also found. The explanation of this changing shape parameter cannot be found easily with deterministic analyses. The reduction of data scatter ("a" larger) at decreasing transverse ply thickness can be explained as follows: Formation of cracks starting from defects in the transverse ply in a range (of unknown depth) at the 0°/90° interphase is suppressed by the presence of a stiff surface ply. If the transverse ply is thick, the influence of this range is negligible, whereas in the case of a thin transverse ply, the influence is large. In the latter case the material apparently contains fewer defects, although in principle the defect distribution can be expected to be similar per unit volume of material. These apparently smaller and fewer defects lead to this second aspect of constraint: the increasing shape parameter at decreasing ply thickness.

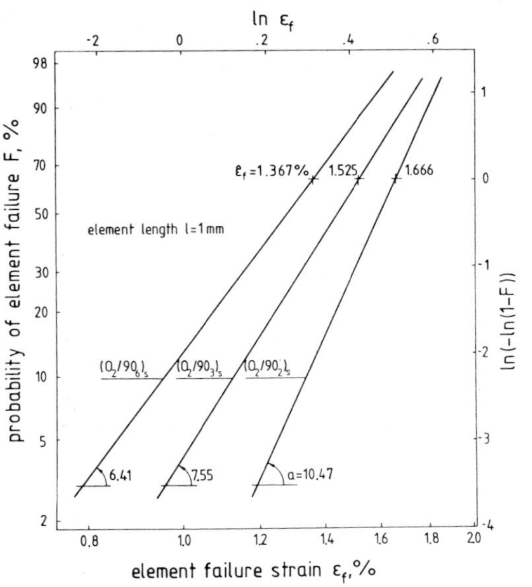

Figure 4.7 Weibull fracture strain distributions of CFRP transverse plies with different thicknesses (T300/914C) [699].

Further, it is clear that the three Weibull distributions intersect in the right hand upper corner of the graph, indicating that a fracture strain is reached which is independent of transverse ply thickness. This is the so-called strain at interphase failure. If this strain at interphase failure is determined according to the procedure outlined in the last section (= fracture strain of strongest element), then a strain at interphase failure of $\varepsilon_{2,int}$ = 1.75% to 1.92% is found for the 90_4, 90_6 and 90_{12} plies. These values deviate only 10% in the range of investigated transverse ply thicknesses from 90_4 to 90_{12}.

CFRP system	Prepreg producer	Fibre	Fibre producer
T300/914C	Ciba-Geigy	T300	Toray
T800/PEEK (APC-2)	ICI	T800	Toray
T800/R6376	Ciba-Geigy	T800	Toray
ST3/MY720	DLR	ST3	Toho-Beslon
ST3/MY720, LY556	DLR	ST3	Toho-Beslon
CG43-750/HG9106	Courtaulds	CG43-750	Courtaulds

Table 4.1 Carbon fibre reinforced plastic systems treated in this section

The effect of constraint is not similar in all CFRP materials. The results given in Figure 4.7 were produced with T300/914C, the matrix system of which is rather brittle (fracture strain 1.4%). Investigations have shown that the effect of constraint is less for more ductile materials [713], like T800/PEEK (APC-2). It can be concluded that the constraining effect gives the material quasi-ductility (which makes the material less sensitive to defects). If the resin system, however, is more ductile, the constraining effect is smaller.

At this point the difference in the constraining effects of different fibre reinforced materials has to be mentioned. The model developed is valid for CFRP, however not for GFRP. The effect of constraint in GFRP is of a different nature. [706]. In contrast to CFRP the Weibull distributions for the different ply thicknesses show a parallel shift and no rotation. Thus the constraining effect acts differently in GFRP. Typical, however, for the constraining effect is that the thickness effect can be described with the aid of a Weibull shape parameter a_t (t=thickness), which is clearly smaller than the Weibull shape parameter "a". Crucial in this respect is the influence of defects, which are reduced in effectiveness as a result of the constraining effect. Defects are primarily voids, which have shown to reduce the transverse fracture strain in CFRP dramatically [707]. Voids, however, are less critical in GFRP. The reason is probably the interaction of void-induced stress fields with the strain-magnified stress fields in the matrix. The simple models discussed before [696, 697] predict a strain magnification factor (SMF) locally in the matrix on transverse loading of SMF=6 for GFRP, but only SMF=3 for CFRP (V_f=0.60). The stress concentration factor of a cylindrical void in an isotropic homogeneous material is 3. Thus in GFRP the strain magnification factor is much larger and thus probably dominates the fracture process, whereas in CFRP the stress concentration factor due to strain magnification and voids is comparable. This is no explanation, but only an indication of which aspect is important in terms of transverse cracking and in what direction future investigations should be performed.

Effect of Increased Temperature on Transverse Fracture: Recently, cracking of the transverse ply in a $0/90_4/0$ laminate of T800/R6376 was investigated in the temperature range of -100 °C to +100 °C [704]. Before presenting the results, let us first consider which aspects play a role in this investigation. As the specimens were cut from a single plate, matrix and fibre volume and defect distribution can be considered to be similar for all specimens. Testing at different temperatures thus only leads to a change of the following parameters. At increasing temperature:

- the matrix ductility increases (the modulus decreases)
- the fibre matrix bond strength changes (probably decreases).

The influence of these parameters on the Weibull distributions is given schematically in Figure 4.8 for the extreme temperatures of -100 °C and +100 °C. It is clear that the onset of cracking occurs at 100 °C at higher fracture strains, whereas at larger crack densities the fracture strain is smaller. This can be explained with the model developed before. Transverse cracking is in this case positively influenced by the increase in matrix ductility, but negatively by the reduction of interphase strength at increasing temperature. This is shown schematically in Figure 4.8. Increased ductility especially causes increased fracture strain for the first occurring cracks (defect-controlled failure). Near the top of the Weibull distribution the increase in strength is small (as shown by the intermittent curve), because in this range the fibre/matrix bond strength dominates over the relative small defects. For this reason a decrease in bond strength changes the Weibull distribution particularly in this top range, finally leading to the Weibull distribution as indicated for the test temperature of +100 °C.

The Weibull distributions at all testing temperatures were evaluated in terms of strain at first failure (FPF, first ply failure) and in terms of the strain at interphase failure (= the fracture strain of the strongest element). These results as well as the thermal strain in the transverse ply are presented in Figure 4.9. From this figure it is clear that up to RT the positive effect of increasing matrix ductility (increasing matrix fracture strain) leads to an increased strain at first ply failure. From RT to 60 °C, however, there is a strong decrease in interphase strength which overrules the positive influence of increased ductility leading to a reduction of the strain at first ply failure. Thus the model presented is able to separate the conflicting influences of transverse fracture strain increase through increased matrix ductility and transverse fracture strain decrease through interphase strength decrease.

Influence of Fibre Surface Treatment on Transverse Cracking: At the end of the production process the carbon fibre is given an oxidative treatment in order to improve the fibre/matrix bond strength. The influence of such a fibre surface treatment on surface physics and chemistry and on mechanical properties has been

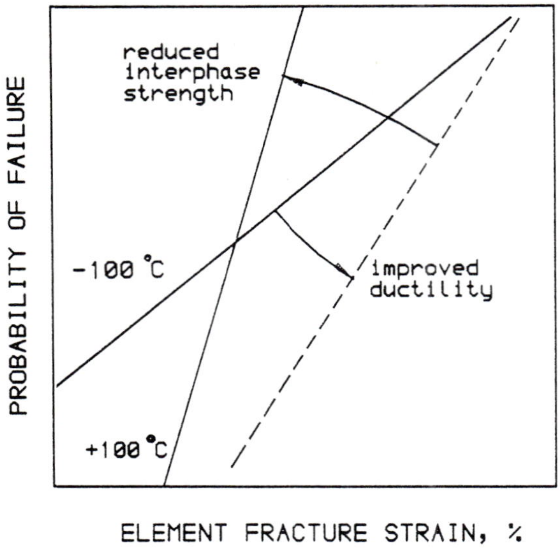

Figure 4.8 Influence of increased test temperature on Weibull distribution for transverse cracking through improvement of ductility and decrease of strength of interphase.

investigated. The fibres (the intermediate modulus fibre CG43-750) were delivered by Courtaulds with four different levels of fibre surface treatment designated 0% (untreated), 10%, 50% and 100% in four different rolls of prepregs with the resin system HG9106. Laminates with the stacking sequence $0/90_4/0$ were produced according to the advised cure cycle. Transverse cracking was measured with the technique described above [714]. The specimens with untreated fibres showed transverse cracking (and partly delamination between 0° and 90°) only as a result of the thermal strain, which was measured to be $\varepsilon_2^t = 0.45\%$. On mechanical loading further cracking occurs, as is presented in Figure 4.10a. A state of saturation of transverse cracking is, however, about to be reached for the 0% treated material. The materials with 10% and 50% treated fibres show at larger fracture strains a small deviation from the determined Weibull distributions (straight lines), indicating that the interlaminar shear strength is being approached. The initial Weibull distribution for the untreated material (cracking due to thermal strain) is estimated roughly with the aid of results from specimens cut from plates cured at 100 °C (which had a lower number of thermally induced cracks because of the lower thermal strain). The specimens with 10%, 50% and 100% treated fibres show increased transverse fracture strain. At increasing fibre surface treatment the Weibull distributions are shifted about parallel to larger fracture strains. Thus not

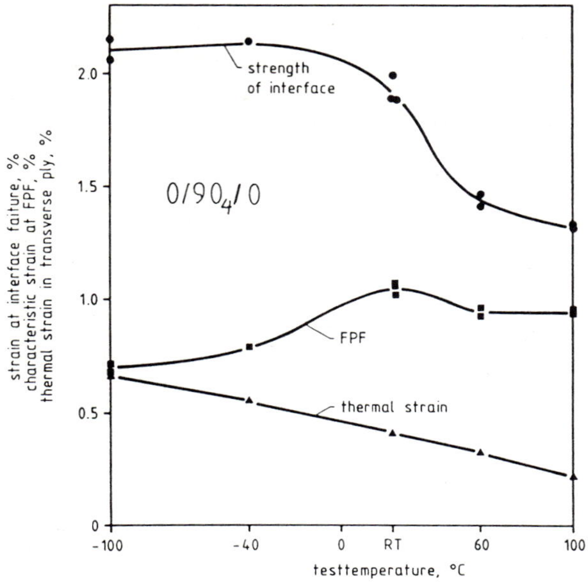

Figure 4.9 Characteristic fracture strain at first ply failure (FPF), strain at interphase failure and thermal strain in the transverse ply of a 0/90$_4$/0 laminate as a function of test temperature.

only is the strain at interphase failure increased with increasing surface treatment level, but also the strain at first ply failure. This indicates that due to the surface treatment defect size and number also are reduced. The reason for this is the improved wettability of a fibre as a result of the fibre surface treatment. The void content as a result of improved wetting was reduced from 1.2% measured on laminates produced from the 0% treated material to a value of 0.5% for the 100% treated material. Thus according to the model described above the influence of interphase strength increases, and the influence of reduction in defect size and number can be explained according to Figure 4.10b. The reduced defect size and number lead to an increase in transverse fracture strain in the tail of the distribution, but not in the top, since in this region the strength of the interphase dominates. Improved bond strength thus mainly influences fracture strain in the top of the distribution, whereas in the tail the effect of interphase strength increase is smaller, since fracture in this range is defect dominated. The overall effect is a shift of the total Weibull distribution with the shape parameter as being effected by the actual defect distribution.

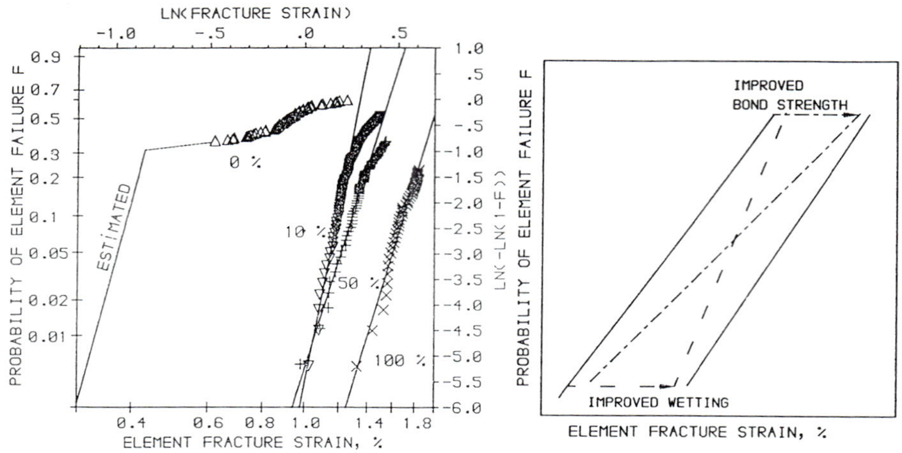

Figure 4.10 a) Transverse fracture data and resulting Weibull distributions for CFRP with four different levels of fibre surface treatment (0%, 10%, 50%, 100%) [714].
b) Influence of fibre surface treatment on the Weibull distribution for transverse fracture through improved wetting and through improved bond strength [714].

4.2 Fatigue Strength of Polymer Matrix Composites

Fibre reinforced composites have a superior behaviour when compared to metals, especially under tension-tension fatigue loading. However, fatigue failure has as well to be expected. In metals the endurance limit is in the range of 40% of the static fracture stress, while in composites about 60-80% of the static fracture stress have to be reached in each consecutive load cycle before fatigue failure can occur.

In Figure 4.11 the various factors influencing the fatigue behaviour of composite laminates are schematically summarized. The fatigue stress is normalized with the static fracture stress σ_0 and plotted versus the number of load cycles.

Fatigue behaviour is improved with increasing anisotropy of the laminate. That means the larger the relative amount of 0° plies, the less fatigue sensitivity has to be expected (Figure 4.11a) [715, 716]. However, the composite constituents (fibre, matrix, interphase) also influence fatigue behaviour. The higher the Young's modulus E_f of fibres, the lower the fatigue sensitivity [717]. An increasing fracture strain, ε_F, of the fibres has the same effect. The higher the fibre fracture strain, the higher also the fatigue life. But matrix properties can also have a positive effect

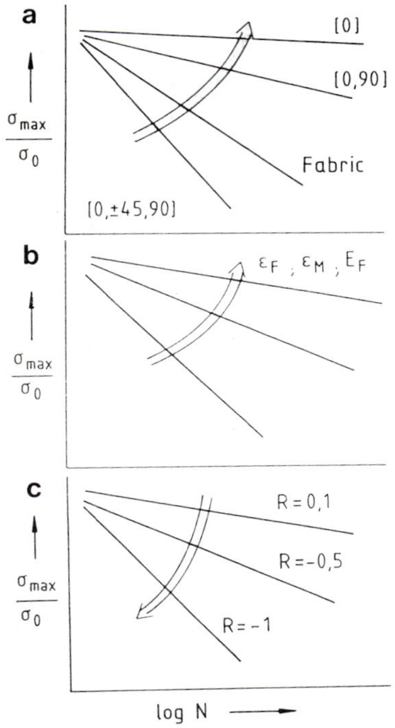

Figure 4.11 Schematic of the various influencing factors to the fatigue behaviour of fibre reinforced composites.
a) Influence of increasing anisotropy towards advanced composites.
b) Composite constituents, increasing ductility of fibre and/or matrix. Increase of fibre Young's modulus.
c) With increasing compressive stresses in load cycle, fatigue life is drastically reduced.

on fatigue behaviour; especially higher matrix fracture strains lead to an increase in fatigue life [717]. The influence of the fatigue load sequence is shown in Figure 4.11c. Due to portions of compressive loads ($R< 0$), fatigue life can drastically be reduced [718, 719].

Fatigue Damage Mechanisms: In the previous sections the mechanical properties and especially transverse ply cracking under static loading conditions have been discussed. Under fatigue loading conditions similar damage as well as other mechanisms occur. The various damage mechanisms, caused by the fatigue loading, are summarized in Figure 4.12 using the example of a cross-ply laminate. The most common damage mechanisms are [715, 720]:

- the formation of transverse cracks
- longitudinal cracks along load carrying 0° fibres
- edge delamination
- internal delaminations at intersections of transverse and longitudinal cracks
- fibre fracture of 0° fibres due to kinking and bridging of longitudinal cracks
- fracture of 0° fibres in the vicinity of transverse crack tips
- fibre/matrix debonding as a result of shear stresses.

The parameters influencing transverse cracking under static conditions have been discussed before. Under fatigue loading conditions the number of transverse cracks is additionally dependent on the number of fatigue load cycles.

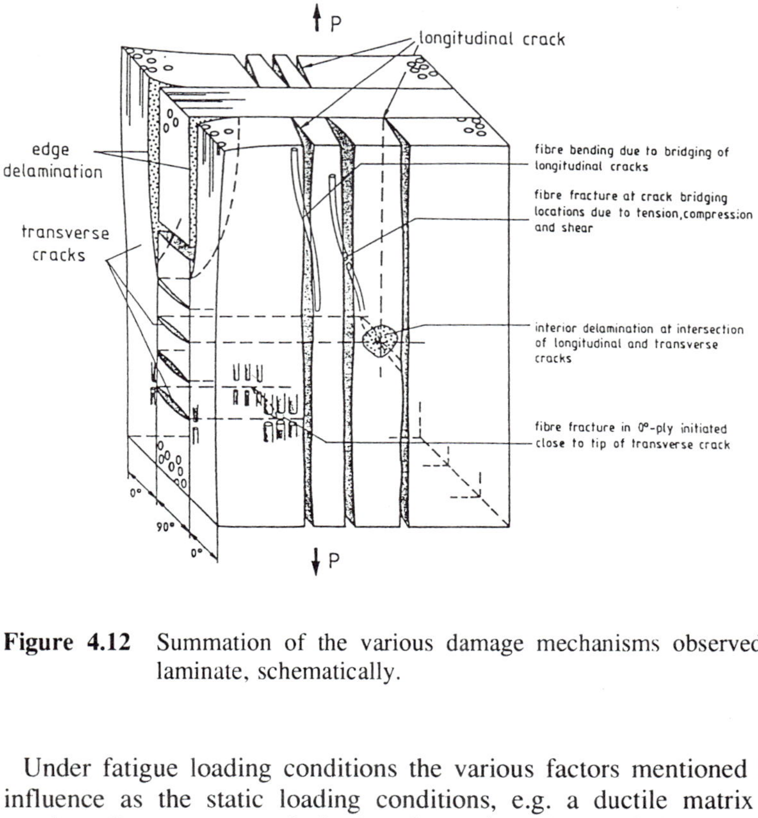

Figure 4.12 Summation of the various damage mechanisms observed in a cross-ply laminate, schematically.

Under fatigue loading conditions the various factors mentioned have the same influence as the static loading conditions, e.g. a ductile matrix decreases the number of transverse cracks both under static and under fatigue loading conditions. Using ductile matrix systems [717], increasing the fibre/matrix bonding and influencing the internal stresses of a laminate [721] are all appropriate measures to suppress transverse ply cracking. It has especially to be mentioned that using thick laminate plies is in all cases extremely bad, as they tend to form cracks in the early load stages. The fact that thick laminate plies form fewer transverse cracks than do thin plies is irrelevant, because the few but broader cracks more easily lead to secondary effects, such as the formation of delamination or the initiation of fibre fracture.

Sensitivity to delamination is mainly dependent on geometric factors; the laminate stacking sequence determines the amount of edge stresses (especially tensile stresses in the thickness direction, peal stresses). The laminate stacking sequence can be chosen such that under an applied tensile stress, the stresses in thickness direction at the free edges are compressive, which means that under static as well as fatigue loading conditions delaminations cannot be initiated. Under fatigue loading the sensitivity to delamination is additionally dependent on the maximum load level in each load cycle. Further factors which influence the delamination growth are matrix properties and fibre/matrix bonding [717]. A ductile matrix and good fibre/matrix bonding can successfully suppress the initiation of delaminations. Also, the formation of longitudinal cracks is mainly dependent on matrix properties, fibre/matrix bonding and laminate stacking sequence. Under fatigue loading conditions again the fatigue load level is the main influencing factor. Under static loading conditions longitudinal cracking can only very seldom be observed [722].

Fibre cracking (fracture of 0° fibres) can normally only be observed if previous transverse or longitudinal cracking has occurred. The dependence of 0°-ply fibre fracture on the formation of angle ply cracks seems to be more important than the hitherto assumption of fibre fracture being dependent on the statistical strength distribution of the single fibres. Failure of fibre reinforced coupons is, therefore, to be expected, if locally a big number of fibres which are close together, or fibre bundles, fail, so that the neighbouring fibres can no longer carry the load being transferred. This means that fibre fracture is mainly dependent on the previous longitudinal and transverse cracking. Longitudinal and transverse cracking, however, is dependent on those factors which were previously described. Therefore, the main factor influencing fibre fracture is the sensitivity to formation of longitudinal and transverse cracks [715, 722, 723].

This leads to an interesting remark. Under long term loading conditions, one could suppose that because of the low loading level, only minor damage can be expected (which allows a test coupon longer life), but in reality the total amount of damage is higher than at loading on a high loading level, which means lower fatigue life [724]. This can be corroborated by two facts. The residual stress and the remaining stiffness in a specimen both have much lower values than in those cases where the specimens were loaded with shorter numbers of load cycles at higher load levels. The largest damage of a specimen is reached at or after the largest possible fatigue life. An example is given in Figure 4.13, which shows an X-ray radiograph of a cross-ply laminate fatigued close to final failure.

Figure 4.13 shows an extremely high damage density with a big number of transverse and longitudinal cracks, edge and interior delaminations.

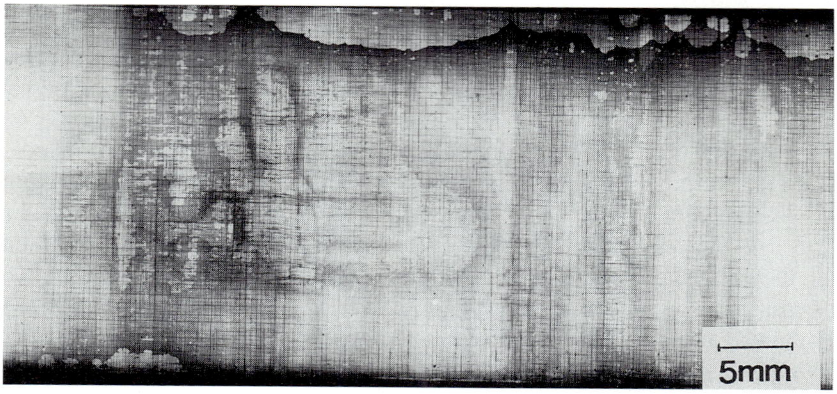

Figure 4.13 X-ray radiograph of a cross-ply laminate. Radiograph taken a few cycles before final failure [715].

Extreme is the comparison of specimens which failed under tensile static, respectively fatigue loading. The specimen which failed under tensile loading has a relatively smooth crack appearance, what means that damage is concentrated in a small area alongside the crack. The specimen which failed under fatigue loading conditions has lost its inner structure along the entirely loaded specimen length, therefore, after fracture it has a brush-like appearance.

Comparing the fatigue life of laminates with a ductile vs. brittle matrix systems, increasing matrix ductility leads to an increase in fatigue life (see Figure 4.14). The main reason for the increase in fatigue life of laminates with ductile matrix systems can be explained by the delay in damage development. Also, the damage density is less pronounced than in the case of brittle matrix systems. Why laminates with ductile matrix systems fail will be explained in the following.

In Figure 4.15 is given the stress condition ahead of a transverse crack tip. σ_1 is the stress in loading direction and σ_2 the stress transverse to loading direction. τ_{12} is the shear stress between the 0° and 90° ply of the laminate. In the case of a ductile matrix, no branching and splitting of the transverse crack into the loading direction occurs. Fibres in loading direction directly ahead of the crack tip will, therefore, be loaded, additionally to σ_1 stresses, by the stress concentrations. In the case of a brittle matrix system the shear stresses cannot be transferred from the 90° ply into the 0° ply, and longitudinal splitting occurs. This splitting favours longitudinal crack growth. The stress concentration ahead of the crack tip is

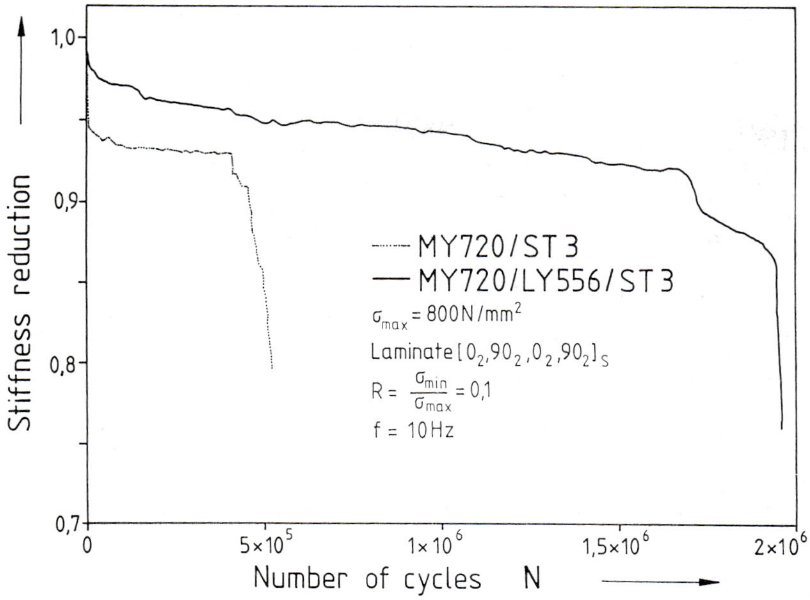

Figure 4.14 Stiffness reduction of cross-ply laminates. Comparison of a brittle (MY720) and a ductile (MY720/LY556) matrix system [717].

reduced. The result is that fibre fracture occurs in a ductile matrix system just ahead of the crack tip. In a brittle matrix system fibre fracture more often occurs some distance away from the transverse crack tip [725].

Figure 4.15 also suggests an explanation for the influence of good, or bad, fibre/matrix bonding on damage development. In the case of good fibre/matrix bonding, stress concentrations ahead of the crack tip cannot be reduced by crack branching and the crack propagates straight through the specimen. In the case of poor fibre/matrix bonding a decohesion of the matrix from the fibre occurs and the stress concentrations due to the transverse crack cannot immediately act on the fibre. It is, therefore, misleading to believe that a good composite material also has extremely good fibre/matrix bonding. In reality a compromise has to be found between good and poor bonding, so that in the case of damage, stress concentrations can be reduced due to fibre/matrix debonding.

Fibre reinforced composites have very complex damage development essentially different from that in metallic materials. While in composites various damage mechanisms such as matrix cracking, delaminations, fibre fracture, and

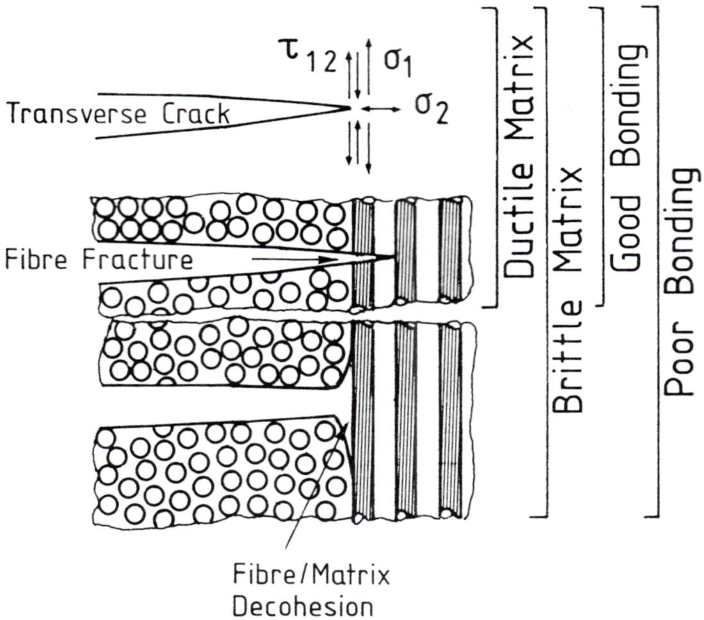

Figure 4.15 Stress distribution at a transverse crack tip and the resulting damage in neighbouring 0° plies. Comparison of ductile and brittle matrix systems and, in analogy, good and poor fibre/matrix bonding.

fibre/matrix debonding occur simultaneously, in metallic materials damage is mainly the initiation and propagation of single cracks. This is schematically shown in Figure 4.16 [726]. During loading, both in composites and in metals, damage increase can be observed. In fibre reinforced composites transverse and longitudinal cracks initiate relatively early in fatigue life. Their number increases with increasing fatigue loading and additionally delaminations and fibre fracture can be observed, distributed over the entire specimen. In metallic materials damage initiates relatively late. First, the dislocation density has to be increased before crack initiation can be observed. Very soon a single crack forms which propagates throughout the entire specimen. During this crack propagation the damage rapidly increases, leading to final failure. This means, due to this single crack formation damage is concentrated only on a small area of a metallic specimen. Due to this single crack propagation in a metal the total damage is lower than in a composite material, in which damage is distributed over the entire specimen.

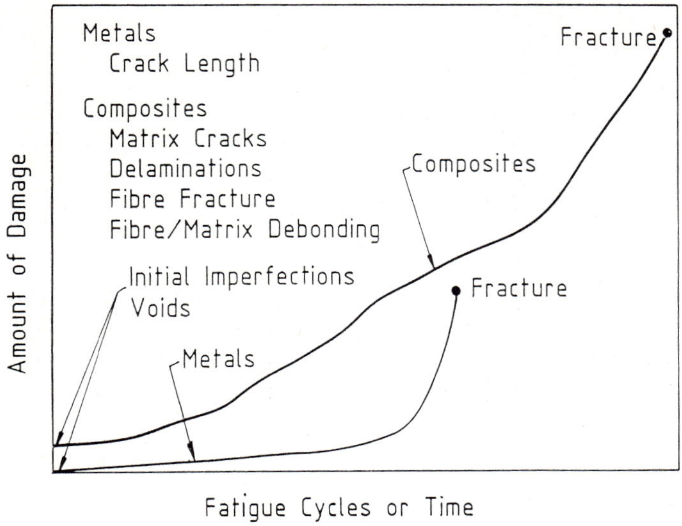

Figure 4.16 Comparison of damage development in metals and composites [726].

5. Materials Characterization and Life Prediction

5.1 Microstructural and Microanalytic Methods in Materials Science

H.J. Dudek

A short review of microscopic, diffractometric and spectroscopic methods and their application in investigating the microstructure of materials is presented, and the physical background, the kind of informations obtainable on materials and the limitations of the microstructural and microanalytic methods are discussed.

Principles of Measurement

The microstructure of materials is always determined according to the same principle: the material is "illuminated" by a primary excitation, the primary excitation experiences an interaction with the elements of the microstructure, in doing so the primary excitation is modified and a secondary radiation is emitted. Information on the microstructure of the material will be obtained by a suitable analysis of the modified primary and secondary radiation, Figure 5.1. The microarea of the materials will be of primary interest; therefore it will be necessary to influence the incoming or emitted radiation by lenses so that an enlargement of the object under investigation is obtained. The microstructure will only be visible inasmuch, as information due to interactions in the radiation coming from the object are included. Structures of the material not undergoing interactions with the primary radiation cannot be seen and will not be recorded by the experiment.

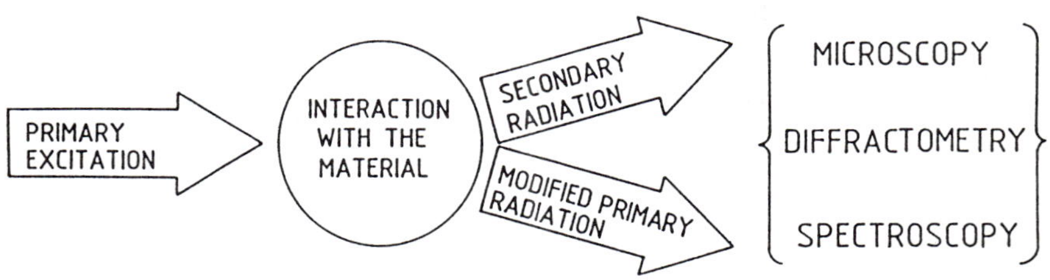

Figure 5.1 Schematic representation of the principles of investigation of the materials using microanalytic and microstructural methods

Primary Excitations

The kind of primary excitation has to be adjusted to the dimensions of the analyzed elements of the microstructure. In order to identify a chemical element in the material, e.g., the energy levels of the atoms have to be analyzed. In this case, primary radiation has to be chosen which has a high interaction probability with electrons located at atomic levels. This will be successful only by a primary excitation which has dimensions (wave length) comparable to those of the atoms. As primary excitation for an analysis of the microstructure of materials, electron beams, ion beams, electromagnetic waves (light and X-rays), as well as statically and dynamically electric and magnetic fields are used.

For reasons of particle-wave duality one is forced to include in the discussion both images of radiation-matter interaction. The particle image of radiation is successful in spectroscopical considerations, the wave image is indispensable for diffractometry. In the particle image primary excitation is defined by the intensity distribution function of energy E and momentum m: $I = I(E,m)$, in the wave image by the wave amplitude A, frequency ν and phase Φ: $I = I(A,\nu,\Phi)$. In the following discussion, for simplicity the particle image will mainly be used.

During interaction with the matter the properties of the incoming radiation are modified. For interpretating the microstructure this modification has to be measured as accurately as possible. This will be possible with sufficient accuracy only if the primary excitation has a high intensity, a small energetic and angular distribution width and a high stability in space and time. Such quality primary radiation is presently obtained by high technical expense, ambitious sources, high electronic stabilization or the application of monochromators.

The following discussion concerns mainly electron beams as primary excitation sources. If another primary excitation is involved, the interaction of the radiation with matter will be different, but the argumentation is similar.

Interaction of the Primary Excitation with Matter

Interactions: The primary excitation is able to interact with the components of the microstructure in different ways (for electron-matter interaction see [727-733], ion-matter interaction [734, 735] and X-ray-matter interaction [736-738]). In the particle image one has roughly to distinguish between elastic and inelastic interaction. Elastic interaction leads to a change in the momentum of the primary radiation; inelastic interaction is connected with a change in energy. The possible interactions are determined by the structure of the material and the kind of primary radiation. The architecture of atoms, with electrons being localized at energy levels, can be probed by excitations and ionizations of these levels. The energy for an excitation or ionization is withdrawn from the primary radiation. The primary particle with energy losses can be identified in the energy distribution of the modified primary radiation as intensity maxima; the more interactions take place the higher the intensity of this "loss-line". The difference between the energy of exciting radiation E_0 and the energy of the particle after an inelastic interaction E_1 is the amount ΔE transferred to the atom. If the surplus of energy ΔE is transferred to an electron of the atom, then ΔE consists of portions which the electron needs to leave the atom, the material and the amount of kinetic energy carried. The latter is continuous and not quantized. The maxima of intensity losses in the energy distribution of the modified primary radiation are therefore only sharp at the low loss side, which relates to the lowest amount needed to ionize the atom, while at the high loss side due to the kinetic energy part a continuous decrease of intensity is observed (absorption edge), Figure 5.2a. The position of the sharp edge in the kinetic energy distribution relates to the energy level of the atom and is used for its identification.

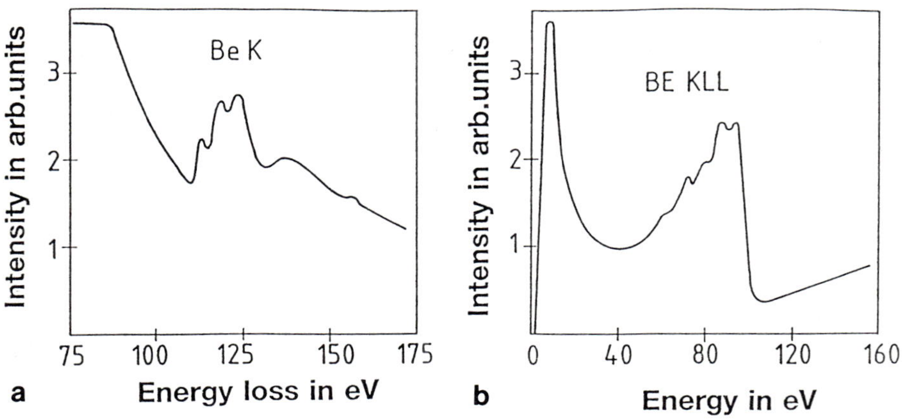

Figure 5.2 Examples of spectra for the modified primary (a) and secondary radiation (b).
a) Absorption K edge of a 50 nm beryllium foil measured in the TEM using EELS at a primary excitation of 120 keV with a beam divergence of 10 mrad (according [739]).
b) Be KLL Auger electron spectrum obtained with a CMA with a primary excitation of 2 keV and a resolution of 0.1 % [740]. The high energy side of the Auger line is sharp due to the characteristic energy levels of the atom; the low energy side is modulated by the interaction of the Auger electrons with plasmons (for the Auger line shape interpretation, see [741]) In the energy range between 0 and 50 eV the emission of "true" secondary electrons is observed.

In a similar way the excitation of energy levels of atoms, electron waves (plasmons) and lattice vibrations (phonons) by the primary beam will be identified in the energy distribution of the modified primary radiation by the electron energy loss spectroscopy (EELS) and the high resolution electron energy loss spectroscopy (HREELS).

In case the primary radiation is considered a wave, this wave interacts with the matter by diffraction. The diffracted wave is then the modified primary radiation, the interference phenomena of which follow the Laue and Bragg laws and are used to identify the crystallographic structure of the material.

Interaction Cross Section and Mean Free Path: In the particle image the interaction of the radiation with the components of the material can be described by an interaction cross section. The cross section is the probability of a special particle-matter interaction. It can be considered to be the area of the electron or atom. A primary particle crossing this area will suffer the considered interaction. The total cross section σ_{tot} describes all possible events, the differential cross section σ_{dif} only some of them. The differential scattering cross section, e.g., describes the scattering in a spatial direction $\Omega(\Phi,\Theta)$ and in an increment space angle $d\Omega$, while the total scattering cross section is given by all possible scatterings

$$\sigma_{tot} = \int \sigma_{dif}(\Phi, \Theta)\ d\Omega \tag{5.1}$$

σ depends on the kind of primary particle p, the considered interaction r, the atomic number of the scattering atoms Z of the matter, on the energy E of the primary radiation and the space angle Ω in which the primary particle is scattered $\sigma^r_p = \sigma^r_p(E,Z,\Omega)$. If the number of scattering centres in volume unity is N, then for the intensity of the primary radiation I(E,m) the number of scattering events in volume dV in the direction $\Omega = (\Phi,\Theta)$ will be

$$n(E,Z,m,N) = I(E,m) \cdot N \cdot \sigma^r_p(E,Z,\Omega)\ d\Omega\ dV \tag{5.2}$$

In a similar way the ionization cross section $\sigma^i_p(E,Z)$ depends on the primary energy E of particle p and the atomic number Z of the ionized atoms. The number of ionizations in volume unity is given by

$$n_i = I(E,p) \cdot N \cdot \sigma^i_p(E,Z) \tag{5.3}$$

When the interaction cross section is known, the number of ionizations (5.3) and the intensities of scattered radiation (5.2) can be determined. This is important for quantifying experiments. The interaction cross sections are determined experimentally and by computations and are summarized in tables (for EELS quantification, see [742]).

After an interaction according to (5.2) or (5.3), the movement of the primary particle progresses in a new direction or with a new energy, respectively. When the density N of the scattering centres is high enough, after passing a way λ the scattered particle can be scattered again. The mean value λ is called the mean free path of the particle and can be determined from the total scattering cross section according to

$$\lambda(E,N) = \frac{1}{\sigma_{tot}(E,Z) \cdot N} \tag{5.4}$$

Multiple Scattering: Computations (5.2) and (5.3) rarely lead to the desirable quantification of measured intensity, because the single scattering situation is realized only in a few cases. Commonly, the density N of the scattering centres is high and the sample under investigation thick enough, so that the primary particle will suffer many interactions before leaving the sample. In every interaction the particle will lose a different amount of energy. Leaving the sample, a particle with few interactions has lost the information on the structure of the material and contributes only to the background of the signal. The information from equations (5.1) to (5.4) can be applied in the case of very thin objects, as are used in transmission electron microscopy (TEM).

Relaxations in the Material: The energy transferred to the atom or crystal results in a relaxation process which differs with the kind of excitation. In the case of ionization of an atomic core level, the acquired energy can by emitted in two ways: a) by emission of an X-ray quantum or b) the energy is transferred to another electron of the atom which then is emitted as an Auger electron. The X-ray quantums and the Auger electron have a characteristic energy determined by the energy levels taking part in this emission, Figure 5.2b. Especially, it is independent of the kind of exciting primary radiation (electrons, ions, X-rays). When X-rays are used for primary excitation, the electron leaving the core level of the atom has an energy determined by the atom level taking part in the emission. This electron is called a photoelectron, and by determining its energy the emitting atom is identified.

The radiations emitted as a result of the primary excitation of the atom - such as the characteristic X-rays, the Auger electrons and the photoelectrons - are called secondary radiations. The modified primary radiation and the secondary radiation are the most important information carriers on the microstructure of the material.

Influence of Neighbour Atoms: The modified primary radiation and the secondary radiation are influenced by the neighbourhood of the emitting atom. The bonding state with the surrounding atoms has an influence on the valence and core level energy. When the bonding energy of an electron in the core level is higher due to chemical bonding with neighbouring atoms, the ionization of this level will be connected with a higher energy loss of the primary radiation, and therefore the appropriate peak in the energy distribution will be found at higher energy losses. Similarly, the relaxation process in the atom following ionization will be influenced by the energy shift caused by the bonding, and the emitted secondary radiation will have an energy displaced according to that energy shift. After calibration both effects can be used to identify of the chemical bonding.

There is another effect of neighbour atoms on the emitted radiation. The secondary radiation emitted from the atom is partly reflected from neighbouring atoms, and the reflected and emitted waves interfere. The modified primary and the secondary radiation are therefore influenced also by the number, arrangement and distances of the neighbouring atoms. These interference effects influence the cross section for ionization and appear in some energy distributions of the modified primary radiations as an intensity oscillation. They can be used to identify the atomic distances and the coordination numbers (see EXAFS, EXELFS).

Energy Transfer to the Lattice: The interaction of the primary radiation with the crystallographic lattice results in a generation of lattice vibrations, which leads to a temperature rise in the material [743]. In many cases this effect causes artifacts in the sample and has to be avoided, but it may also be used for materials characterization, e.g. in acoustic microscopy [744].

Fundamental Methods of Measurement

Decoding Information: The information included in the modified primary radiation and in the secondary radiation can be used for an understanding of the structure of the material when the energy and momentum distribution of the radiation will be converted into a spatial or time distribution. A localization of the interaction region in the material is possible, if the microstructure is enlarged by optical processing of the radiation. Optical enlargement and conversion of the energy and momentum distribution into a spatial or time distribution is carried out by an application of electrical and magnetic fields. The enlargement of the microstructure is called microscopy, the decodation of the energy distribution, spectroscopy and the conversion of momentum distribution, diffractometry. (In most cases a conversion of the energy and momentum distribution into a spatial distribution is performed; examples for a conversion into a time distribution are energy dispersive X-ray spectroscopy (EDX) and time-of-flight spectroscopy (TOF).) Microscopy, spectroscopy and diffractometry are therefore fundamental methods for investigating materials.

Optical enlargement of the microstructure can be achieved in two different ways: a) in the integral method by parallel "microscopical" imaging and b) by a sequential scanning of the object with a focused primary beam and an electronic correlation of the emitted signal from every point of the object with its position on the sample. Similarly, spectroscopical and diffractometrical analysis can also be performed sequentially or in parallel. In nearly all cases where the parallel method can be applied it is much more effective than the sequential one. This is due to the possibility to collect the signal simultaneously from a large spatial, angular or energy distribution, at the same time and results in a larger signal intensity acquired during a time interval compared to that obtained using sequential detection. The introduction of electronic parallel detection and the processing of signals by a computer has therefore in many regions of science resulted in a considerable extension of the detection limit and an increase in spatial resolution [745, 746].

The present paper concentrates on the microregion in the sense that the properties of a small region of the material are considered. X-ray diffraction is, e.g., a method which analyzes the microstructure, but in general not in the microregion (for X-ray microscopy, see [747]). Optical enlargement - microscopy - is therefore one of the criteria of the microstructural and microanalytical methods. The interpretation of the energy distribution and the momentum distribution of the signals from the microarea is the domain of microspectroscopy and microdiffractometry, respectively.

According to the discussed assignment into the three fundamental methods, the most important microanalytic and microstructural methods for determinating the microstructure of materials are arranged, see Figure 5.3. Some of them will be discussed briefly in the following sections.

	SPECTROSCOPY	DIFFRACTOMETRY	MICROSCOPY
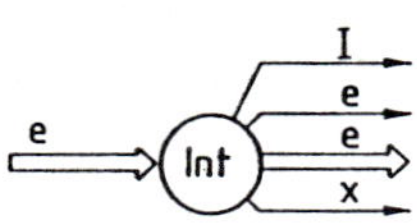	ESID AES EELS, HREELS EPMA	ED, CBED, RHEED	SAM, SEM CTEM, STEM, HREM
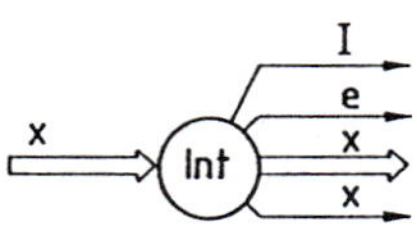	LAMMA XPS, UPS EXAFS	XPD, ARXPS XD	LM, XM
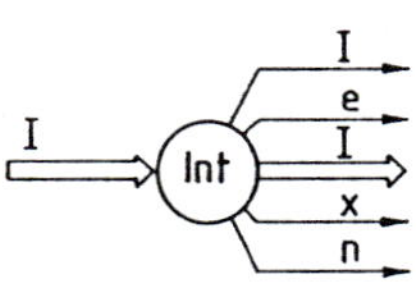	SIMS RBS, ISS PIXE SNMS		IM, SIM

Figure 5.3 Microanalytical and microstructural methods applied in materials science. The acronyms are explained in the text. e - electrons, X - light and X-rays, I - ions, and n - neutrals.

Kinds of Information: The aim of applying of microstructural and microanalytic methods is to gain as complete an image of the microstructure of the material as possible. The information obtainable by these methods can be summarized as follows:

- Structure: For every phase in the material the dimensions and angles in the crystallographic unit cell, its space group, the atomic number of the atoms, their coordinates and their chemical bonding conditions,
- Morphology: The spatial distribution of matter, especially the structure and distribution of segregations, lattice defects in every phase and the grain and phase boundary distribution and
- Chemistry: The chemical composition of the bulk and surface, relative arrangement and frequency of the phases, grain, phase and interface segregation of chemical elements.

Criteria for Application in Materials Science: The above information can be provided only with methods which fulfil the following "analytical" criteria: a) Spatial resolution: The structures which have to be analyzed are very small, the spatial resolution of the microscopical method must be very high. It has to be distinguished between lateral and depth resolution. Surface analytical methods have good depth resolution, in many cases however only moderate lateral resolution. b) Angular resolution: In diffractometry, the angular resolution in intensity distribution determines the precision of the lattice parameters evaluation. c) Energy resolution: In spectroscopy, energy resolution determines the separability of neighbourhood lines in the spectrum and therefore the separability of elements and chemical states. d) Detection limit: The intensity of spots in a diffractogram or the energy lines in a spectrum compared with the background intensity determines the smallest detectable amount of the element being considered in the sample region analyzed. Finally, the criteria of the natural sciences, reproducibility, accuracy and correctness, decide on the applicability of a method. The methods discussed in this section meet the above criteria to different extents, so that in many investigations of the microstructure of materials more than one method has to be applied.

Present microstructural and microanalytic methods require costly techniques and equipment. To achieve a high signal-to-noise ratio with high spatial resolution, sources with high brightness and small energy distribution are necessary. Field emission sources for electrons, liquid metal ion sources for ions, and synchrotron radiation for X-rays are of increasing importance. For the design of lenses and deflection fields, new concepts, increasing the acceptance angle of the emitted radiation, high stable currents and voltages, and exact mechanical treatment are applied. Signal registration by parallel detection is used together with computer control and recording.

Microscopy

Imaging of microscopically small structures in the materials using electron or ion beams is performed similarly to the methods applied in light microscopy [748]. Some of the important microscopic methods applied in material science are discussed in this section.

Conventional Transmission Electron Microscopy (CTEM) [730, 749-751]: The intensity distribution in the transmission image of a thin object is given by the distribution of the scattering matter in the object (the scattering or absorption contrast). The regions with a larger scattering cross section and higher density in (5.2) and (5.3) result in darker regions in the image. The scattered part of radiation is faded out by an aperture of the microscope limiting the angle of rays contributing to the image. Similarly, in imaging crystalline objects the diffracted beams are stopped by an aperture which results in dark regions in the image, with lattices being in reflection conditions for the primary beam (diffraction contrast). Lattice distortions, such as dislocations, grain and phase boundaries can be made visible by diffraction contrast. The spatial resolution of the CTEM is limited by lens errors to approximately 0.2 nm (for a discussion of the lateral resolution in the CTEM and HREM, see [752]).

Scanning Transmission Electron Microscopy (STEM): Electron beams focused into a small diameter are scanned over a small area of a thin sample and the transmitted intensity is measured by a detector. The detector signal is used to control the brightness of a TV image with a synchron scanned electron beam. The magnification is given by the ratio of the TV area to the area on the specimen scanned. STEM contrast formation is similar to that of CTEM with some modifications due to differences in the ray path and areas of the STEM detector and CTEM aperture. Lateral resolution is limited by the beam diameter whose smallest dimensions are restricted by lens errors. STEM has a large area of application in connection with the spectroscopy of the characteristic X-ray emission (EDX = energy dispersive X-ray analysis) and electron energy loss spectroscopy EELS in the analytical transmission electron microscope (ATEM) [731, 752]. Using field emission sources, beam diameters as small as 1.7 nm (FWHM) with beam currents of 1 nA are possible [753]. With high brightness focused electron beams submonolayer segregations can be detected at grain and phase boundaries using EDX and EELS [754, 755].

High Resolution Electron Microscopy (HREM): Improved lateral resolution in transmission electron microscopy to point resolutions of approximately 0.2 nm [445] allows the imaging of the atomic lattices of the materials [756-761] and of single heavy atoms [762]. Presently, for lattice imaging mostly CTEM is used, but STEM is also applicable [763]. Within the point resolution lattice structures, grain and phase boundaries, point defects and interfaces can be imaged on atomic scale and interpreted using computer simulation of image formation [756, 761].

Imaging of Surfaces: Using scanning electron microscopy (SEM) imaging of surfaces is performed using the same imaging principle as in STEM, however collecting secondary or reflected electrons [764-766]. The lateral resolution of SEM is in the region of 10 nm and is limited mainly by the beam diameter and the multiple scattering of primary electrons in the sample.

The scanning principle is also realized in scanning tunnelling microscopy (STM) and in atomic force microscopy (AFM). In the STM a tip with nearly atomic dimensions is scanned on a surface controlled by piezoelectric crystals and the tunnelling current is measured. In AFM the tip is positioned on the surface by a spring and the deflection of the tip caused by Van der Waales, electrostatic, magnetic and friction forces is measured. The piezoelectric crystals allow horizontal movements with an accuracy of approximately 0.2 nm and vertical movements with 0.01 nm [767] resolution; tunnelling currents are in the region of 1 nA [768]. The movements of the spring in AFM can be measured with an accuracy of 0.001 nm and the forces are in the range of 10^{-13} to 10^{-6} N [769, 770]. Although STM and AFM are very new methods they already have applications in materials science [771].

The imaging of surfaces on the atomic scale is also realized in low energy electron microscopy (LEEM) [772], using CTEM [773-776], in reflection electron microscopy REM [777-779], using STEM [780] and in field emission ion microscopy (FIM) [781].

Ion Beam Imaging: Using ion beams for primary excitation, the imaging of surfaces can be performed applying the scanning principle [782] or the integral imaging principle [734]. As imaging signals, often particular mass numbers m/e are used. In this case the imaging of the surface is connected with information on the distribution of chemical elements, which corresponds to spectroscopy of the surface. As the imaging accompanies surface erosion, taking images at different times results therefore in a three dimensional surface analysis. The application of liquid metal ion sources [783] allows surface imaging with a lateral resolution definitely better than 0.1 μm [734, 735, 782].

Diffractometry

Diffractometry of Modified Primary Radiation: The most powerful primary excitation for diffractometry in a microarea is the electron beam mainly due to its excellent focusing possibilities and, in comparison to ions, small interaction cross section with the matter [784, 785]. Applying parallel illumination, selected area diffraction (SAD) allows diffractometry at microregions not smaller than approximately 0.5 μm with a resolution of $\Delta d/d = 5 \cdot 10^{-3}$ (d=lattice spacing, [727]). In the new generation of transmission electron microscopes, however, convergent electron beams are applied (CBED = convergent beam electron diffraction [731, 752, 786]) with a lateral resolution comparable to the dimensions of the crystallographic unit cell [752]. Using beams with different divergency in relation to the diffraction angle or applying beams "rocking" in different angles a great variety of diffraction diagrams results with information on lattice parameters, space group and sample thickness [752, 786, 787]. The resolution of convergent electron diffraction ($\Delta d/d = \pm 3 \cdot 10^{-5}$ [727]) is comparable with that of X-ray diffraction [752].

In thin samples, inelastic scattered electrons may be reflected at lattice planes, if the Bragg condition is fulfilled, resulting in lines in the diffraction diagram of thin specimens in the TEM. The lines are called Kikuchi lines and can also be used for determinating crystal structure, crystal orientation and texture [727, 752].

In the scanning electron microscope the scanning of the electron beams in an interval (similar to the rocking beam method in CBED), and a correlation of the emitted intensity of reflected and secondary electrons to the angle of the primary beam in relation to the crystal orientation result in an electron channelling pattern diagram (SECP = selected area channelling pattern) [765, 766, 788]. The preferential reflection of the primary electron beam in a low indiced crystallographic direction is caused by the development of standing electron waves (Bloch waves) in the crystal lattice. Standing electron waves with maxima at the position of the atom rows of the lattice develop, if the incident primary radiation has an angle to the lattice plane a little lower than the Bragg angle. Maxima of the Bloch waves between the atomic rows originate, when the incident angle is a little larger then the Bragg angle. Optimum reflection of electrons occurs with Bloch waves at positions at the atomic rows, and channelling diagrams in reflection are observed.

Using slow electrons (a few hundred electron volts) the crystallography of the surface can be analyzed by normal incidence of the primary beam (LEED = low energy electron diffraction) [789]. The crystallography of the surface is analyzed with highly energetic electrons (10 to 100 keV) using convergent beams in the grazing angle incidence with reflection high energy electron diffraction (RHEED) [790-793]. In both cases only near surface regions of the material are investigated. The crystallographic structure of surfaces can be significantly different from the bulk, what is of importance for the interaction of atoms with surfaces, e.g. in the development of interfaces in composite materials (see section "Interfaces"). The surface structures are sensitive to contaminations and therefore diffraction experiments using LEED and REM have to be performed in UHV [790]. The high surface sensitivity of this diffraction method is especially visible in the oscillation of RHEED intensities during molecular beam epitaxy [794].

Diffraction of the Secondary Radiation: The secondary radiation generated in the material may also experience diffraction at the periodic structures of the near surface atoms before leaving the material. The measured secondary radiation shows a modulation of intensity with the emission angle to the crystallographic orientation. Important for the investigation of the crystallographic and valence band structure of the surface near regions [795, 796] is photoelectron diffraction (XPD = X-ray photoelectron diffraction), the angle resolved X-ray photoelectron spectroscopy (ARXPS) [797], and the angular distribution Auger microscopy (ADAM) [798, 799].

Spectroscopy of the Modified Primary Radiation

X-ray Absorption Spectroscopy: X-ray absorption spectroscopy is an example of a method in which the energy of the primary radiation is varied and the modified primary radiation or a secondary radiation is measured after interaction with the material. Using an electron synchrotron high intensity X-ray radiation is obtained, which can be energy modulated and focused into a small area. A classical application of this method in materials science is the determinating the neighbourhood conditions of atoms in the material by EXAFS (extended X-ray absorption fine structure). The EXAFS structures are visible on the high energy side of the absorption edge as an intensity oscillation caused by the interference effects (see section on interactions) [800-804]. EXAFS can be used also for surface characterization when grazing angle incidence geometry is used or surface sensitive signals such as Auger electrons are detected. The method is called SEXAFS (surface-sensitive EXAFS) in this case [805].

Electron Energy Loss Spectroscopy (EELS): The electron energy loss spectroscopy (EELS) performed in the TEM is one of the typical spectroscopies for modified primary radiation [731-733, 752, 806,807]. Thin layers with thicknesses of the order of the mean free path, Eq. (5.4), are investigated. The electrons suffer one or a few interactions in the material. The electron spectrometers are located between the objective and the projective lenses [808], or below the projective plane [732, 733, 806]. Applying field emission sources, the primary beam can be focused into a diameter of 1 to 2 nm [753]. Theoretical considerations show that the lateral resolution for EELS is limited to approximately 0.3 to 0.4 nm [809].

The modulation of intensity found in X-ray absorption spectroscopy at the higher side of the absorption edge (EXAFS) is also visible in the EELS-spectrum. The determination of the neighbourhood using EELS is called EXELFS (extended energy loss fine structure) [810]. Using EXELFS the distances to the surrounding atoms can be determined with an accuracy of 0.02 nm [752], which is important for amorphous materials in which electron diffraction cannot be applied with sufficient accuracy. Structures observed at the lower energies of the absorption edge in the EELS spectrum are called ELNES (energy loss near edge structure). They contain information about the energy density distribution of unoccupied states above the Fermi level [739, 810, 811].

*High Resolution Electron Energy Loss Spectroscopy (HREELS):*The vibrational states of surfaces near atoms can be analyzed by applying low energy electrons of a few hundred electron volt as a primary radiation and electron spectrometers with high energy resolution. The interaction between submonolayers of atoms and surfaces can be analyzed and the initial states of oxidation, nucleation and growth of interface layers can be determined [812].

Spectroscopy of the Secondary Radiation

Electron Probe Microbeam Analysis (EPMA): The electron beam is focused on the surface of a bulk in a SEM [764-766] or on thin layers in the TEM [731, 752] and the characteristic X-rays emitted due to the ionizations in the material are measured using crystal spectrometers (WDX = wave length dispersive X-ray analysis) or energy dispersive detectors (EDX = energy dispersive X-ray analysis). WDX is a sequential method in which energy distribution is measured at one energy interval after the other. In the EDX detectors the incoming X-rays are sorted according to their energy during small time increments. Advantages of the WDX systems are their smaller energy width of the characteristic X-ray lines (a few eV in WDX compared with 140 eV in the EDX system) and an higher signal compared to background intensity.

The discussion of EPMA demonstrates the physical limitations of spatial resolution and the need for mathematical computations to gather information on the sample composition from intensities measured by a spectroscopical method. Primary electrons suffer in the material many scattering events before they are absorbed or emitted from the surface as backscattered electrons. During their zigzag way through the sample the electrons ionize the material in an excitation volume, the diameter of which is independent of the primary beam diameter (for diameters smaller than a few 0.1 μm) and depends on the primary energy and the sample composition. The spatial resolution of the EPMA is therefore physically determined by the radius of the excitation volume down to a few micrometers, Figure 5.4.

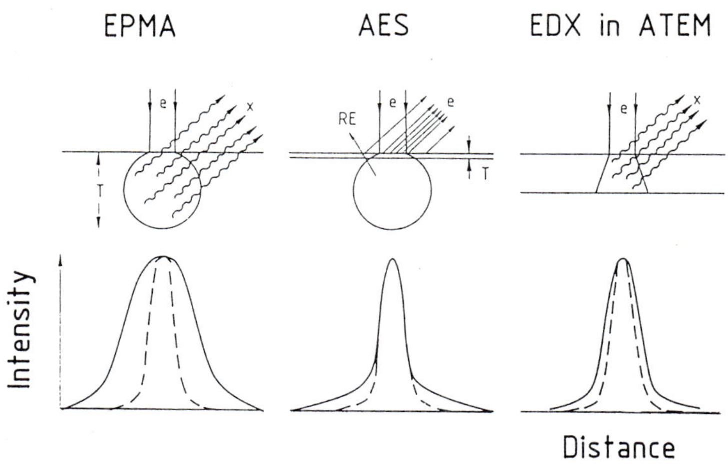

Figure 5.4 Schematic representation of the lateral resolution of three analytical methods using electron beams for primary excitation: EPMA, AES and EDX in the ATEM.

Three corrections have to be applied in the EPMA to determine the concentration of the elements from the measured intensity:

1. Atomic number correction: In a relaxation process the ionizations in the excitation volume lead to the emission of characteristic X-rays. The depth

of X-ray generation depends on the element concentration, which is described by the atomic numbers Z.

2. Absorption correction: The X-ray intensity is absorbed on the way to the surface according to the exponential absorption law $e^{-r \cdot \rho/\chi_x}$ (supposing isotropy of the material, ρ is the density of the material and χ_x the X-ray absorption factor). The intensity generated can be obtained by measuring X-ray intensity correcting for absorption.
3. Fluorescence correction: The X-rays of one element can excite characteristic X-rays of another element. The whole correction procedure is called ZAF correction and has to be performed iteratively.

In the analysis of thin specimens in the TEM using EDX no (or only absorption) correction has to be performed. The lack of an excitation volume will also improve lateral resolution, so resolution is mainly limited by the electron beam diameter (Figure 5.4).

Auger Electron Spectroscopy (AES)[740, 813, 814] : As mentioned above, beside characteristic X-rays in the excitation volume, Auger electrons are also generated. The Auger electrons will also undergo absorption on their way to the surface by an exponential absorption law $e^{-r \cdot \rho/\chi_A}$ (considered an isotropy of material). But the Auger electron absorption is much larger than the absorption for X-rays: $\chi_A \ll \chi_X$ Therefore the mean free path (5.4) is very short, only a few nanometres. The Auger electrons generated inside the material cannot leave the surface without suffering more than one inelastic scattering. After leaving the surface the Auger electrons generated inside the material have lost an undefined amount of energy and contribute only to the background. Only Auger electrons generated near the surface will show characteristic energies for the emitting atoms. Auger electrons are therefore characteristic for the surface composition; the AES is a surface analytical method. In combination with an ion sputtering of the surface a characterization of thin surface layers by so called AES depth profile analysis is possible [815].

X-ray Photoelectron Spectroscopy (XPS)[734, 735, 784, 785, 816] : The energy of photoelectrons is comparable to that of Auger electrons. They also originate from the topmost atomic layers of the surface. XPS is applied in materials science for the analysis of the bonding states of surface atoms, especially for determinating the chemical state of elements on powders [817], surface oxidation [818], corrosion [819, 820], and surface passivation [821] in interface analysis, and for the analysis of polymer surfaces [822]. Imaging of surfaces illuminated by X-rays using photoelectrons [823, 824] or focusing the primary X-rays [825, 826] allows analyzing and imaging small areas by XPS. Lateral resolution of a few μm is achieved.

Ion Spectroscopy: Applying ions as primary radiation, the most successful spectroscopical methods in materials science are secondary ion mass spectroscopy (SIMS) [734, 735], ion scattering spectroscopy (ISS) [735, 827] and secondary neutrals mass spectroscopy (SNMS) [828]. In Rutherford backscattering spectroscopy (RBS) [735] and in particle induced X-ray emission spectroscopy (PIXE) [735] high energetic ions (in the MeV region) are applied. ISS and RBS are spectroscopies of the modified primary radiations. For the generation of secondary ions (and for a post ionization of emitted neutrals) laser beams can be used (LAMMA = laser ablation microprobe mass spectrometry). Ion spectroscopic methods have mostly a very low detection limit, allowing the detection of elements in the ppm and even in the sub-ppm region.

Combination of Different Analytical Methods

In some experiments special information is obtained only when more than one interaction effect acts in combination. An example is X-ray photoelectron diffraction (XPD) [797] where the generated X-rays suffer diffraction at the near surface crystallographic structures. Another example is the ALCHEMI (atom localisation by channelling-enhanced microanalysis) method [829, 830]. The localization of the standing waves (channelling) at special atomic rows of the lattice causes an enhanced ionization of this atom row and an increased emission of X-rays. Using this effect atom rows in the lattice can be identified also in cases when the concentration of this element is low.

In many materials science problems, one analytical method alone does not sufficiently characterize the material. In this case more than one method has to be applied. The analytical transmission electron microscope (ATEM) is an example of equipment which combines microscopic (CTEM, STEM), spectroscopic (EDX, EELS) and diffractometric (CBED) methods to investigate the same sample position. In surface science often more than one method is applied (AES, XPS, SIMS, ISS, etc.). Combining methods sometimes may have the disadvantage that the geometry doing not allow optimal analysis conditions.

In many cases, especially using surface analytical methods, a materials science experiment has to be performed "in situ" under the UHV conditions needed for the analytical method [831]. The "in situ" experiments allow a study of the fundamental processes of oxidation, corrosion [821], grain boundary segregation, the importance of grain surfaces in the powder metallurgy and the conditions of adhesion and reaction in the interface of composite materials.

Limits for the Extraction of Informations

The reproducibility, accuracy and correctness of the microstructural and microanalytic methods are limited by a number of factors, (Figure 5.5) [832]. The primary excitation I(E,m) is influenced by the instabilities of the source. The focusing lenses modify the primary radiation and reduce the intensity due to lens errors. In consequence, spot size and thus spatial resolution is limited.

Every preparation method modifies the sample being investigated. The preparation of a surface by fracture, grinding, polishing, or ion etching changes signif-

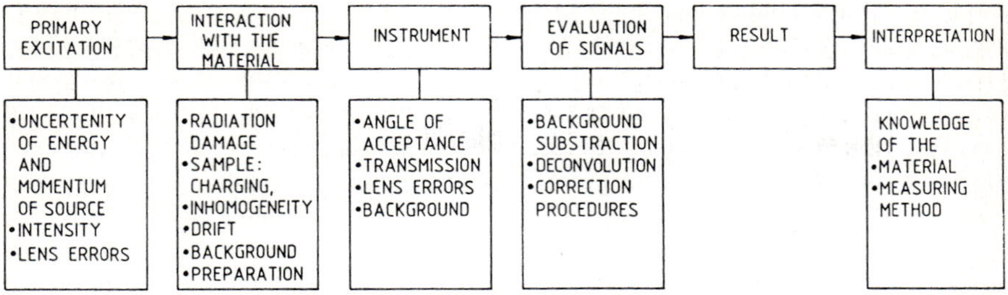

Figure 5.5 Limitations to the acquisition of information using microstructural and microanalytic methods

icantly the state of the surface, which is quite different to the state of the bulk material. The interaction of the primary excitation with the material modifies not only the primary radiation, but also its introducing energy, momentum and charge into the sample, which results in heating, charging up and radiation damage of the analyzed volume. Secondary effects are introduced by diffusion, precipitation and phase transformations. As a consequence unwanted deviation from the state of the material on investigation, sample drift and distortions of the signal can occur.

The instruments - the microscope, spectroscope and diffractometer - record only a limited range of the emission angle, only a part of the incoming radiation is transferred from the entering aperture to the escape aperture (transmission of the instrument). The transformation of the energy and momentum distribution into a spatial or time distribution is performed with electric or magnetic fields having errors, which results in a spread of the signal and an increase of the background radiation.

For the evaluation of results, theoretical assumptions have to be made which are only an approximation of reality and introduce further errors. Finally, the interpretation depends on knowledge of the material and the analytical method applied.

Comparison of Methods

The investigation of the microstructure of materials in the micro- and nano-area is always carried out according to the same principle: the material is illuminated by a primary excitation and the modified primary radiation as well as the secondary radiation are analyzed. The evaluation of the emitted radiation is performed by fundamental methods for determining the microstructure of materials, that is by microscopy, spectroscopy and diffractometry. Information can only be obtained from those elements of the microstructure which interact with the primary radiation. Information on the microstructure in the modified primary radiation and in the secondary radiation is coded as an energy and momentum distribution. This is converted into a spatial or time distribution by spectrometers or diffractometers. Using different primary excitations different analysis methods are obtained. They differ by the different interaction cross sections and according to the differences of the instruments which are adjusted to the requirements of the primary radiation. The information gained includes the parameters and the space group of the crystallographic unit cell, the atomic number of the elements and their bonding state in every phase, the lattice defects and interfaces and the phase distribution. The different analysis methods deliver this information with different accuracy, reproducibility and correctness, with different energy, angular and spatial resolution and with different signal-to-noise ratios. As an example the spatial resolution for different methods is compared in Figure 5.6. The success of mathematical quantification of the analysis methods and the error sources are different.

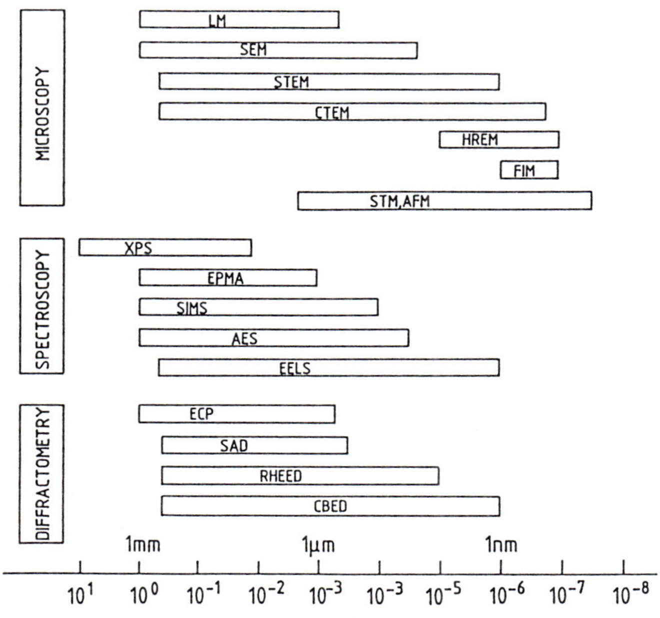

Figure 5.6 Comparison of the lateral resolution of different microscopic, spectroscopic and diffractometric methods

Microstructural and microanalytic methods are in a rapid state of development. New methods have been introduced such as STM or AFM, and established methods are being improved by new sources of primary excitation (electron field emission, liquid metal ion sources), new analysis methods, e.g. time-of-flight mass spectroscopy, and by application of parallel recording and computer signal acquisition. The methods presently available are able to solve nearly every materials science problem, the limitations are mainly personnel and financial resources.

5.2 Fatigue and Fracture of Metallic Materials

5.2.1 Random Load Fatigue and Life Prediction

H. Nowack and H. Döker

Structural components of aircraft are subjected to variable amplitude loading. In most cases it is not possible to keep the stresses in the complete structure so low that they fall below the endurance limit of the construction materials, and it has to be proved that the component will endure the expected service loading in a safe and reliable manner. Such a proof can be performed experimentally in special tests or analytically by the application of cumulative damage hypotheses [834]. These two directions do not actually represent alternatives, but rather complement each other.

Another essential viewpoint regarding primary components of aircraft is whether they can be routine inspected during service life or whether they are internal parts of structures, as for example of a wing, and are concealed by other structural elements. In the first case the design can be "damage tolerant". That means that it has only to be assured that macrocracks which are present from the beginning will not become critical during service. This can be proved by a macrocrack propagation analysis. In case of structural components, which cannot be routine inspected during service life, it has to be ensured by a "safe life design" that the component endures the service life without significant damage. In the majority of cases, e.g. for a highly loaded wing root structure, it is appropriate to require that the technical crack initiation stage is not exceeded during the design life of the component. This contribution focusses on methods which consider either the total life or the technical crack initiation life. In order to increase safety, safe life components are often realized as redundant structures. However, this approach is not always possible, as for example for pivot bearings of wings or for discs of aircraft turbine engines.

As mentioned before, experimental and analytical methods are used to prove the fatigue design of components. Because tests are costly and time consuming, there is a strong tendency to prefer life calculation hypotheses based on cumulative damage. In the past various types of cumulative damage hypotheses have been developed. They can be subdivided into three categories: category A type hypotheses, which consider the total life, category B type hypotheses, which consider the technical crack initiation, and category C type hypotheses, which consider the formation and the propagation of small cracks separately.

A most essential problem is that damage accumulation under random loading occurs nonlinearly. Despite the extremely large amount of research work which has already been invested, the physical and mechanical reasons for nonlinearities in damage accumulation are still not widely understood [834]. That is a main reason for the large number of cumulative damage hypotheses which presently exist. Some typical hypotheses will be briefly described in order to facilitate the selection of a suitable hypothesis for a special practical problem.

Component Development: Significance of Cumulative Damage Hypotheses

As indicated in Figure 5.7, the development process for fatigue critical components passes through several stages [835]. The first stage involves the design. Several groups of input data and boundary conditions have to be considered: the expected loading, the design principle, existing design rules, the integration of the component into its structural environment, governmental regulations, etc. Then a material with suitable monotonic and cyclic properties has to be selected. Afterwards FEM calculations are performed in order to identify the critical areas of the component or structure. After a brief check whether the behaviour of the component under static loads is acceptable, fatigue performance is evaluated using a cumulative damage hypothesis. If the fatigue properties are not sufficient, the design has to be improved on the basis of fatigue life calculations. Then a prototype is built and tested. If it turns out that fatigue performance is not sufficient, modifications become necessary and the process is repeated. The damage hypotheses are normally improved according to the experience from the prototype testing. After the component has passed the laboratory tests, further tests, where the component is integrated into larger structural units, may become necessary. At the end, the complete structure is tested (full scale test). If the observed behaviour is satisfactory, production of the component can be started. During the first in-service phase further deficiencies may become visible. Very fast modifications need again cumulative damage hypotheses. The effect of changes during the in-service operation of components can also be evaluated by means of cumulative damage hypotheses.

Nonlinearity of Damage Accumulation Under Random Loading

As already mentioned, fatigue damage under variable loading accumulates in a nonlinear way. Jacoby showed about 20 years ago that the fatigue lives from random tests and from program tests, which were derived from the same aircraft loading spectrum, differed by a factor of more than 8, with the program tests on the unconservative side [836]. In order to show load sequence influences on cumulative damage, it is worthwhile to select simple histories as two-step block loading or sequences with peak overloads and/or underloads, and to compare the fatigue life as observed in these tests with those lives which are calculated assuming linear damage accumulation. Further it is important to consider the total fatigue life not only as one stage, but to specially account for the individual stages, such as the technical crack initation stage or substages as the precrack stage and

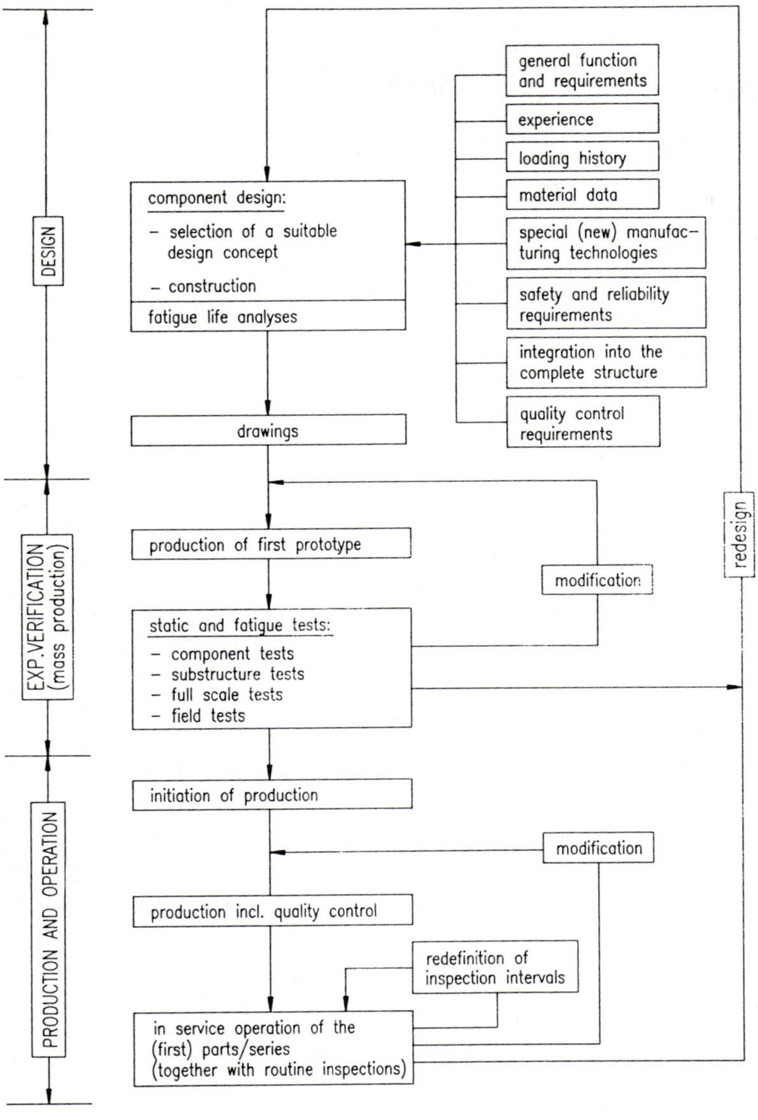

Figure 5.7 Development process of fatigue critical components.

the small crack stage. Figure 5.8 shows corresponding results [837]. The linear damage calculation differs significantly from the actual behaviour.

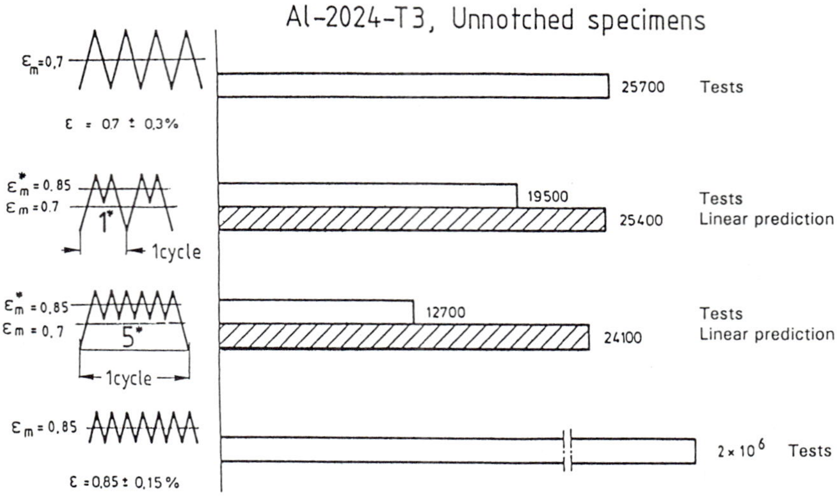

Figure 5.8 Number of cycles to crack initiation (crack length of 50 μm) compared to linear predictions.

Different Categories of Prediction Methods

Figure 5.9 gives a survey of presently available cumulative damage hypotheses, and which stages of the total fatigue life they consider. Three main categories of cumulative damage hypotheses can be distinguished.

Category A Type Hypotheses for Total Fatigue Life: The eldest and simpliest cumulative damage hypothesis, the linear cumulative damage hypothesis of Palmgren and Miner, does not account for load sequence effects. Its prediction capability is rather limited [838, 839]. Nevertheless, engineers still use it because of its simplicity. Another reason is that a linear damage calculation represents a well defined calculation basis, where modifications which become necessary to improve the prediction of cumulative hypotheses, can be related to. Many attempts have been made to improve the prediction capability of the simple linear hypothesis. One way is to modify the input data for the calculations. These are normally constant amplitude SN-curves from tests with the components themselves or with specimens with equivalent notches. The SN-curves are prolonged into the range below the endurance limit with the same or a modified slope. An important possibility to improve the predictions is to include trends which are observed in actual fatigue tests into the predictions in the form of relative damage calculations. The sum of damage increments reached at the end of life is no longer 1, but chosen according to the test results. Relative damage hypotheses have attained the greatest

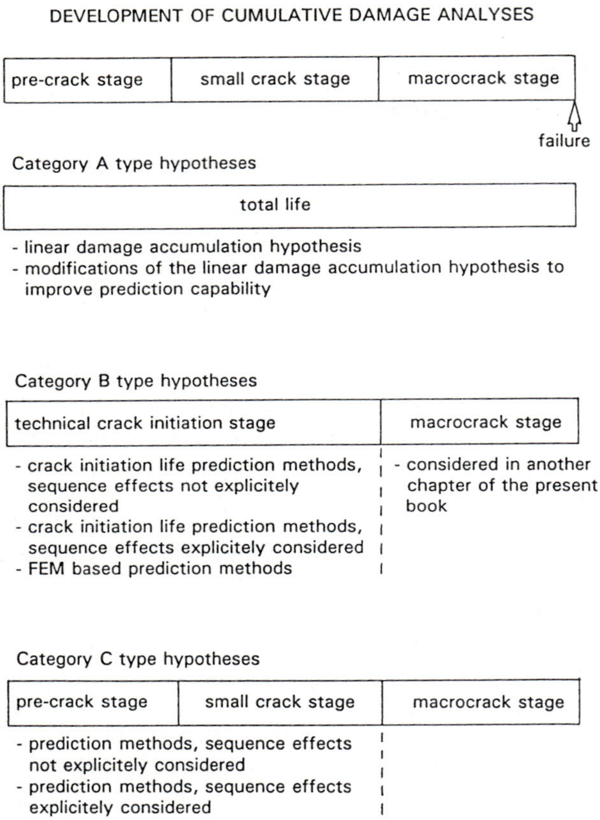

Figure 5.9 Different categories of cumulative damage hypotheses.

practical significance. Meanwhile, relative cumulative cycle ratios for a large number of practical cases have been evaluated. It is further important that industrial design engineers are normally concerned with a limited number of components, so empirical knowledge can develop rather quickly.

Other further developments of the linear hypotheses that try to specially account for load sequence effects are (ordered according to expense for the generation of the basic input data): the double-linear damage rule, where systematic two-step tests have to be performed [840]; damage calculations using Gaßner lines (Gaßner lines represent life curves which are derived with standard random sequences, as for example TWIST) and the U_0-concept, where the results from standard random sequences are used and where the most damaging level of the loading spectrum is specially accounted for [841, 842]; and, further on, specialities based on the material behaviour and on constructional details of the individual component. The accuracy of the predictions can be significantly improved by these engineering prediction methods.

Independent of progress made in the further development of category A type hypotheses, there remain serious objections: the expenses for the generation of SN-curves with actual components are extremely high. From a physical viewpoint, it is further not acceptable that all the individual stages of fatigue life are treated as one, irrespectively of the fact that the physical and mechanical mechanisms in the individual stages are different.

Category B Type Hypotheses for the Technical Crack Initiation Stage: To overcome some of the physical and mechanical shortcomings of category A type hypotheses, category B type hypotheses have been developed, especially also at DLR [843]. These hypotheses are mainly based on the companion specimen concept of Crews and Hardrath [844], where it is assumed that the fatigue behaviour at stress risers (notches) can be evaluated from the fatigue behaviour of unnotched specimens of the same material, if the same local stress/strain sequence is applied. The Crews and Hardrath concept has been transferred into an analytical analysis scheme, which comprises three modules, one, where the material stress/strain response is analytically evaluated using basic material input data from data banks, a second one for the consideration of notches in the component, and a third one for calculating crack initiation life from strain-crack initiation life curves from strain controlled constant amplitude tests with unnotched specimens. The hysteresis loops which close during variable amplitude loading are taken to determine individual damage increments. These are summed up until a damage sum of 1 is reached. In contrast to category A type hypotheses, these "notch analysis concepts" need basic monotonic and cyclic material data from unnotched specimens, and the relevant notch factor. Meanwhile, more than 1000 sets of material data are available in data banks. DLR developed computer programs to apply the notch analysis concepts both for medium size and personal computers.

Because the crack initiation life prediction methods described above are highly attractive from an engineering viewpoint, they have been extensively studied during recent years. However, it turned out that the deviations between predictions and actual test results were not significantly smaller than those for category A type hypotheses. One reason is seen in the fact that the stress gradient at notches is not considered. Therefore, fatigue notch factors are often used instead of elastic stress concentration factors. Another important reason for the differences observed between actual and predicted crack initiation life data has again to be seen in the fact that the damage increments as caused by the individual hysteresis loops are linearly summed up. That means that sequence effects are not specially accounted for.

In order to improve the situation, special "damage parameters" have been developed (compare [834]). At DLR Hanschmann developed a semiempirical damage parameter which led to improved crack initiation life prediction [843].

For some time, multiaxial load inputs at the fatigue critical locations of structural components have explicitly been considered. Two of these new analysis concepts shall be briefly mentioned. They both incorporate FEM calculations. Hanschmann et al. [840] proposed a prediction method for such highly complex components as the wing root structure of aircraft. An (elastic) FEM code is used to determine how the external load at the component is transferred to the integration points of the FEM mesh at the fatigue critical location in terms of σ_x, σ_y, τ_{xy} stresses. Multiple load inputs at a component are considered by superposition. The resulting variations of the σ_x, σ_y, τ_{xy} stress components are then combined to equivalent stresses by an equivalent stress criterion, which is specially derived for fatigue stresses. For a consideration of σ_x, σ_y, τ_{xy} stress components, which may all occur at variable frequencies and amplitudes, a special algorithm has been developed. This new concept is pragmatic and straightforward.

The other method of crack initiation life prediction for arbitrarily loaded complex aircraft structures has been developed by Ott [845]. The evaluation of fatigue damage is based on the plastic energy accumulated at the fatigue critical location. Plastic energy and fatigue damage are incrementally determined. In contrast to the method mentioned before, which remains still flexible in essential points, Ott's method is determined in every detail.

If the present development state of category B type methods is seen in more general terms, those methods have become highly attractive from an engineering viewpoint. Artificial residual stresses which may be introduced at the fatigue critical locations of components by shot peening, rolling, etc., or two-phase materials, for example, with a thermally harded surface layer, can also be considered [834].

Various aircraft engine components exposed to elevated temperatures operate in the low cycle fatigue (LCF) range, as for example, compressor and turbine disks. Especially at those locations where the buckets are later placed, a significant notch effect is present. Category B type hypotheses can also be applied to *fatigue at elevated temperatures*; however, changes have to be made especially within that module where the stress/strain path is evaluated and in the crack initiation life evaluation modules. Suitable constitutive equations, which consider the time dependent changes of the stress/strain path, have to be used. It has been shown that separate treatment of the damaging effect of time independent and time dependent parts of stress/strain hysteresis curves (one based on conventional strain-crack initiation life curves and one based on creep curves) to predict fatigue behaviour does not lead to acceptable results. Various procedures have been proposed to consider the simultaneous action of the time independent and time dependent effects on fatigue damage, such as the strain range partitioning (SRP) method of Manson and Halford, the frequency separation method of Coffin, and a method based on plastic work in the positive stress range proposed by Ostergren [846]. Today the SRP method has attained the greatest significance. Recently Agatonovic et al. suggested a revised version of the SRP method [847]. This modified SRP method has to be integrated into the crack initiation life evaluation module of the convenient notch analysis concepts. In case of high temperatures a careful examination is necessary to deermine whether the microstructural mechanisms remain unchanged between the case for which the prediction is made and the tests from which the basic input data are derived.

Category C Type Methods for the Precrack and the Small Crack Stage: It is reasonable to assume that the category B type methods as they were considered in the previous sections should generally yield realistic results if their use is limited to the precrack stage.

During the past decade much emphasis has been placed on considering of the behaviour of the small crack stage. The initiation and propagation of small cracks is highly controlled by the microstructure. Figure 5.10 gives a survey of the most important influences on small crack behaviour. If unnotched specimens are considered first, small cracks can start at extrusions and intrusions at the specimen surface and then follow persistant slip bands. They also may initiate at incoherent particles which fail as the amount of slip accumulated at these particles exceeds a certain level, or they may start at grain boundaries at the specimen surface or at triple points in the interior. If the cracks then grow transgranularly, the next phase or grain boundary represents an obstacle, which may cause the crack to grow slower or even stop completely. Those cracks which overcome these obstacles grow further and reach, later on, the status of macrocracks. Small cracks usually grow faster than macrocracks. An important reason is that crack closure behaviour differs for small and large cracks. Small cracks are assumed to be fully open at the beginning.

In Figure 5.10 the behaviour of small cracks at notches is also indicated. In this case the cracks usually start in a highly elastically and plastically deformed notch field. For realistic predictions of small crack behaviour, all microstructural and micromechanical details important for crack extension should be accounted for. These are tilting and bifurcation of the cracks, interaction with phase and grain boundaries, multiple cracking within a grain, crack closure effects, initiation of several small cracks which interact as colinear or parallel cracks, coalescence of small cracks, etc. Some of these influences. can already be quantitatively estimated [834]. However, the presently available approaches are far from being applicable in engineering practice. As long as the stresses and propagation rates of the small cracks remain small, linear elastic fracture mechanics principles can be applied. El Haddad et al. proposed correcting the instantaneous crack length by l_0 to account for the small crack behaviour derived from the threshold stress intensity factor of the material and from its endurance limit. The l_0 concept is quite practical in application [848].

As plasticity effects become important, elastic-plastic fracture mechanics based methods have to be applied. The J-integral is normally preferred. Dowling extended the use of the J-integral to cyclic conditions [849]. He proposed cyclic J-integral equations for unnotched and notched specimens. In recent times Vorm-

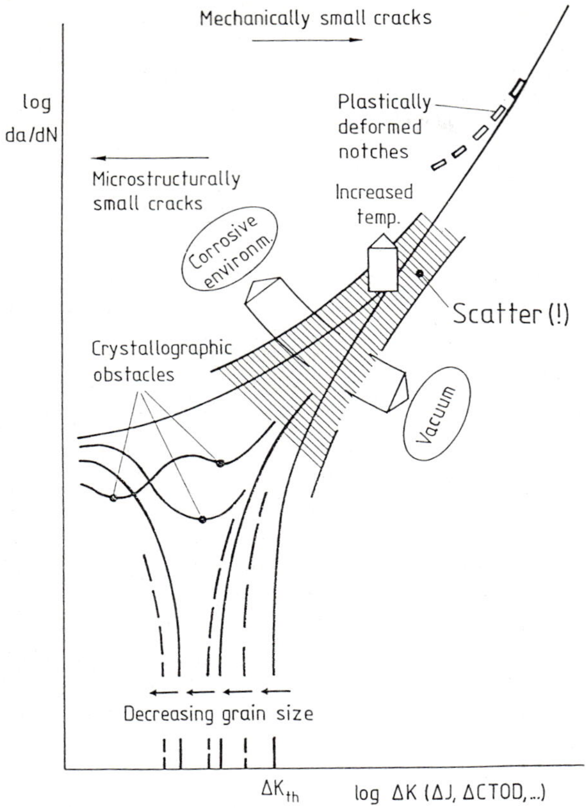

Figure 5.10 Important characteristics of small crack behaviour.

wald suggested a further refined J-integral approach for small cracks and transformed it into the shape of a "damage parameter" which can be introduced into category B type methods [850]. This new concept is applicable to random loading and led to satisfactory results. At DLR further developments were performed in two directions. The behaviour of cracks in a macroplastically deformed environment - a situation which is similar to that for small cracks at notches - was investigated and correlated to a simplified type of cyclic J-integral, which was derived from the gross section stress-COD curves and which correlated the observed crack behaviour in a linear manner [851]. The goal of the second development was to establish a bridge between the cyclic J-integral and convenient macrocrack propagation prediction methods for variable loading conditions as based on the linear elastic stress intensity factor. This was achieved by the deri-

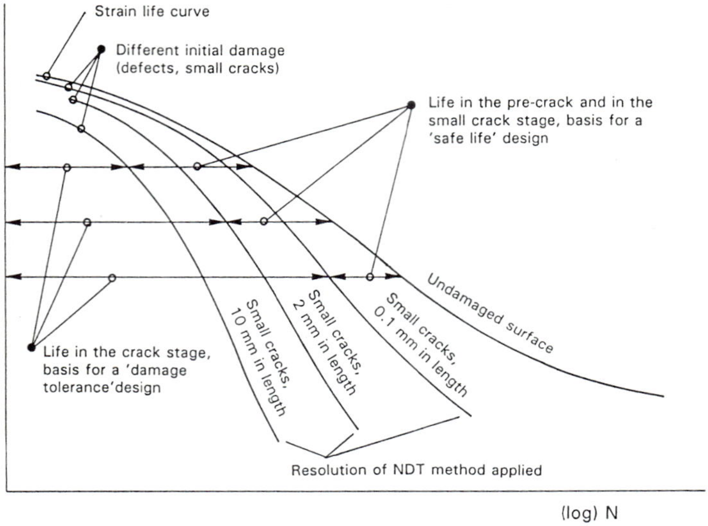

Figure 5.11 Relevance of the individual stages of the fatigue life for the design of aircraft components.

vation of "equivalent stress intensity factors" at the end of every half cycle [852].

Concluding Remarks

Random loading leads to significant nonlinearities in fatigue damage accumulation in all stages of the fatigue process. Conventional category A type cumulative damage hypotheses do not specially account for the individual stages of the fatigue process and their specific microstructural and mechanical mechanisms. Independent of that, the relative linear cumulative damage hypothesis is presently widely used. The category B type hypotheses consider the technical crack initiation stage and the macrocrack stage separately. The significance of the hypotheses for the technical crack initiation stage is rapidly expanding. One reason is that the recent developments operate in combination with FEM codes and enable a direct evaluation of stress/strain and the fatigue response at fatigue critical locations of components. Category C type hypotheses try to consider the physical and mechanical mechanisms in the precrack and in the small crack stage in a detailed manner, however, their development has not yet come to an end. In Figure 5.11 a brief survey of the relevance of category B and C type hypotheses for the structural design of aircraft is given. It can be seen that the initial crack size, which can reliably be detected by the applied NDT methods, is of predominant importance.

5.2.2 Physical Reasons for the Existence of ΔK_{eff}

G. Marci

Ever since Elber [853] proposed his closure concept, there were numerous investigations which led to doubts about closure as reason for the existence of the so-called ΔK_{eff}, namely the reduction of the applied ΔK to an effective one with respect to fatigue crack propagation (FCP). For clarification, closure comprises all phenomena associated with the fracture surfaces of a propagating fatigue crack which in concert act as a wedge between the fracture surfaces. Closure thereby modifies the loading condition. On the assumption that a fatigue crack propagates only if tensile stresses act on the crack front, Elber postulated that the end of the wedging action of closure would be the start of the effective part of the loading cycle ΔK. This partitioning point of ΔK is commonly denoted as K_{op}.

$$\Delta K = K_{max} - K_{min} \tag{5.5}$$

$$\Delta K_{eff} = K_{max} - K_{op} \qquad \text{if} \quad K_{op} > K_{min} \tag{5.6}$$

$$\Delta K_{eff} = K_{max} - K_{min} \qquad \text{if} \quad K_{min} > K_{op} \tag{5.7}$$

Equations (5.6) and (5.7) are not quite correct, if Elber's preposition

$$\Delta K_{eff} = 0 \qquad \text{no FCP} \tag{5.8}$$

$$\Delta K_{eff} > 0 \qquad \text{FCP} \tag{5.9}$$

is accepted. This is due to the existence of a closure-free threshold ΔK_T in the stress intensity range K_{max} - K_{op} [854]. Then, Eqs. (5.6) and (5.7) have to be written as:

$$\Delta K_{eff} = (K_{max} - K_{op}) - \Delta K_T \tag{5.10}$$

$$\Delta K_{eff} = (K_{max} - K_{min}) - \Delta K_T \tag{5.11}$$

In the scientific literature, the compressive stress fields in the cyclic plastic zone and associated phenomena are often referred to as closure, too. Schijve and Broek [855] explained their results from overload experiments on Alclad 2024-T3 by compressive stress fields produced in the plastic zone. Yet, it is not clear by which phenomena cyclic plastic deformations should be prevented from producing additional damage during the lower part of a loading cycle. The assumption that fatigue cracks cannot grow under compressive stress fields around their crack front does not suffice.

Such doubts about the closure concept by Hertzberg et al. [856] were based on the fact that "artificial" closure was not reflected in the appropriate way by FCP rates. Similarly, determination of the partitioning point K_{op} by means of a "growth versus no-growth" criterion [857, 858] furnished values agreeing with semi-empirically determined values, but not with measured closure values.

The present work proposes a plastic zone model for the plane strain region of parts or specimens to account for the ΔK_{eff} and FCP behaviour. The model of the plane strain region of a fatigue crack is essentially identical to that proposed originally by Schijve [859] to explain striation formation. Since the existence of ΔK_{eff} is based on different physical reasons (compressive hydrostatic stress fields) than those for closure, the partitioning point equivalent to Elber's K_{op} is denoted as "K_G".

Plastic Zone of a Fatigue Crack under Plane Strain Condition

Consider the simplified model of a plastic zone associated with a propagating fatigue crack as shown in Figure 5.12. The area ABCCBA is taken to model the active plastic zone of a propagating fatigue crack; the areas ABDE are the regions plastically deformed due to passage of the active plastic zone during crack propagation (plastic wake zone). The contour EAAE is the crack profile. The distance "X" between the crack tip and the end of the active plastic zone is to be on the order of the propagation rate per cycle. Let it be assumed that this plastic zone model is loaded under K_{max} loading.

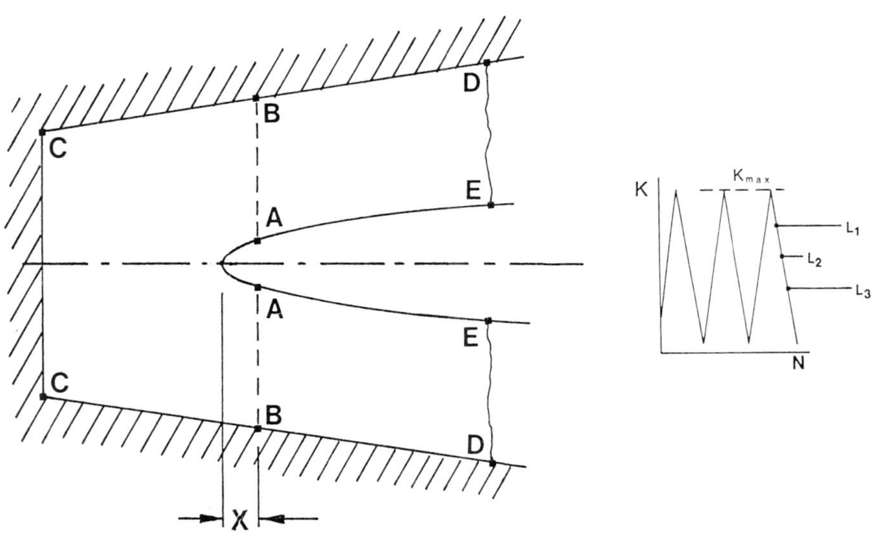

Figure 5.12 Simplified model of the plastic zone of a propagating fatigue crack

Such a model, as shown in Figure 5.12 was fabricated as follows. Three rigid wooden bars were hinged as shown in Figure 5.13 at the positions indicated by the arrow heads. Artificial rubber was poured between the bars and two parallel glass plates, and cured in a "soft" condition so as to model the active plastic zone (black region in Figure 5.14). On top of it, artificial rubber was poured and cured to a "hard" condition (grey region in Figure 5.14). The later layer was to model the wake zone. After curing and removal of the glass plates, the artificial crack was inserted by cutting out the crack profile. A slice (80 mm wide) of the plastic zone model as shown in Figure 5.12 was produced.

Figure 5.13 Mechanical model using rubber to simulate the effect of specimen unloading on deformation behaviour of the plastic zone

Per definition of the active plastic zone, plastic deformations across the lines AB cannot occur. This was modeled by choosing the "hard" cured condition of the artificial rubber for the wake zone. The fabricated plastic zone model was clamped between two thick, greased PMMC plates. As was mentioned earlier, the fabricated model should correspond to the plastic zone under K_{max} cyclic loading conditions. During unloading, the active plastic zone is squeezed between the elastic

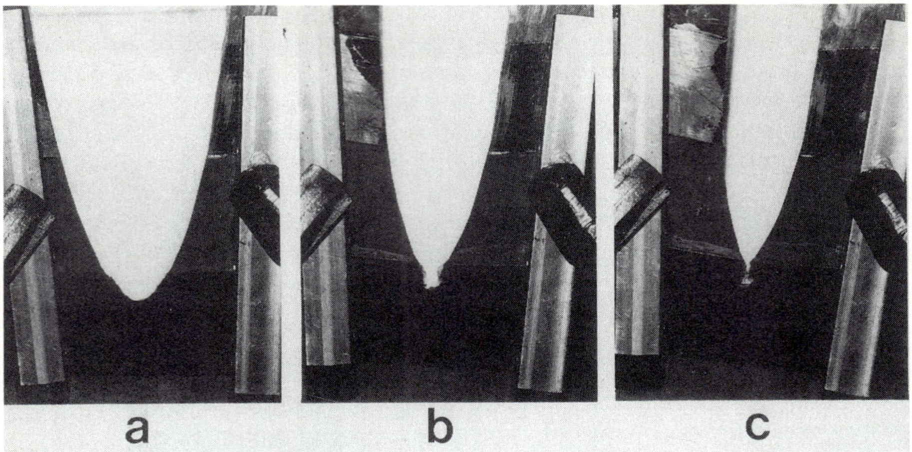

Figure 5.14 The response of simulated fatigue plastic zone (see Figure 5.12) during unloading (a) through (c), respectively

region (modeled by the wooden bars) and the wake zone (hard cured rubber). The unloading was modeled by applying a compressive load "L" at the free ends of the wooden bars (Figure 5.13). Figure 5.14 shows three pictures of the fabricated plastic zone model under increasing compressive load modelling of the unloading cycles. The pictures were photographed through one of the PMMC plates. Since the fabricated model could deform only between the rigidly held, parallel PMMC plates, plane strain conditions were modeled. The compressive loads relative to K_{max} are schematically shown in the insert of Figure 5.12. Figure 5.14a - c correspond to compressive loads L1, L2, and L3. As can be seen from Figure 5.14, the crack profile in the active plastic zone collapses symmetrically to the macroscopic fracture plane. This is due to the volume constancy of plastic deformations. The material in the active plastic zone - stretched during the loading cycle - has to be squeezed out into these area where no constraining material prevents it. The elastic region around the active plastic zone, as well as the boundaries to the plastic wake zone, prevent (constrain) the active plastic zone from relieving the volume constancy requirement across these interfaces. Therefore, material is squeezed into the crack opening, as can be clearly seen in Figure 5.14b and c.

The question arises, what does this particular deformation behaviour mean in terms of FCP? First, such reverse plastic deformation during the unloading part of a fatigue cycle in the plane strain region (plane stress is dealt with subsequently) can materialize only under high compressive hydrostatic stress states. Furthermore, the material plastically deforms "practically" toward one point, namely the crack tip (the distance "X" is minute in comparison to the dimensions of the active plastic zone). Therefore, these reversed plastic deformations are occurring under highly oriented deviatoric stress fields. Both conditions - high compressive hydrostatic stress states and highly oriented deviatoric stress fields - prevent any additional damage of the type well known from extrusions. Looking at the sequence of events occurring in the cyclic plastic zone during unloading, elastic unloading occurs first. Then the material is squeezed in the crack opening. First, the immediate vicinity of the crack tip material element is filled out. Figure 5.14a does not show it clearly because of the inadequacy of the mechanical model. With further unloading, the crack opening around the crack front fills out. As soon as the material element ahead of the crack front is completely enclosed, no plastic deformation (only translations) can occur on further unloading in this material element. Only hydrostatic compressive stress fields are acting on this crack tip material element during further unloading. But behind it, more and more material is squeezed into the crack opening (Figure 5.14b and c). These sequences of plastic deformations are reversed during the lower part of the loading cycle. The point at which the crack tip material element is relieved of the compressive hydrostatic stress state is - in terms of stress intensity factor - the partitioning point K_G. Only if the partitioning point K_G is reached and surpassed during the loading cycle, can fatigue damage and FCP occur. The following conclusions can be drawn from the above for the plane strain region of the plastic zone associated with a propagating fatigue crack:

Volume constancy of plastic deformation in the plastic zone leads to high compressive hydrostatic stress states and highly oriented deviatoric stress fields during the lower part of the unloading cycle. In the lower part of the loading cycle, the plastic deformations are reversed and compressive hydrostatic stress states persist in the crack tip material element until the partitioning point K_G *is reached. These compressive hydrostatic stress states in the lower part of fatigue cycles prevent accumulation of fatigue damage, i.e., are the cause for the existence of* ΔK_{eff} *in the plane strain region.*

Plastic Zone of a Fatigue Crack under Plane Stress Conditions

It is general praxis to apply the ΔK_{eff} concept indiscriminately to thin and thick specimens, i.e., independent of the stress state. Yet, the model proposed depends on plane strain conditions. Thus, what happens in the plane stress region, i.e., in thin specimens? One has to realize that semi-finished products with the same material denomination have to be considered different materials if the thickness is substantially different. This is due to the different metallurgical conditions during fabrication. It is a known experimental fact that the FCP in the surface regions (plane stress) lags behind. The resistance to fatigue damage is obviously higher in the plane stress region. On the other hand, in thin specimens two different propagation mechanisms are usually encountered, namely the flat (mode I) and slanted (mixed mode) fracture. In addition, every specimen showing "mode I" FCP has to have a symmetry plane on which plane strain conditions have to exist. Considering the topics cited above and the pertinent scientific information available, it is not at all certain whether ΔK_{eff} exists in the plane stress region or that - if it exists - it is comparable to that found in the plane strain region. The author's "opinion" is that the plastic deformation (under mode I) in the plane stress regions of the plastic zone are not so different, in particular due to the coupling between the plane strain and plane stress regions, and a similar mechanism should operate in the plane stress region as in the plane strain region. Arguments are presented to show that this could be possible, i.e., that the crack stress field itself and/or environmental atoms cause a similar behaviour in the plane stress regions as persists in the plane strain region.

The interference fringe pattern around a fatigue crack shows that practically no deformations have occurred across the original surface of the specimen. The fringe patterns around the fatigue crack correspond nearly ideally to the transverse displacement imposed by Poisson's contraction. The relatively small deformations across the original surfaces can be explained by a surface layer of the plastic zone acting as a membrane - from a continuum mechanics standpoint - which prevent deformations perpendicular to itself. This might be a consequence of the crack stress field itself or be caused by transport of environmental atoms (H, O, ... [860]) into the surface region of the material by dislocations. The later process starts as soon as a material element in the surface layer would be crossed by the elastic-plastic boundary. At this point the plastic deformation amplitude is still small. As the material element is approached by the crack front, where higher deformation amplitudes are encountered, the slip systems in the surface layers with displacement vector perpendicular to the surface are already blocked. Only a zone with width of one to several grains, where the crack actually passes through or passes by, is highly deformed across the original surface. The later zone can be excluded from continuum mechanics considerations. If the explanation given is correct, then a "quasi" plane strain condition is obtained for the plastic zone of a fatigue crack even for the plane stess region. Therefore, a similar mechanism as that which causes the existence of ΔK_{eff} in the plane strain region could act in the plane stress region.

Discussion of the Proposed Plastic Zone Model

The model of cyclic plastic deformation in the plane strain region of a fatigue crack is not a new idea [859], but the states of stress involved justify the existence of ΔK_{eff}. For the plane stress regions of the plastic zone in commercially pure titanium Sugano et. al. [861] came to the same conclusions as were obtained from interference fringe analysis. The investigations of Wei and Shih [862] on 5 mm thick Ti6Al4V sheet material hint at the potency of the processes by which foreign atoms are trapped by surface layers. They found that constant amplitude FCP rates were the same in air and dehumidified argon, while the FCP rates were considerably lower in argon dehumidified and further purified by TSP. Their overload experiment in air and dehumidified argon furnished similar results (overload tests in dehumidified and further purified argon were not made). Therefore, only traces of foreign atoms capable of transport by dislocation are necessary for blocking the slip systems perpendicular to the surface.

Ranganathan et al. [863, 864] found for their overload tests in air and vacuum stonger retardation in vacuum than in air. They explained the - qualitatively identical - growth behaviour by the lesser fatigue damage under vacuum conditions. The 10 mm thick specimens of Al alloy 2024-T3 they used are to be considered "mainly plane strain" under the usual FCP test conditions.

Investigations of thickness effects on retardation behaviour after overloads on Al 2024-T3 [865] in air and on Al-alloy 5083 and HT 80 steel [866] in air and vacuum showed that retardation increased with decreasing specimen thickness. This experimental fact suggests the necessity of a retardation model for the plane stress regions. Yet, the thickness effect on FCP rates for overload tests is not a clear cut matter, as surface removals after overloading showed [866]. As will be shown later, closure mechanisms and ΔK_{eff} changes can intermingle.

The proposed model for the plane stress regions of specimens, namely that the surface layers of the fatigue plastic zone act as a membrane which is practically inpenetrable for dislocations with displacement vector perpendicular to the surface, could be due either to the crack stress field itself or could be aided by the entrapment of foreign atoms in the surface region. In the latter case, very thin specimens should show retardation after overload in air, but under identical test conditions in vacuum, no retardation should occur. These comparative tests are presently in preparation in our laboratory.

Discussion of FCP Behaviour

R-Ratio Effect: The closure model, i.e., the wedging open of the fatigue crack, cannot account for ΔK_{eff} under high R-ratio conditions. In order to check the present model, cyclic loading conditions and the associated plastic deformations are qualitatively analysed.

The compliance force squeezes the plastic zone together during unloading. For larger plastic zone sizes (more plastic tensile stretching at higher K_{max}), more reversed plastic deformations are necessary to comply with the "constant-volume" requirement for a given degree of unloading. Therefore, the amount of reversed plastic deformation for a given degree of unloading is a function of K_{max} which produced the plastic zone.

For a given plastic zone size, i.e., a given K_{max}, the amount of reversed plastic deformation increases the more the specimen is unloaded, that is, the larger ΔK becomes. Therefore, the amount of reversed plastic deformation for a given K_{max} is a function of ΔK. It follows that the reversed plastic deformation is a function of K_{max} as well as of ΔK.

These statements are in agreement with the plastic zone model proposed for the plane strain region as shown in Figure 5.12 and Figure 5.14. The partitioning point K_G must be a function of the total amount of reversed plastic deformation. For fatigue cracks loaded under certain K_{max} and ΔK conditions, such that the plastic zones are in equilibrium with the cyclic loading, the partitioning point K_G must then be a function of K_{max} and ΔK. Since $\Delta K = (1\text{-}R)\ K_{max}$, K_G is a function "f" of K_{max} and R.

$$K_G = f(K_{\max}, R) \tag{5.12}$$

Equation (8) written as:

$$\frac{K_G}{K_{\max}} = a_1 + a_2 R + a_3 R^2 + a_4 R^3 \ldots \tag{5.13}$$

$$h = a_1 + a_2 R + a_3 R^2 + a_4 R^3 \ldots \tag{5.14}$$

is the usual form to express the results of a semi-empirical determination of the partitioning point from da/dN data [853, 867-872]. The same functional relation as Eq. (5.13) was obtained by K_G measurement via a "FCP versus no FCP" criterion [857, 858, 873]. Therefore, the proposed model accounts for the existence of ΔK_{eff} under high R-ratios.

Load History Effects: The proposed plastic zone model should be analysed as to its ability to account for known FCP behaviour. Let FCP behaviour in an overload experiment be split up in stages, as shown in Figure 5.15.

FCP Rates During Single and Multiple Overloads: Let the FCP rates be a function of Eq. (5.10) or (5.11) under constant amplitude fatiguing, i.e., da/dN=f(ΔK_{eff}). In overload experiments, FCP rates under constant amplitude with a certain K_{max} (ΔK) and R are considered reference for the overload FCP behaviour. The ΔK_{eff} for the reference loading is:

$$\begin{aligned} \Delta K_{eff}{}^{*} &= [(K_{\max})\ reference - (K_G)\ reference] - \Delta K_T \\ &= [(K_{max}\ reference) \cdot (1 - h)] - \Delta K_T \end{aligned} \tag{5.15}$$

where h is the function defined by Eq. (5.14). For constant amplitude cyclic loading corresponding to the overload cycle, Eq. (5.10) gives:

$$\begin{aligned} \Delta K_{eff}{}^{**} &= [(K_{\max})\ overload - (K_G)\ overload] - \Delta K_T \\ &= [(K_{max}\ overload) \cdot (1 - h)] - \Delta K_T \end{aligned} \tag{5.16}$$

For the first overload cycle, K_G is fixed by the reference cycle, i.e.,

$$\begin{aligned} \Delta K_{eff}{}^{***} &= [(K_{\max})\ overload - (K_G)\ reference] - K_T \\ &= [K_{max}\ overload - (h \cdot K_{\max})\ reference] - \Delta K_T \end{aligned} \tag{5.17}$$

Therefore,

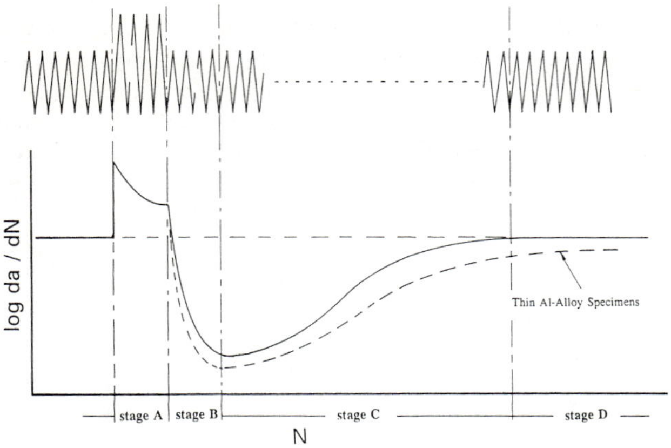

Figure 5.15 Schematic division of an overload test into four stages. The individual stages are denoted as:

stage A:	FCP rates during overload cycles
stage B:	Delayed retardation
stage C:	Maximum retardation and subsequent retarded FCP rates
stage D:	Decreased FCP rates after the fatigue crack passed through the overload plastic zone

$$\Delta K_{eff}^{*} < \Delta K_{eff}^{**} < \Delta K_{eff}^{***} \tag{5.18}$$

and the highest FCP rates should be found for the first overload cycle.

The cyclic plastic deformations cannot be brought into equilibrium conditions during one overload cycle. More overload cycles are necessary to achieve the same condition as persists under constant amplitude fatiguing. As more and more overload cycles are applied, K_G approaches K_G-overload and therewith ΔK_{eff}*** decreases toward ΔK_{eff}**. That means that the FCP rates decrease after the first overload and approach that FCP rate of constant amplitude cycling under overload conditions. The number of overload cycles required to establish equilibrium conditions certainly depends on the respective material and the overload conditions themselves. The numbers 5 to 30 cycles, obtained from the results of striation measurements and multiple overload tests on Ti alloys and steels, can be taken as a general guide. Delay as a function of number of overload cycles in tests on thin aluminum alloy specimens with reference cycles of low R-ratios should not be considered due to the influence of closure, as will be shown later. FCP behaviour under single and multiple overloads can be easily documented today by measurements of striation spacings in a SEM.

Delayed Retardation: Delayed retardation is defined as decreasing FCP rates immediately after the application of overloads up to the point where the FCP rates reach a minimum. There is no common agreement yet as to the general growth behaviour in this stage of an overload experiment. The reasons are the present limits in measuring capability of FCP rates, the influence of specimen thickness, and in connection with the thickness, the lack of knowledge about general FCP behaviour under plane strain versus plane stress for identical materials. With relatively thick specimens under mostly plane strain conditions the fatigue crack tends to advance in the plane strain region unproportionally under overload cycles, i.e., the crack fronts in the surface regions stay behind. As a consequence the stress intensity factor is increased unproportionally for the surface regions. In addition, if the plane stress regions are made into "quasi" plane strain regions by blockage of slip systems by foreign atoms, one would not expect this mechanism to be as stable as that caused by geometric constraint. This explains the start of FCP in the surface regions after overloads [874, 875]. After overloads, the crack extends in the surface region and the relatively high stress intensity factor there is reduced. FCP rates decrease up to the point when FCP starts at low rates in the plane strain region.

The FCP rates occurring during a change of ΔK (no over- or underload) or equivalently by changing ΔK_{eff} are commonly thought to be represented by the respective change in ΔK_{eff}. That means that the FCP rates should correlate to the respective ΔK_{eff}. But this assumption is probably not correct; there seems to be an influence of the damage zone which cannot be taken care of by continuum mechanics parameters. This damage zone, or more precise, the width over which FCP rates do not correlate with ΔK_{eff}, extends up to approximately 0.1 mm. It should be noted in Figure 5.16[876] that after a decrease of ΔK (K_{max} was kept constant) the FCP rates were much higher than when measured under constant amplitude tests (shaded band in Figure 5.16). Similar findings was reported by Schijve [877]. The deviation from that expected from fracture mechanics considerations depends on the magnitude of change in cyclic loading conditions. Hertzberg [878] measured the striation spacing after overloads. For roughly 15 cycles after the overload cycle the FCP rates were higher than for the reference cycles, while they should have been lower. Cyclic loading below the partitioning point K_G[872] or smaller than the FCP threshold [879] produce FCP rates in the range of 10^{-8} mm/cycle over a distance of less than 0.1 mm.

It should be clear from the above that "delayed retardation" is probably a very complex process involving more than one phenomenon. Furthermore, it might not be primarily connected to changes in ΔK_{eff}.

FCP Retardation and its Dependence on Overloading: It is a generally accepted experimental fact that retardation increases with the relative increase of overload-

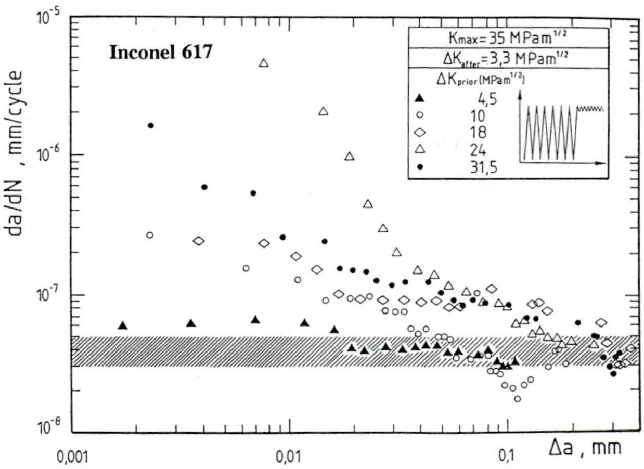

Figure 5.16 FCP rates after a sudden change in amplitudes for Inconel 617

ing with respect to reference fatigue conditions [877, 880-887]. As K_{max} of the overload cycle increases relative to the K_{max} of the reference cycles, the partitioning point K_G increases and therewith retardation increases. It should be noted that the amount of retardation depends on the difference between the partitioning point K_G after overloading and the K_{max} of the reference cycles, i.e.,

$$\Delta K_{eff} = [(K_{max})\ reference - (K_G)\ after\ overload] - \Delta K_T \tag{5.19}$$

As was pointed out before, equilibrium cyclic plastic deformation conditions cannot be established during a single overload cycle. For a given K_{max} overload condition, the amount of retardation increases with increasing number of overload cycles [877, 880, 882, 884-887] up to saturation. The magnitude of ΔK for an overload condition with a given K_{max} is in its effect equivalent to the number of overload cycles. The larger the ΔK of this overloading condition, the fewer the overload cycles required to achieve saturation, i.e., the maximum amount of retardation. The maximum retardation for a given overload cycle is reached when the number of overload cycles is sufficient to increase K_G to its value under constant amplitude cycling with overload conditions, i.e.,

$$(K_G)\text{ after overload} = (h \cdot K_{max})\text{ overload} \tag{5.20}$$

The above results agree with expectations from the proposed model. It should be clear from Eq. (5.20) that crack arrest can occur [888, 889] if $\Delta K_{eff} = 0$ after overload. Quite freqently, it is found that much higher overload conditions are necessary to produce crack arrest than would be expected based on K_G defined by Eq. (5.13) and ΔK_{eff} defined by Eq. (5.19). As with delayed retardation, the surface regions can be responsible for this. If the crack advances under overloading unproportionately in the mid-section of the specimen (plane strain), then the stress intensity factor in the surface region is increased considerably. The fatigue crack starts growing in the surface region (delayed retardation). Depending on specimen thickness, crack growth in the surface region (quasi plane strain) could destroy the retarding mechanism in the plane strain region by gradual growth from the surface into the interior of the specimen. Therefore, overloads which otherwise lead to crack arrrest result only in strong FCP retardation.

Lower FCP Rates After the Crack Passed Through the Overload Plastic Zone: Closure, that is the wedging action of the plastic wake zone, of oxide layers, or of corrosion products in the fracture surfaces, is an experimental fact. The question here is "has it anything to do with ΔK_{eff}"? From experimental measurement it is known that the wedging action of closure is active only in fatigue cycles with R-ratios less than 0.35. Analysis of "load - displacement" curves of fatigue cycles to determine closure was made by Elber [853] with the intent to determine the last point in the loading cycle at which the crack front experiences compressive stresses. Equivalently, the first point in the unloading cycle at which the crack front experiences compressive stresses had to be determined. Elber argued that for an ideal crack - a crack without closure - a linear relation between the load and the displacement (COD or CMOD) should exist. Since a propagating fatigue crack had to be open at least in the upper part of the cycle, the first deviation from the linear upper part was taken as the point sought.

Agreeing with Elber that closure exists and that in the ideal crack there is a linear relation between the cyclic load and the associated displacement (COD or CMOD), a quite different analysis can be made. Figure 5.17 shows a load-COD curve of a fatigue cycle schematically. The broken line is the extension of the linear upper part of the loading cycle. In accordance with Elber's analysis F_{op} and COD_{op} were determined; this is shown in Figure 5.17 for reference.

The analysis [890, 891] goes as follows: if there is a linear relation between load and COD, then the load is conveyed appropriately to the crack front. In the lower part of the loading cycle it comes to a deviation between the linear and the actual relation. Under the minimum load F_{min} the COD_{min} should have been reached in an ideal crack. But, the lower limit reached by the actual COD is COD_{LL}. COD_{LL} corresponds for an ideal crack to a load denoted as F_{LL}. Therefore, under COD_{LL} the load conveyed appropriately to the crack front is F_{LL}. Let the loads be converted to stress intensity factors, then a fatigue cycle $K_{max} - K_{min} = \Delta K$ is externally applied. Yet, only a reduced fatigue cycle $K_{max} - K_{LL} = \Delta K_{red}$ is conveyed appropriately to the crack front. This is graphically shown in Figure 5.18.

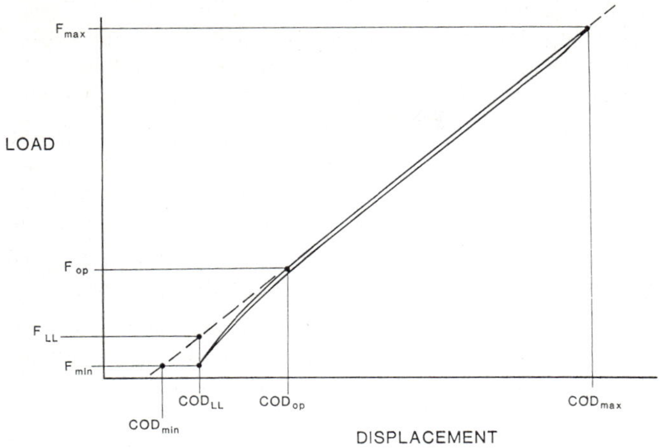

Figure 5.17 The load versus displacement (COD) curve of a fatigue crack, schematically

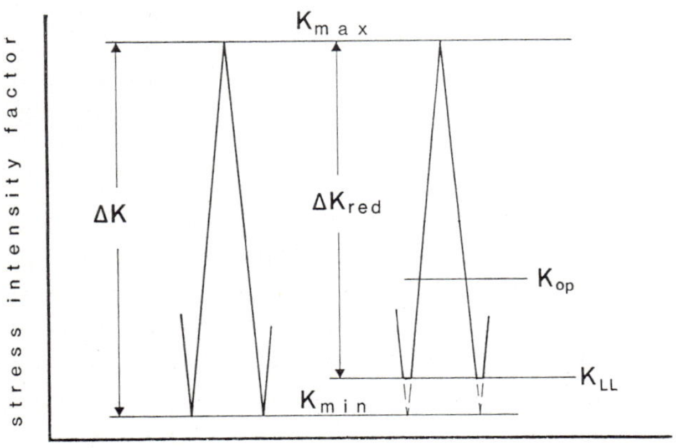

Figure 5.18 The loading conditions of a fatigue crack in fracture mechanics terms

Measurements [891] in accordance with Figure 5.17 and Figure 5.18 were made on two materials which can be considered extreme cases as far as closure is concerned. The materials are the quenched and tempered steel with German designation 42CrMo4 and the Al-Li alloy 8090-T6. Figure 5.19 and Figure 5.20 show the experimentally determined ΔK_{red} versus the calculated ΔK for the 42CrMo4 steel and the Al-Li alloy 8090-T6 respectively. It can be seen from Figure 5.19 that for the fine grained steel the calculated ΔK and the ΔK_{red} conveyed to the crack front are the same; that means that closure has no affect on ΔK_{eff} or equivalently on FCP behaviour. On the other hand, the difference between calculated ΔK and ΔK_{red} can be higher than 40% for the Al-Li alloy (Figure 5.20) with its rugged fracture surfaces[9]). This agrees with previous measurements on an Al alloy [892]. It should be noted that the closure effects as shown in Figure 5.20 become more pronounced as K_{max} of the loading cycle decreases. Furthermore, since ΔK is reduced and K_{max} is not altered, the actual R-ratio is increased. Eq. (5.14) and Eq. (5.15) it follows that ΔK_{eff} is reduced.

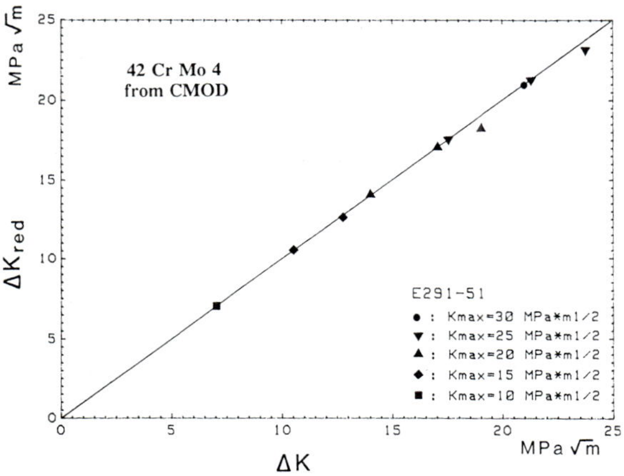

Figure 5.19 Comparison between ΔK externally applied and ΔK_{red} (Figure 5.18) transferred to the crack front for 42CrMo4

Now we can return to the topic of this section, namely retardation beyond the extent of the overload plastic zone.

---

9) This experimental analysis can be extended to fatigue cycles with negative R-ratios, too [893].

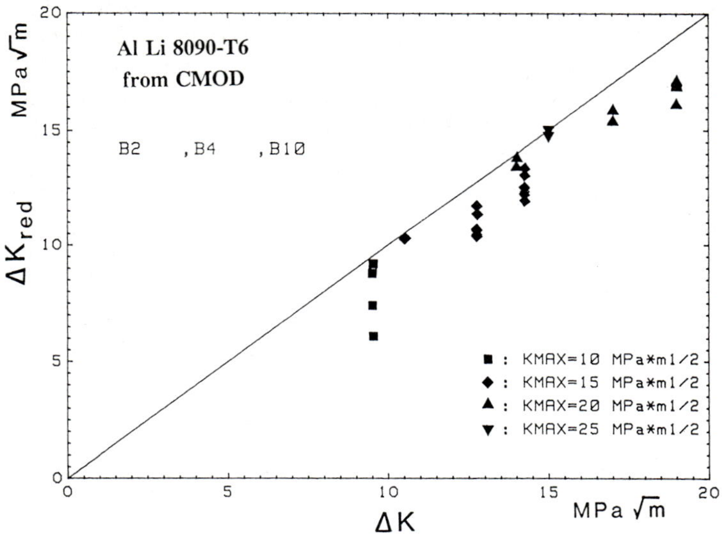

Figure 5.20 Comparison between ΔK externally applied and ΔK_{red} (Figure 5.18) transferred to the crack front for AlLi 8090-T6

Overloads produce humps of excessively stretched materials in the fracture surfaces as the crack grows through the overload plastic zone. The humps can affect the reference cycles up to a value corresponding to 0.35 K_{max} of the overload cycles. Therefore, one would expect for reference cycles with low R-ratios and low K_{max}, for certain materials and for thin specimens to find a pronounced effect of closure on the FCP rates if overloads were applied. This was documented by the investigation of Mills and Hertzberg [865], who showed that reduced FCP rates can persist far beyond the extent of the overload plastic zone. The reduction in FCP rates after the crack passed through the overload plastic zone in Al 2024-T3 was found to be the highest in thin specimens (B = 1.6 mm); no reduction was found in the thickest specimen (B = 26 mm).

Effects of Combinations of Overloads and Underloads: As shown in Figure 5.21, two types of conditions have to be considered, namely underloads before overloading (Figure 5.21a) and underloads after overloading (Figure 5.21b). It should be noted that the unloading part of the last cycle determines the position of the partitioning point K_G for the next cycle. In connection with the R-ratio effect on ΔK_{eff}, it was stated that the partitioning point K_G was a function of the reversed plastic deformation range. From that assumption function "h" was obtained. For the unloading range AA in Figure 5.21a, which leads to underloading before overloading, it is clear that the reversed plastic deformation range is increased. Furthermore, this unloading corresponds to a lower R-ratio than that of the reference cycle. Due to transient effects the partitioning point K_G would not correspond to an unloading cycle AA in constant amplitude fatiguing. Therefore, Eq. (5.14) does not apply directly, but it gives the direction and an estimate of the magnitude in which K_G is changed by the underload. K_G is decreased relative to its position in the reference cycle. The FCP rate for the first overload should increase. The overload primarily acts on the monotonic plastic zone size and should therefore be little affected by the previous reversed plastic deformation. Therefore, prior underloads should have little effect on the retardation caused by the overload. Only the FCP rates of the first few overload cycles should be increased.

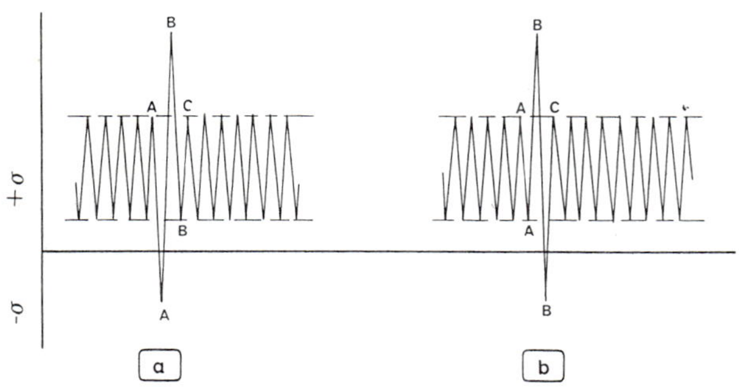

Figure 5.21 Sequences of overload versus underload, schematically

Comparing the unloading ranges BB in Figure 5.21a and b and applying the same argument as for the unloading range AA in Figure 5.21a, K_G for the loading range BC in Figure 5.21a must be higher (same K_{max}, higher R, see Eq. (5.14)) than K_G for the loading range BC in Figure 5.21b. Let it be assumed that the reference cycle would establish a certain K_G*. After a certain overload the unloading range BB in Figure 5.21b might be chosen such that K_G for the loading range BC is equal or less than the K_G* of the reference cycle. For such a case, despite the overload no retardation would occur. Therefore, underloading after overloading lowers K_G and therewith reduces the retardation effects of the overloads.

5.2.3 Crack Growth Life Prediction

G. Marci

The goal of fatigue research is to predict failure of structural parts subject to fluctuating service loads, and thereby design parts, or limit the service life such that structural failure is reliably prevented. Since most engineering structures contain manufacturing flaws, their initiation does not determine service life, but rather the growth of the - mostly minute - flaws already present at the start of structural service. That is the main reason for the success of fracture mechanics in safe-life analysis. Initially, the capability of fracture mechanics as a tool for safe-life predictions was demonstrated for relatively simple mechanical loading conditions. Yet, the initial success prompted the ongoing effort to extend fracture mechanics analysis capabilities to more complex loading conditions.

Presently there are approximately one dozen computer codes operating which are able to estimate the fatigue crack growth life of parts or specimens in a cycle by a cycle way of calculation. These codes vary considerably even in the basic correlation between crack propagation rates and fracture mechanics loading parameter, i.e. in the equations correlating da/dN with ΔK. If it comes to load interaction effects, the modeling is limited only by the researcher's gusto and experience.

In the remaining part of the paper, the equations correlating da/dN and ΔK are presented which are usually the basis on which a computer code is built. Then some recent developments in the area of experimental data acquisition are presented. Certain generalizations concerning modeling of service type loading conditions are made. Finally, some results from a DLR fatigue crack propagation life prediction code are presented.

Equations Correlating Fatigue Crack Propagation and the Fracture Mechanics Loading Parameter ΔK

Paris et al. [894] were the first who tried to correlate fatigue crack propagation (FCP) rates with fracture mechanics parameters, K_{min} and K_{max}, describing the loading conditions in the near-region of the crack front. They proposed the following equation for predicting FCP rates:

$$\frac{da}{dN} = C_1 \cdot (\Delta K)^{n_1} \tag{5.21}$$

where $\Delta K = K_{max} - K_{min}$. Equation (5.21) constituted an enormous advance in fatigue life prediction capability at that time. Yet, Eq. (5.21) by itself was not the only reason for the relative fast adaptation of the fracture mechanics approach to FCP. At the same time Forsyth [895] was able to show by TEM of replicas from a FCP fracture surface that the magnitude of crack advance per cycle, i.e. the FCP rate, depends on the magnitude of the load excursion. Equation (5.21) correlates well with the experimentally observed FCP rates only in the region denoted as "Stage II". Foreman et al. [896] fitted Eq. (5.21) to the experimental FCP rates at the high ΔK end by introducing a scaling factor which they obtained via limit arguments. They modified Eq. (5.21) as follows:

$$\frac{da}{dN} = \frac{C_2 \cdot (\Delta K)^{n_2}}{(1-R)K_C - \Delta K} \tag{5.22}$$

or equivalently:

$$\frac{da}{dN} = \overline{C}_2 \cdot \frac{1}{K_C - K_{\max}} \cdot \Delta K^{n_2} \tag{5.23}$$

where K_C stands for the fracture toughness of the material and $\overline{C}_2 = C_2/(1-R)$.

Lui [897] modified Eq. (5.21) by subtracting from ΔK a threshold range ΔK_{th}. The excursion of the stress intensity factor at the lower end of a da/dN ΔK curve at which no FCP is observed is denoted as ΔK_{th}. With this later modification Eq. (5.21) becomes:

$$\frac{da}{dN} = C_3 \cdot (\Delta K - \Delta K_{th})^{n_3} \tag{5.24}$$

Combining both modifications of Eq. (5.21) the following equation is obtained:

$$\frac{da}{dN} = C_4 \cdot \frac{1}{K_C - K_{\max}} \cdot (\Delta K - \Delta K_{th})^{n_4} \tag{5.25}$$

A further advance in FCP rate prediction capability brought Elber's discovery [898], that the tip region of a fatigue crack was in a state of compressive stresses over a considerable part of a fatigue loading cycle when the loading cycle was entirely in the tensile range. Due to the plastic tensile stretching of the material in the presently active plastic zone, i.e. the plastic zone ahead of the crack front, this plastically stretched material constitutes a wedge between the fracture surfaces as the crack is propagated by fatigue cycling. At zero load with constant amplitude fatiguing in the tensile range, part of the active plastic zone and the material at some distance behind the crack front are under high compressive stresses. With increasing load the compressive stresses are reduced till they become zero and change into tensile stresses. This transition point for the stresses coincides with the condition at which a fatigue crack opens at the crack tip. The transition point is, in fracture mechanics terms, the stress intensity factor K_{op} at which the fatigue crack is open. Elber denoted that part of ΔK as the effective stress intensity range which constitutes the excursions of the stress intensity factor between K_{op} and K_{max} i.e.

$$\Delta K_{eff} = K_{\max} - K_{op} \tag{5.26}$$

ΔK_{eff} characterizes that part of ΔK where FCP can commence. Elber has chosen to make ΔK_{eff} the prime correlation factor for the FCP rates, i.e.

$$\frac{da}{dN} = C_5 \cdot (\Delta K_{eff})^{n_5} \tag{5.27}$$

He determined ΔK_{eff} semi-empirically for 2024-T3 from da/dN versus ΔK data.

$$\Delta K_{eff} = U \cdot \Delta K \tag{5.28}$$

and

$$U = 0.5 + 0.4 \cdot R \tag{5.29}$$

The assumption was made that for high R-ratios ($R \geq 0.7$)

$$\Delta K = \Delta K_{eff} \tag{5.30}$$

This assumption requires (5.26) to have the counterpart

$$\Delta K_{eff} = K_{max} - K_{min} \quad \text{for} \quad K_{min} > K_{op} \tag{5.31}$$

Experimental results from FCP threshold determination [899-901], which show the existence of a closure-free FCP threshold ΔK_T, require modification of Eqs. (5.25) and (5.31)

$$\Delta K_{eff} = (K_{max} - K_{op}) - \Delta K_T \quad \text{for} \quad K_{op} > K_{min} \tag{5.32}$$

$$\Delta K_{eff} = (K_{max} - K_{min}) - \Delta K_T \quad \text{for} \quad K_{min} \geq K_{op} \tag{5.33}$$

The equations presented above are the basis of life prediction models for FCP under service type fatigue loading.

Recent Experimental Developments Relevant to FCP Modeling

Over the past decade an iterative process at DLR between FCP modeling and experimental development has led to novel test procedures for material properties as input into cycle-by-cycle life predictions. Attention was firstly focussed on the FCP threshold [899-901]. It was found that a threshold exists under all fatigue loading conditions and that it has a constant value for some materials (steels and Al alloys), while with other materials it decreases with increasing K_{max}. As far as the threshold is concerned, the following materials can be "roughly" ranked (room temperature):

7475-T7351	$\Delta K_T = 1.0 \ MPa \cdot \sqrt{m}$
Ti-6Al-4V	$\Delta K_T = 2.2 \ MPa \cdot \sqrt{m}$
Ti8Al1V1Mo	$\Delta K_T = 2.4 \ MPa \cdot \sqrt{m}$
Inconel 617, annealed	$\Delta K_T = 2.5 \ MPa \cdot \sqrt{m}$
42CrMo4, quenched + tempered	$\Delta K_T = 2.7 \ MPa \cdot \sqrt{m}$

The FCP threshold ΔK_T does not include any part of the so-called "closure range". Furthermore, it was proven to be applicable to service type loading [901]. Therefore, for physical reasons, threshold should be included in every equation on which a computer code is built. From a practical standpoint, the many modeling schemes included in a cycle-by-cycle calculation code easily compensate for the error made when neglecting threshold. But, for certain spectra - high mean load, small amplitudes - neglecting threshold results in underestimation of life by the code which cannot be overcome by adjustment of other modeling schemes [902].

It had been documented [902] that the computer codes based on the ΔK_{eff} concept were far superior to others not relying on this concept. The usual way to obtain ΔK_{eff} in a code is via an equation:

$$\frac{K_{op}}{K_{\max}} = a_1 + a_2R + a_3R^2 + a_4R^3 ... \tag{5.34}$$

Equation (5.34) can be obtained semi-empirically from da/dN versus ΔK curves under the assumption that for high R-ratios ($R \geq 0.75$) $\Delta K = \Delta K_{eff}$. Elber obtained Eq. (5.29) for 2024-T3 this way, which when converted into the functional form of Eq. (5.34) gives:

$$\frac{K_{op}}{K_{\max}} = 0.5 + 0.1R + 0.4R^2 \tag{5.35}$$

Equation (5.34) can be obtained by "Finite Element Computations", as was shown by Newman [903, 904]. Furthermore, experimental determination of Eq. (5.34) via the criterium "FCP versus no FCP" can be made by the experimental procedures shown in Figure 5.22 [899, 905, 906]. The following equations were obtained experimentally by the latter method:

$$\frac{K_{op}}{K_{\max}} = \frac{1}{1.9 - 0.9R} \tag{5.36}$$

Ref. [905] for 2024-T3, method Figure 5.22a

$$\frac{K_{op}}{K_{\max}} = 0.455 + 0.321R + 0.208R^2 \tag{5.37}$$

Ref. [908] for 7475-T7351, method Figure 5.22b

$$\frac{K_{op}}{K_{\max}} = 0.49 + 0.195R + 0.361R^2 \tag{5.38}$$

Ref. [908], from Ref. [907] for Ti6Al4V, method Figure 5.22c

$$\frac{K_{op}}{K_{\max}} = 0.583 + 0.323R + 0.068R^2 \tag{5.39}$$

Ref. [908] for Inconel 617, method Figure 5.22c

$$\frac{K_{op}}{K_{max}} = 0.653 + 0.18R + 0.16R^2 \tag{5.40}$$

for Ni base alloy Nicrofer 5219 Nb, method Figure 5.22b

Equations (5.37) through (5.40) are graphically shown in Figure 5.23.

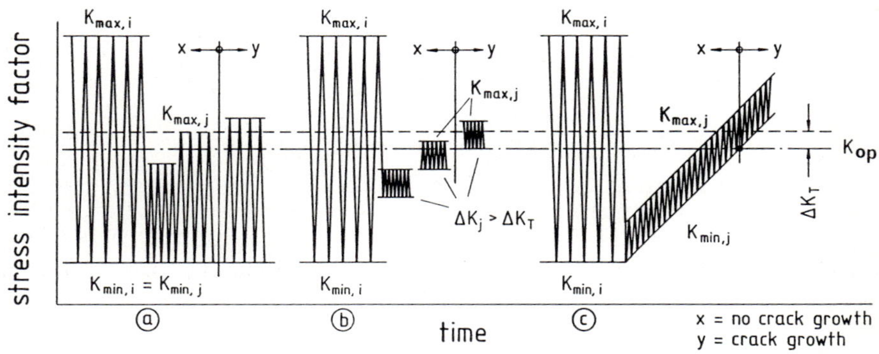

Figure 5.22 Test methods for K_{op} determination via the criterion "crack growth versus no crack growth"

Some Basic Modeling Schemes

Monotonic plastic zone sizes associated with propagating fatigue cracks, as well as cyclic (reverse) plastic zone sizes, depend on cyclic loading conditions and material properties. They are needed for several modeling schemes in a cycle-by-cycle life calculating computer code. Even such important parameters as plastic zone dimensions can be calculated only to rudimentary agreement with reality. Plastic zone sizes are needed to model FCP under plane strain, plane stress and in the transition region between the two states in parts of a given thickness. Furthermore, load history effects - in particular high load excursions (overloads) - are "frozen into the material" as large plastic zone sizes. Therefore, modeling load history effects always involves plastic zone dimensions. Every computer code for FCP life predictions has to calculate the plastic zone dimension R_p.

For many computer codes a different plastic zone size is calculated under identical loading conditions. Newman [903] used in his analytical work the Dugdale approach for calculating the plastic zone dimension,

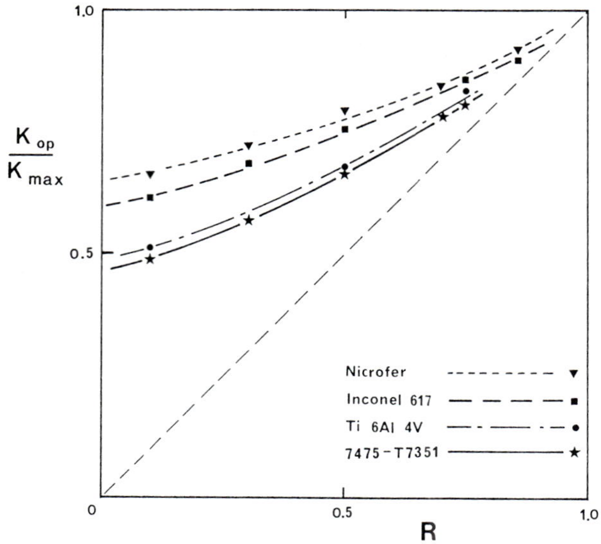

Figure 5.23 Normalized K_{op} versus R determined with test methods shown in Figure 5.22b and Figure 5.22c

$$R_p = a \cdot \left\{ \frac{W}{\pi \cdot a} \cdot \sin^{-1}\left[\sin \pi \cdot \frac{a}{W} \cdot \sec\left(\frac{\pi \cdot S_{max}}{2 \cdot \omega \cdot \sigma_0} \right) \right] - 1 \right\} \tag{5.41}$$

but it is not known whether the same route is taken in his life prediction program. De Koning [909] used the following equation to calculate R_p:

$$R_p = \frac{a}{\lambda} \cdot \left(\frac{S_{max}}{\sigma_0} \right) \cdot \left[1 + \left(\frac{a}{W} \right)^2 + \frac{\left(\frac{S_{max}}{\sigma_0} \right)^2}{\lambda} \right] \tag{5.42}$$

where $\lambda = 1.32$ for plane stress and $\lambda = 9.0$ for plane strain. Baudin and Robert [905] as well as Marci [902, 910] use Irwin's equation

$$R_p = \frac{1}{\lambda} \cdot \left(\frac{K_{max}}{\sigma_0} \right)^2 \tag{5.43}$$

where $\lambda = 2\pi$ is employed by Baudin and Robert while Marci adjusts λ for plane strain 2π to plane stress π transition via an interactive routine involving the thickness of the structure.

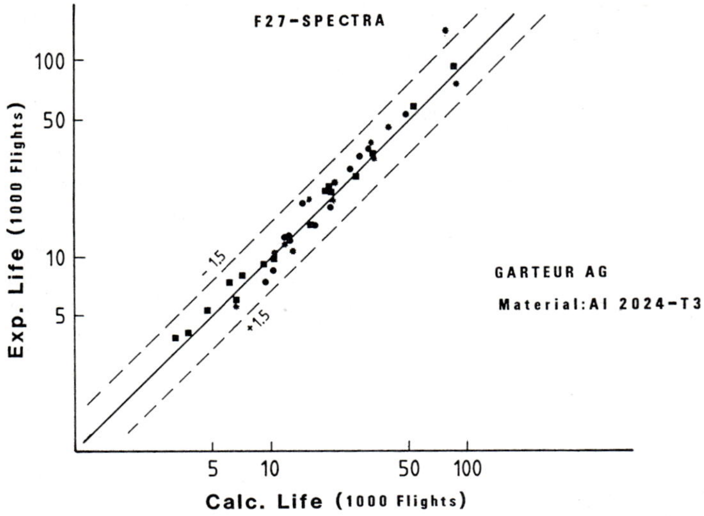

Figure 5.24 Comparison of experimental life and life predicted with the DLR computer code for F27 spectra and Al 2024-T3 material of different specimen thicknesses [902]

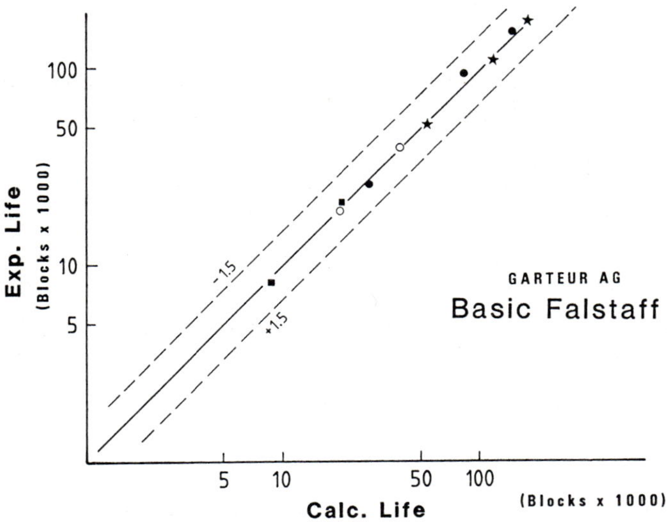

Figure 5.25 Comparison of experimental life and life predicted with the DLR computer code for Basic Falstaff for different materials and specimen thicknesses [902]

Results of FCP Life Prediction with the DLR Computer Code

The computer program to model FCP under spectrum loading relies on the modified Elber equations, i.e. Eqs. (5.32) and (5.33). Overload as well as underload phenomena are modelled by connecting the plastic zone dimension with the partitioning point K_{op}. The effect of a structural part's thickness is taken into account by the dependence of plastic zone dimension on plane stress or plane strain conditions. The crack dimensions primarily determine the influence of compressive loads on K_{op} via the wake zone. Results of life predictions obtained in the frame of a Garteur action group [902] are shown for different F-27 and Falstaff spectra for Al and Ti alloy plates with different thicknesses (Figure 5.24 and Figure 5.25).

5.3 Special Testing Techniques

5.3.1 Stress Corrosion Testing

R. Braun and H. Buhl

In aircraft structures primary design goals are safety, reliability, and lifetime. Premature failure of structural components caused by corrosion processes such as stress corrosion cracking (SCC), corrosion fatigue, or hydrogen embrittlement has to be avoided. The permanent demand for higher efficiency and economy is usually answered by the selection and development of materials with higher strength and/or lower density. Unfortunately, it is a long experience that increasing strength implies increasing SCC susceptibility. The SCC behaviour of commercial alloys depends upon their microstructure and, thus, is influenced by the chemistry, production route, and the heat treatment. More economic machining methods may introduce additional stresses or change surface conditions, resulting in altered SCC behaviour. Therefore, in materials research as well as quality assurance, reliable testing techniques for determining SCC behaviour are indispensable. In relatively short time periods those tests have to predict the service behaviour. This requires more severe testing conditions, like the use of more aggressive environments or higher stress levels than those anticipated in service.

In terms of classical corrosion chemistry, stress corrosion cracking results from the combined action of a tensile stress and a corrosive environment on a susceptible material. With aluminium alloys stress corrosion cracks propagate predominantly along grain boundaries. Therefore, the SCC susceptibility of wrought aluminium products is highly anisotropic due to the directional grain structure (pancake structure) with grains elongated in the direction of flow. The lowest SCC resistance is observed in the short transverse direction and the highest in the rolling direction [911].

With aluminium alloys stress corrosion cracking occurs usually in aqueous environments. Initiation and propagation of stress corrosion cracks are accelerated by the presence of chloride. Most tensile stresses occurring in service originate from residual stresses caused by heat treatment, machining, or assembly and may be simulated with constant deformation tests. Constant load tests are used to simulate operating stresses [912]. In accelerated laboratory tests using smooth statically loaded specimens time-to-failure data are determined, and a threshold stress σ_{th} is evaluated below which SCC should not occur. The value of those threshold stresses depends upon the method of loading. Constant load tests (dead weight loading) yield a lower threshold value than constant deformation tests, and regarding the latter, uniaxial tension stressing is more severe than bending. Several types of SCC specimens such as C-ring, bent beam, tuning fork or tension specimens are used, depending on the size and shape of the products investigated [912, 913].

The SCC behaviour of a material evaluated in accelerated tests depends on environmental factors like species and concentration of ions, the pH-value , dissolved gases, and temperature. Therefore, synthetic environments used in accelerated SCC tests must be carefully chosen leading to predictions in agreement with long term outdoor exposure tests in marine or industrial atmosphere, or other service environments. For aluminium alloys exposure to aqueous 3.5% NaCl solution is widely used for testing. In the following, SCC test standards and other testing procedures, particularly for Al alloys, will be briefly outlined, with priority given to the Slow Strain Rate (SSR) technique.

- Smooth, tensile loaded specimens are alternately immersed in 3.5% NaCl solution, as recommended in US specification ASTM G44 [914]. Depending upon the alloy group and the direction tested, the exposure time varies from 10 to 90 days. However, this environment may also promote severe pitting and intergranular corrosion, which can interfere with the initiation of stress corrosion cracks, which is the failure criterion for specimens with nonuniform stress throughout the cross section. C-ring or bent beam specimens fall into this category. Furthermore, for longer immersion time periods, corrosion products can deposit on the surface of the specimens and impede the detection of initiated cracks.
- Uniform and pitting corrosion may be eliminated using an inhibitor such as Na_2CrO_4 which is added in a concentration of 0.5% to an aqueous 2% NaCl solution at pH 3, according to German aviation standard LN 65666 [915]. To reduce the incubation period of environmentally induced cracking, specimens are pickled prior to testing. Moreover, this pickling procedure has the advantage of reducing the scatter of the time-to-failure data because the natural oxide film grown between pickling and testing has the same thickness with all specimens. Tests are performed under continuous immersion and constant load conditions using dead-weight or spring loaded test devices. Testing time is limited to 1000 h.
- In service the presence of cracks or flaws in structural components has to be expected. Environmentally induced growth of those cracks is related to stress

intensity at the crack tip. Therefore, SCC tests with smooth specimens have been supplemented with tests based on a fracture mechanics approach. Using pre-cracked specimens crack velocity vs stress-intensity curves are measured. From these curves a stress intensity threshold K_{ISCC}, below which crack growth does not occur, can be obtained. Pre-cracked double-cantilever beam (DCB) specimens are widely used for this SCC testing technique [916, 917].

However, K_{ISCC} as well as σ_{th} should not be overvalued or used as design parameters , but rather for rating alloys. In service, structural components are frequently subjected to fluctuating stresses and a great variety of environments. Therefore, the concept of constant influence parameters cannot be maintained. Furthermore, K_{ISCC} and σ_{th} values may in some cases depend upon the patience of the observer or the maximum length of exposure.

Although standardized SCC tests useful for indicating the behaviour of materials in natural environments have been developed, there are still problems with regard to quantitative rating of the degree of SCC susceptibilty, estimation of risks, and life time expectancy. Thus, reliable accelerated tests are further needed to provide data which can be used to evaluate the service performance of engineering structures. A new accelerated testing technique, the Breaking Load Method, has been developed and explored by Sprowls et al. [918, 919]. In this testing method smooth specimens are exposed to a corrosive environment for various periods and at various sustained tensile stresses. After exposure, specimens are tensile tested under inert conditions, and the residual strength compared to the tensile strength of specimens not exposed is the criterion for the environmentally induced damage. Tensile testing of unstressed exposed specimens enables a distinction between degradation caused on the one hand by SCC and on the other hand by local corrosion attack such as pitting and intergranular corrosion. The Breaking Load Method is claimed to require shorter exposure times and to reduce the number of specimens tested compared to conventional testing procedures. Since small stress corrosion cracks can be detected, this test method rates SCC susceptibility particularly among more resistant materials.

Slow Strain Rate Method

In the early 1970s it was shown for pipe line steels that the classical picture of SCC has to be extended by the strain rate $\dot{\varepsilon}$ as mechanically influencing parameter, documented for the first time by R.N. Parkins [920, 921].

The prevailing majority, if not all, of material-environment systems where SCC is observed are passive systems, i.e. the metal surface is covered by an oxide or hydroxide layer. One characteristic property of such a passive system is its possibility to repassivate after local damage has occurred. However, such a local breakdown of the passive layer is a necessary precondition to release SCC, independent of the SCC mechanism. It can occur

- electrochemically, due to pitting or intergranular attack at weak points of the passive film,
- or mechanically, due to slip steps at the surface, caused by creep processes under sustained load or deformation.

The first case is the aforementioned classical one with $\dot{\varepsilon} = 0$. In the second case a strain rate $\dot{\varepsilon} > 0$ is necessary to induce SCC. Under sustained load conditions, strain rate (in this case caused by creep processes) decreases with time. Therefore, if the controlling parameter for SCC is the strain rate, crack initiation has to occur before all creep processes are terminated. This could be shown by the following experiment performed by Parkins with a C-Mn steel under constant load and potentiostatic conditions: After loading the potential was maintained outside the cracking range. After termination of all creeping processes ($\dot{\varepsilon} = 0$) it was moved inside the SCC range. This experiment proved the material to be not susceptible to SCC. However, if the potential was held in the SCC range from the beginning of the experiment, fast crack initiation and crack growth occurred [921]. This observation led to the development of simple devices for tensile testing specimens with well defined strain rates in the range of 10^{-2} to 10^{-8} s^{-1}, and this special type of SCC was called "strain rate induced SCC".

To find the critical strain rate range within which SCC occurs, the specimens exposed to a certain environment are tensile tested at different strain rates over several decades. This range depends upon the repassivation kinetic of the material, which is strongly related to the environment. If the strain rate is too high specimens will fail by ductile fracture before the necessary corrosion reactions can take place. With decreasing strain rate the influence of corrosion increases. However, strain rate may become too low, when film repair or repassivation is fast enough to keep pace with the formation rate of bare metal, and reactions leading to cracking are suppressed. This influence of the strain rate upon SCC susceptibility is illustrated in Figure 5.26 [922].

The strain rates that trigger SCC have to be determined for each system. For titanium alloys which repassivate very fast in aqueous environments (see section "Titanium Alloys"), SSR tests have to be performed at strain rates in the range of $10^{-4} < \dot{\varepsilon} < 10^{-2}$ s^{-1} (see Figure 2.19). Aluminium alloys with a much lower repassivation rate in aqueous chloride solutions exhibit possible SCC susceptibility at strain rates $\dot{\varepsilon} < 10^{-4}$ s^{-1}. Strain rate induced SCC as schematically illustrated in Figure 5.26 is not found in all susceptible systems. The majority of alloy/environment combinations shows no maximum of SCC susceptibility vs strain rate, but SCC resistance decreasing continuously with decreasing strain rate, identifying classically behaving systems. Most aluminium alloys in aqueous chloride solutions fall into this category. This is shown exemplarily in Figure 5.27 for a sensitized Al-5Mg alloy. Nevertheless, for aluminium alloys strain rate induced SCC could be established, too (e.g. [923-925]).

It was demonstrated that the SSR method can also be applied with Al alloys exhibiting classical SCC behaviour, and that results of this testing technique are

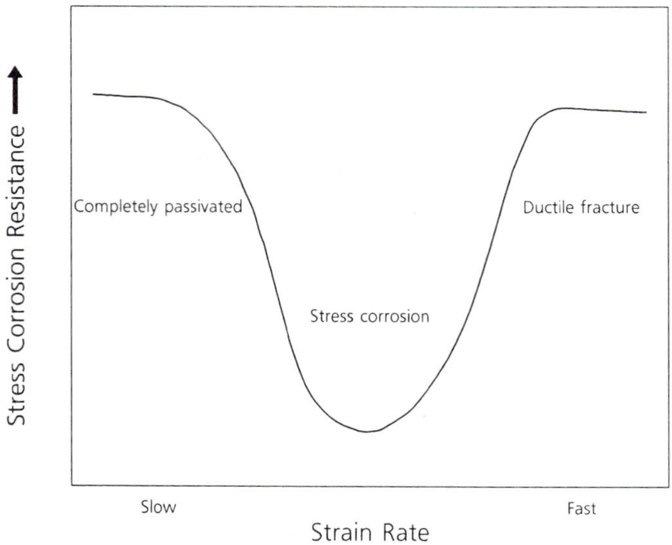

Figure 5.26 Schematic illustration of the variation of SCC susceptibility with strain rate

in accordance with those from static loading tests. Moreover, conclusions on the mechanism of SCC are supported more evidently by SSR tests than by conventional testing techniques.

In static loading tests using smooth specimens, times-to-failure are measured, and the threshold stress determined from these data is used as a rating parameter for SCC susceptibility. In SSR tests specimens are continuously strained until fracture. SCC behaviour is indicated by changes of parameters of the stress-strain curve obtained in a corrosive medium compared to those in an inert environment at the same strain rate. With aluminium alloys, dry laboratory air can be used as the inert environment generated by embedding specimens in $Mg(ClO_4)_2$ powder. The maximum load achieved or the elongation to fracture can be used for assessing SCC susceptibility, but the fracture energy, which is the area beneath the stress-strain curve and a combination of both parameters, may be more adequate. The relative fracture energy (RFE), a normalized value with reference to fracture energy in inert environments, is an appropriate parameter for ranking alloys. An increasing deviation from unity (or 100%) indicates increasing corrosion influence.

With lower strain rates, e.g. longer exposure times, the tensile properties of the specimens can be deteriorated by corrosion independent of the applied stress, e.g. by pitting, intergranular corrosion, or hydrogen embrittlement. Therefore, decrease

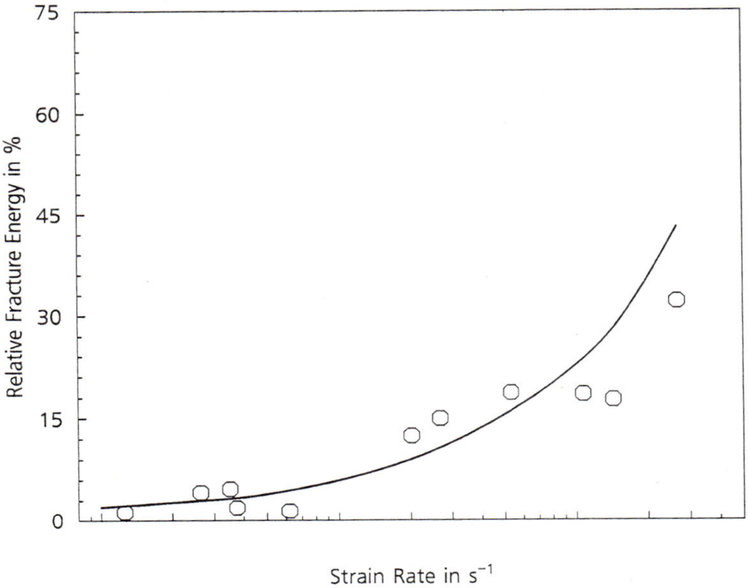

Figure 5.27 Relative fracture energy vs strain rate for a sensitized, highly SCC susceptible Al-5Mg alloy. Short transverse tensile specimens were tested in aqueous 3.5% NaCl solution [924].

of the RFE with decreasing strain rate is not necessarily an indication of SCC susceptibility. The reduction in fracture energy due to corrosion damage not associated with stress can be determined by pre-exposure tests. Unstressed specimens are continuously immersed in the corrosive environment. The exposure periods correspond to the immersion times of the specimens during the SSR tests. After pre-exposure the specimens are tensile tested in an inert environment at the same strain rates as in the corrosive environment (see column B in Table 5.1). If pre-exposure does not have any influence on fracture energy, the reduced RFE observed in the SSR tests in corrosive environment is caused by SCC exclusively. Otherwise, if the same decrease of the fracture energy is found with pre-exposed specimens as in the SSR tests, the degradation is not dependent upon stress and is related to pitting, intergranular corrosion, or hydrogen embrittlement. Metallographic examinations and fractography can assist in finding the causes of the degradation.

Embrittlement by hydrogen results primarily in a deterioration of ductility. The damage observed is related to the transport mechanisms by which hydrogen, absorbed near the surface of the metal, reaches microstructural locations in the bulk material where it leads to premature fracture. Penetration of hydrogen into the metal can occur by diffusion, particularly along paths with enhanced activation energy such as grain boundaries. However, hydrogen can be transported by dislocations at a rate appreciably in excess of that for diffusion [926]. Since a substantial density of mobile dislocations exists only after yielding, hydrogen embrittlement is usually not observed before plastic deformation occurs. Thus, susceptibility to hydrogen embrittlement, not detected in static loading tests applying stresses below the yield strength, can be revealed by SSR testing.

A	B	C
SSR test at the strain rate $\dot{\varepsilon}_1$ with a duration t_1 in a corrosive environment.	• Pre-exposure in the same corrosive environment used in A for the time t_1 without applied stress. • SSR test at the strain rate $\dot{\varepsilon}_1$ in an inert environment.	• Pre-exposure in the same corrosive environment used in A for the time t_1 without applied stress. • Storage of the pre-exposed specimens under vacuum at room temperature or slightly elevated temperatures. • SSR test at the strain rate $\dot{\varepsilon}_1$ in an inert environment.

Table 5.1 Sequence of investigations supplementing SSR tests.
A) SSR test in a corrosive environment.
B) Pre-exposure and SSR test in inert environment to evaluate effects of corrosion damage, not promoted by stress but resulting in a decrease of RFE.
C) By an intermediate storage reduction of RFE caused by pre-exposure can be attributed to hydrogen embrittlement or to corrosive attack due to pitting and/or intergranular corrosion.

With Al alloys pre-exposure embrittlement caused by hydrogen is frequently reversible. This was shown for the alloys Al-5Zn-3Mg and 7049-T651 by the work of Ratke and Gruhl [927] and of Holroyd and Hardie [928], respectively. Storage in a desiccator or in vacuum at room temperature or at slightly elevated temperatures after pre-exposure can lead to a recovery of ductility as indicated in subsequent tensile tests in an inert environment. These recovery effects may be attributed to redistribution of hydrogen including effusion or diffusion to innocuous traps. Such a storage gives the possibility to distinguish whether pre-exposure embrittlement is caused by local corrosion or by hydrogen embrittlement (see column C in Table 5.1).

The significance of pre-exposure tests and intermediate storage for completing SSR tests in corrosive environments is demonstrated below with the alloy 7010 in the two tempers T73651 and T7651.

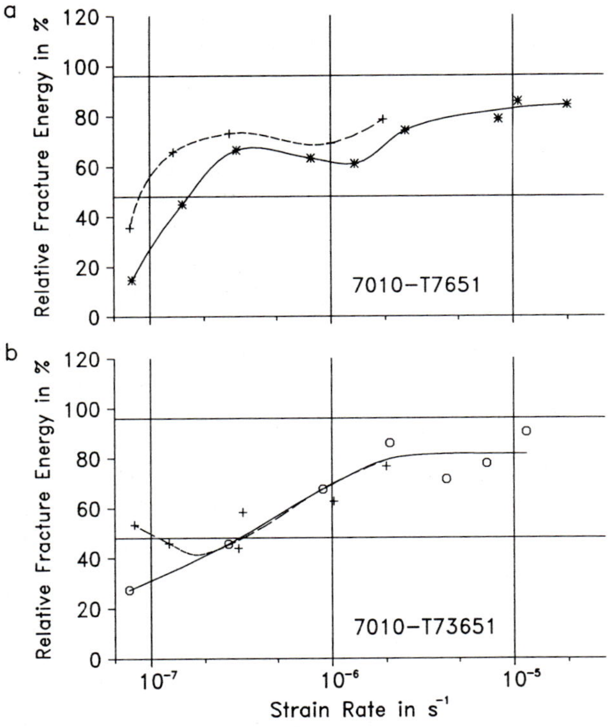

Figure 5.28 Relative fracture energy of the alloy 7010 in the tempers T7651 and T73651 vs strain rate. At least triplicate tensile specimens were tested in ST direction in aqueous 3.5% NaCl solution. The dotted curves represent results of pre-exposure tests [929].

As found in alternate immersion tests according to ASTM G44, 7010-T73651 is virtually immune to SCC in the ST direction (σ_{th} = 96% of $\sigma_{0.2}$), whereas in the T7651 heat treatment the SCC resistance is slightly reduced (σ_{th} = 77% of $\sigma_{0.2}$) [929]. Figure 5.28 illustrates the SSR results in 3.5% NaCl solution for the alloy 7010 in both tempers T7651 and T73651 showing decreasing RFE with decreasing strain rate. As proved by pre-exposure tests (dotted curve, evaluated as described in Table 5.1 column B) the reduction of RFE with the alloy 7010-T73651 is primarily caused by stress independent corrosion processes. Only at the slowest strain rate used is a deviation of both curves, e.g. pre-exposure tests and SSR tests in

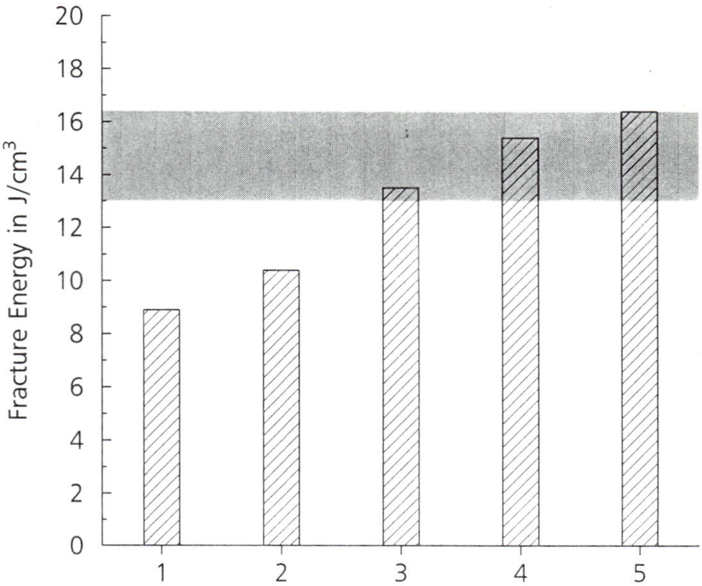

Figure 5.29 Effect of intermediate storage upon the fracture energy of 7010-T73651 specimens which were 75 h pre-exposed in 3.5% NaCl solution and then tensile tested in an inert environment at a strain rate of $3 \cdot 10^{-7}$ s^{-1}. The different storage conditions were:

1. no storage, immediately tensile tested after pre-exposure
2. stored in a desiccator at room temperature for 24 h
3. stored in a desiccator at room temperature for 120 h
4. stored in vacuum ($2 \cdot 10^{-1}$Pa) at room temperature for 120 h
5. stored in vacuum ($2 \cdot 10^{-1}$Pa) at 80 °C for 120 h

The shadowed area refers to the scatter band of the fracture energy of non pre-exposed specimens tested under inert conditions [929]

3.5% NaCl solution, observed. For the alloy 7010-T7651 the decrease of RFE over the whole strain rate range cannot be explained exclusively by pre-exposure effects. To some extent this decrease is caused by SCC. Thus, the slight SCC susceptibility of 7010-T7651 and the high resistance of this alloy in the T73651 temper, as observed in alternate immersion tests, are confirmed by SSR tests supplemented by pre-exposure.

Metallographic examinations of pre-exposed 7010 specimens exhibited only pitting. No intergranular corrosion was observed. Pits may act as possible stress

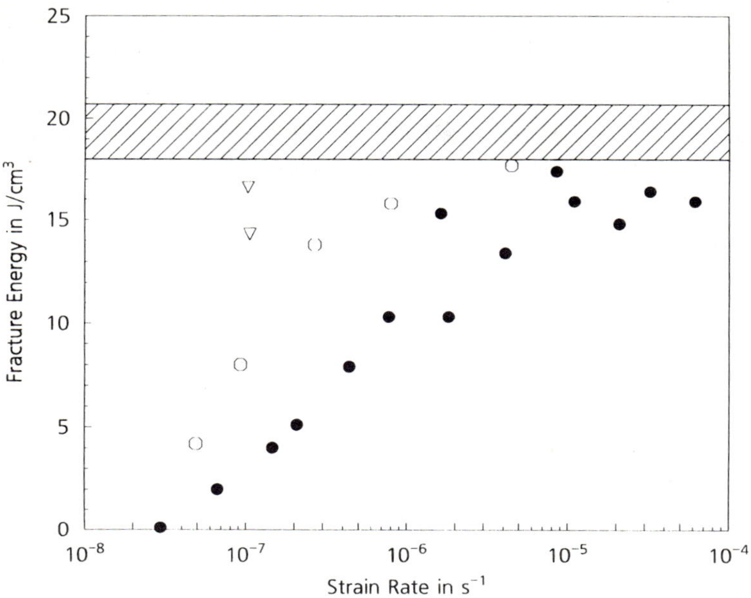

Figure 5.30 Fracture energy vs strain rate for the alloy 2014-T651. Both pickled (•) and degreased (○) tensile specimens were tested in the ST direction in an aqueous solution of 2% NaCl + 0.5% Na_2CrO_4 at pH 3. Pre-exposure tests using pickled specimens are represented by ∇; the shadowed area refers to reference values [935].

concentrators, but it was found that pre-exposure effects were independent of the depth of the pits. Thus, the decrease of the RFE shown in Figure 5.28 is not associated with pitting; the pre-exposure embrittlement of both 7010 alloys is rather caused by hydrogen. This was proved by an intermediate storage of the pre-exposed specimens in a desiccator or in vacuum (column C in Table 5.1), as illustrated in Figure 5.29. The full recovery of the RFE due to this storage confirms that the pre-exposure embrittlement is not related to mechanical damage such as pitting but only to hydrogen. The rise of the RFE of pre-exposed 7010-T73651 specimens at a strain rate of $2 \cdot 10^{-7}$ s^{-1} (Figure 5.28) is associated with recovery effects, too. It can be attributed to the beginning of redistribution or effusion of hydrogen during the relatively long duration of the SSR tests (>100 h) in an inert environment.

Effects of Synthetic Environments

3.5% NaCl Solution: Following ASTM G44, neutral 3.5% NaCl solution was used in most investigations of the SCC behaviour of Al alloys employing the SSR technique. However, when totally immersed in 3.5% NaCl solution 2XXX alloys do not exhibit significant stress corrosion crack growth [916], and this electrolyte does not promote SCC with Al-Li alloys under continuous immersion conditions [930, 931]. Under the more severe loading method of dynamic straining the results of SSR tests are in accordance with those of static loading tests (continuous immersion): with Al-Li alloys no reduction of the fracture energy resulting from SCC was observed [79, 932], and the degradation of 2XXX alloys is mainly caused by pre-exposure effects [933]. For the alloy 7475-T651, which is quite susceptible to SCC in the ST direction, no influence of the 3.5% NaCl solution on fracture energy was found [934]. Therefore, this electrolyte should not be used in SSR tests using smooth specimens under continuous immersion conditions.

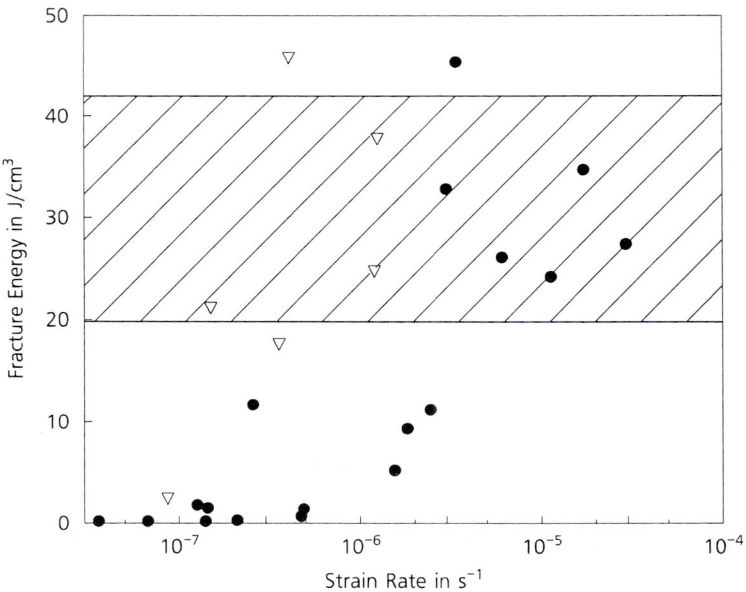

Figure 5.31 Fracture energy vs strain rate for 2091-T8X51 plate material. Short transverse tensile specimens were tested in substitute ocean water. Pre-exposure tests are represented by ∇; the shadowed area refers to reference values [79].

2% NaCl + 0.5% Na_2CrO_4 at pH 3: SSR results with alloy 2014-T651 in acidified 2% NaCl solution inhibited by chromate are illustrated in Figure 5.30 [935]. As proved by pre-exposure tests the decrease of fracture energy with decreasing strain rate is caused by stress corrosion cracking. Thus, the SSR technique indicates the low SCC resistance of 2014-T651 in the ST direction. The influence of surface treatment is illustrated in Figure 5.30, too. With unpickled specimens the decrease of the fracture energy is observed at lower strain rates because the SCC incubation time is shortened by pickling. Nevertheless, the use of chromate inhibited 2% NaCl solution for SSR tests with Al alloys is doubtful. Whereas the low SCC resistance of 2014-T651 and 2091-T8X are indicated in this solution [80, 935], the SCC behaviour of 2091-T8X51 and 7475-T651 plate material could not be evaluated due to scattering data [79, 934], and the SCC susceptibility of 8090-T81 sheet was not confirmed [936].

Substitute Ocean Water According to ASTM D1141: Figure 5.31 shows SSR results of 2091-T8X51 alloy in substitute ocean water [79]. Plate material of this

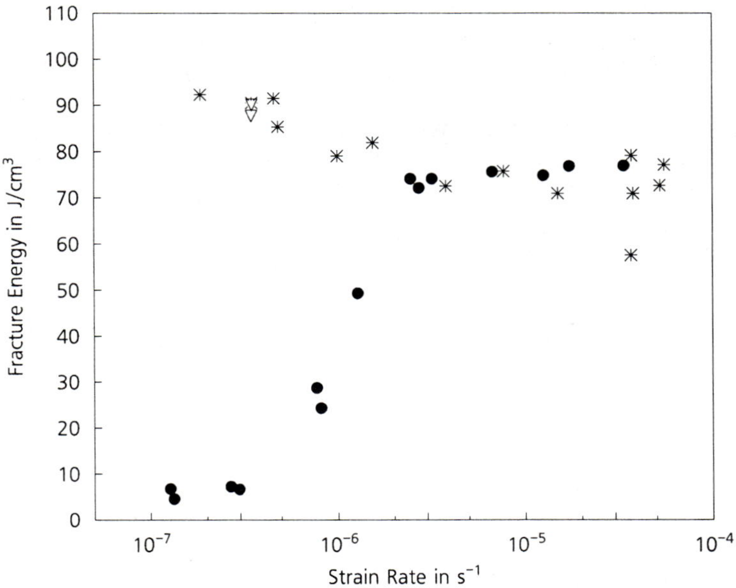

Figure 5.32 Fracture energy vs strain rate for 2091-T8X sheet material. Tensile specimens were tested in L direction in inert environment (∗), in an aqueous solution of 0.5 M NaCl + 0.1 M LiCl + 0.05 M $NaHCO_3$ + 0.05 M Na_2CO_3 (•), and after pre-exposure (∇) [80].

Al-Li alloy is quite susceptible to stress corrosion cracking in the ST direction as found in outdoor exposure as well as in alternate immersion tests. In the SSR tests the low SCC resistance is indicated by the pronounced decrease of fracture energy with decreasing strain rate. Similar plots of the fracture energy vs strain rate were also found with other SCC susceptible 2XXX and 7XXX alloys such as 2014-T651 [935], 2024-T351 and 7475-T651 [934], and with SCC susceptible sheet of recrystallized Al-Li alloys 2091-T8X and 8090-T81 [80, 936]. The high SCC resistance of 7475-T7351 in the ST direction is also predicted by SSR tests in substitute ocean water. Therefore, this electrolyte is a promising synthetic environment for the SSR testing technique. However, as observed with neutral 3.5 NaCl solution, pre-exposure effects interfere with the evaluation of SSR data at lower strain rates.

0.5 M NaCl + 0.1 M LiCl + 0.05 M $NaHCO_3$ + 0.05 M Na_2CO_3: This electrolyte should be appropriate for SCC testing of Al-Li alloys [937]. SSR results of the Al-Li alloy 2091-T8X in an aqueous solution of 0.5 M NaCl + 0.1 M LiCl + 0.05 M $NaHCO_3$ + 0.05 M Na_2CO_3 are illustrated in Figure 5.32. The SCC susceptibility of this sheet is indicated in agreement with results of alternate immersion tests in which a threshold stress below 100 MPa was determined in the rolling direction [59]. Similar results were obtained for 2091-T8X51 plate [79]. Further work is needed to answer the question whether or not an aqueous NaCl/LiCl solution containing carbonate can be used as an effective corrosive environment in accelerated SCC tests for Al-Li alloys.

5.3.2 Biaxial Testing

K.-H. Trautmann, H. Döker and H. Nowack

The stresses and strains in aircraft wing and fuselage structures like in many other plate constructions are normally biaxial. The corresponding elastic and plastic deformations are highly influenced by eventual anisotropy of the plate material as in the case of hybrid or full composite materials. Nevertheless, the design of such structural parts is normally based on data from uniaxial material testing. This applies to deformation behaviour, structural fatigue and fracture. In order to exhaust the properties of advanced materials to their limits, multiaxial design methods based on experimentally determined material properties have to be developed.

Biaxial Testing Equipment

For systematic investigations of multiaxial effects suitable testing equipment, where the loading conditions can be exactly realized and controlled, has to be available. DLR decided to select a biaxial testing system, because the thin structural parts of aerospace structures are predominantly biaxially loaded. Another requirement was that large test pieces could be investigated, especially for the determination of crack resistance curves of precracked structural parts. Figure 5.33 shows the biaxial testing facility with its crosswise arrangement of four hydraulic actuators on their bedplate. It was developed in cooperation with Schenck AG, Darmstadt, which also manufactured the system.

The maximum specimen size which can be tested is 1m x 2m. Maximum load in one direction is 1000 kN, in the other direction 640 kN. Load ratio and phase correlation can be arbitrarily chosen. A special requirement was high stiffness of the testing system in order to allow testing frequencies as high as possible in fatigue testing and to keep the capacity of the oil pumps moderate. The high stiffness of the system was achieved by the four columns which are mounted at each loading direction as support. Between the columns there is still enough space available for observing the specimen and/or for the installation of additional equipment, for example, for high temperature tests or tests under special environments.

Development of Test Specimens

A predominant requirement for a biaxial test specimen is that a certain testing area is achieved where the stresses and strains are uniformly distributed. Besides this, different types of experiments - determination of stress/strain behaviour, crack initiation, crack propagation, determination of crack resistance curves - require different kinds of specimens. Figure 5.34 shows some examples of test specimens.

Figure 5.33 DLR's biaxial testing equipment

The development of specimens for biaxial testing requires optimization by finite element calculations and final verification of the deformation behaviour of the test section of the specimens by strain gage measurements or comparable measuring techniques. A critical point is the design of the loading arms of the specimens. In the grip region no transverse deformation of the specimens is possible. To allow for free deformation of the specimens' test section, the loading arms need a certain length and have to be cut into strips. In order to decrease the notch effect, the radii of the slots at the transition to the test section have to be increased. The cross section of the loading arms and, consequently, the transferable load to the test section are reduced. To realize high plastic deformation of the test section, loading arms of higher strength had to be welded to the test section (Figure 5.34a) [938].

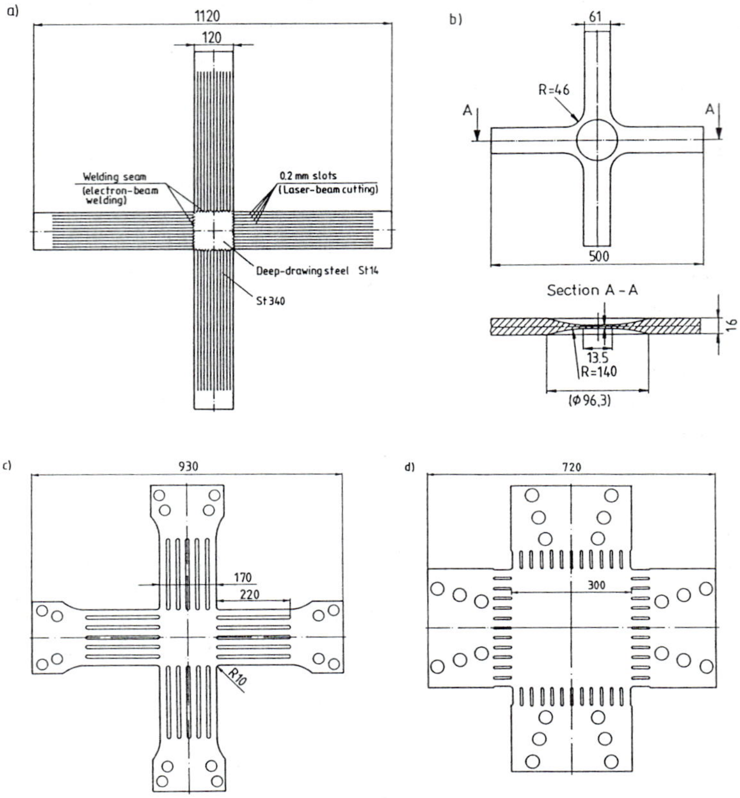

Figure 5.34 Specimen types for biaxial testing:
a) specimen for monotonic loading up to extremely high deformations [938]
b) specimen for fatigue crack initiation tests
c) specimen for fatigue crack growth measurements
d) specimen for determining crack resistance curves

The specimen for the crack initiation investigations (Figure 5.34b) is hourglass shaped and has a flat test section of 13.5 mm diameter. An essential requirement for the specimen design was the possibility to introduce tensile and compressive strains into the test section up to 1%.

For macrocrack propagation the specimen in Figure 5.34c was developed. It has a large observation field with a homogeneous strain distribution. The strip width of the loading arms and the radii of the slots are optimized such that high loads can be transferred to the test section and cracks do not develop at the slots, even for very small crack starter notches. Similar considerations apply to specimens for determining crack resistance curves under monotonically increasing loads (Figure 5.34d).

Stress-Strain Behaviour Under Biaxial Loads

In cases where the deformations remain low ($\leq \pm 2\%$), as under fatigue conditions, and because under fatigue loading other aspects such as permanent changes in the loading direction become important, a multi-yield surface model like the Mroz model may be adequate to describe material behaviour [939, 940]. In this model the yield surfaces are circles in the deviatoric stress space according to the van Mises yield criterion, which can move after certain rules (kinematic hardening) or expand or shrink (isotropic hardening). The ranges between two yield surfaces (Mroz circles) represent ranges with constant hardening parameters. Calculations with the Mroz model together with a suitable FEM code such as Ott's FEMFAT program system [940] require limited numerical expense, provided that the stress increments are not too small. That is the reason why the model is well suited to fatigue problems, where cycle by cycle fatigue life calculations have to be performed. Comparisons between strains as predicted by the FEMFAT program of Ott [940], and as measured experimentally have been performed for a car axle and other components. Both data coincided well.

In cases, where stresses and deformations on structural parts become rather large, as for example during a proof test, the phenomenological plasticity theory as developed by Krawietz [941] forms a most efficient basis for the treatment of this case, because it offers a mathematical model for the material behaviour. If a model like Krawietz' model is available, multiaxial experiments have to be performed only to determine the parameters of the model. Krawietz' model was investigated within a research project together with INPRO, Berlin, on an automotive steel [942].

The experiments to be performed include the determination of the initial yield curve and the evaluation of the three model parameters μ_1, $y_1(\kappa_1)$, $B(\kappa_2)$ from subsequent yield curves, which are generated after specially designed biaxial load sequences have been applied to the biaxial specimens. This requires tests with deformations as high as possible.

For determining of the initial yield curve, biaxial tests with loads with different load ratios into both biaxial loading directions have to be performed until initial yielding occurs. Yielding was defined to occur as soon as a permanent strain of 0.05% was reached. Because the specimens were rather thin, load into compression was impossible. In order to get further points of the yield curve into compression, a special procedure had to be applied. Pieces of a size of 13 x 20 mm were taken from the test section of the biaxial specimens and loaded into compression in a special loading device, which was mounted into a uniaxial servohydraulic testing machine. Figure 5.35 shows an initial yield curve, as determined in the tests.

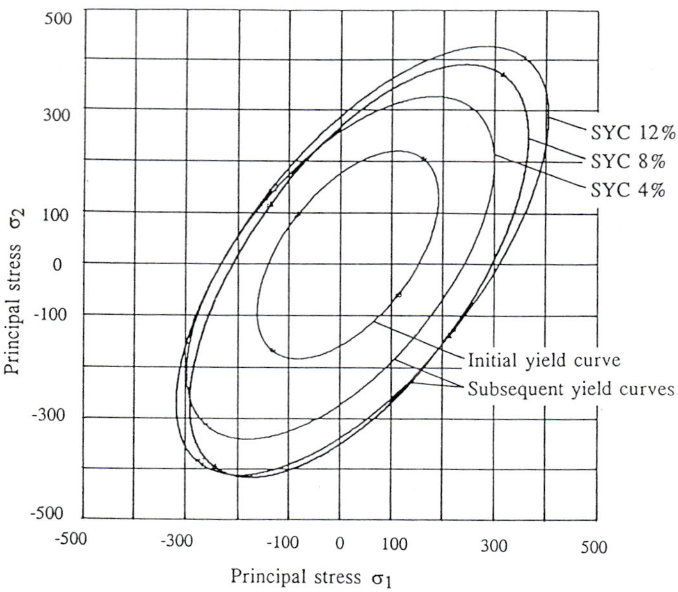

Figure 5.35 Initial and subsequent yield curves as derived from biaxial tests for STE14 material

For determining subsequent yield curves and the parameters which describe material behaviour after initial yielding, biaxial tests, where significant plastic deformations in a certain direction are generated first (by applying a certain biaxial ratio) and where - thereafter - the new (subsequent) yield curve is determined, have to be performed. For a determination of yield points in compression, small specimens were also taken from the test sections of the deformed cruciform specimens and loaded into compression in a separate testing machine. In Figure 5.35 examples of experimentally determined subsequent yield curves are also shown.

The parameters of the material model of Krawietz, μ_1, $y_1(\kappa_1)$, $B(\kappa_2)$, can now be determined in the following manner: the stress and displacement state after some certain biaxial loading is calculated by the material model. Since the parameters of the model are not known at the beginning, some rational assumptions have to be introduced into the model. The predicted results are then compared to the actual stress and displacement states of the biaxial specimen. In subsequent calculations the material parameters are then iteratively modified until the differences between the predicted and the actual stresses and displacements of the biaxial specimens become significantly small.

Fatigue Behaviour Under Biaxial Stresses

Under fatigue loading in the limited life range several individual stages are passed through in the course of the fatigue life, the precrack stage, the small crack stage, and the macrocrack stage [943]. Although it would be worthwhile from a physical viewpoint to consider all individual stages separately, the precrack and the small crack stage and parts of the macrocrack stage are often treated as one part of the total fatigue life, and are denoted as the "technical crack initiation stage", which ends as soon as a crack of an engineering size (normally a crack of a length of 0.5 - 1 mm) is present. The second part of fatigue life is the macrocrack propagation stage.

In the macrocrack stage the shape, orientation and array of the macrocracks are influenced by the multiaxial stresses and further controlled by the anisotropy of the material.

An essential basis for fatigue life predictions is that the multiaxial stress/strain history at the fatigue critical locations is available. For crack initiation (without a dominant crack) the stress/strain behaviour at these locations can be evaluated by methods described above. From these multiaxial stress/strain paths quantities have then to be derived which can be correlated with certain amounts of fatigue damage. From a practical point of view it is highly desirable that the reference data for the calculation of the fatigue damage increments should be available from data from uniaxial tests [943]. Under uniaxial conditions, parameters or quantities correlating with uniaxial fatigue life data can be easily derived from the local stress/strain hysteresis curves.

Under multiaxial conditions stress/strain events like closed hysteresis loops only happen under proportional loading in the stress space. Most of the fatigue prediction methods for multiaxial conditions in the LCF range, as they are known today, are applicable to proportional loading only. Three of them, the equivalent strain hypothesis, the maximum shear hypothesis, and a hypothesis by Brown and Miller, which consider the combined action of shear and of normal stresses in the plane with maximum shear displacement, will be specially considered later on. Details of these hypotheses are given in [943, 944]. They all use a strain life relationship like the well-known Morrow-Manson strain life equation as a basis for the evaluation of fatigue damage.

In the case of multiaxial loading with constant amplitude loading in each direction, however, with different frequencies or in cases where the loading histories in the different directions behave more or less irregularly, closed curves are no longer formed in the stress space. Under these circumstances "incremental models" like the incremental plastic work concept as developed by Ott [940] may be applied. (However, there are also other procedures under development (compare, for example, [945]). In Ott's model it is assumed that fatigue damage is a function of the plastic work a material element consumes under the action of the multiaxial stresses or strains.

The stress/strain path of the material element, from where the plastic work is derived, is determined using the material model by Mroz. Every transition through a range with constant hardening parameters between two yield surfaces in the Mroz model causes a certain amount of damage. The damage per unit plastic work in a range with constant hardening parameters is derived from uniaxial tests. The stress/strain hysteresis curves from these tests are subdivided into segments which correspond to the ranges with constant hardening parameters in the multiaxial case. Every segment of the stress/strain hysteresis curve causes an individual contribution to damage.

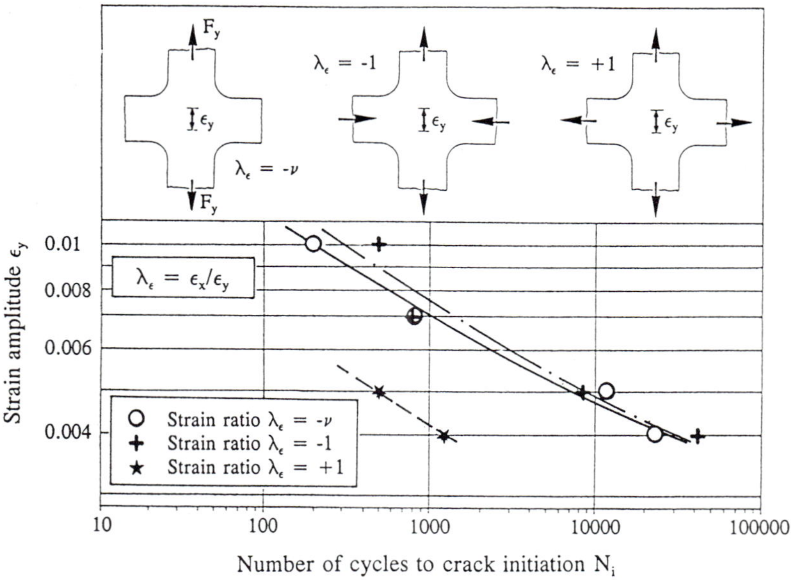

Figure 5.36 Crack initiation life behaviour under different types of biaxial loading

In Figure 5.36 the results of cracks initiation life tests are shown. The initiation of a crack was assumed to be finished after a crack of a length of 0.05 mm was present. The scope of the investigations was to determine crack initiation life behaviour under biaxial loading with different but constant phase angles between the two loading directions and, further on, for the case where one of the biaxial directions remained unloaded during the tests. The fatigue process was monitored using the foil replication technique. Of interest were form, orientation and morphology of the cracks. The initiation lives were compared to the results of two types of prediction methods, one type which is applicable to proportional loading only and which uses uniaxial strain amplitude crack initiation life curves as a basis, and secondly, the incremental plastic work analysis by Ott, which is applicable to arbitrary loading histories. These investigations were performed in close cooperation with LUK, Indonesia [944]. The material investigated was Al 7475-T7351.

In biaxial tests with an hourglass specimen, loads and stresses are parameters not well suited to describe the stress/strain situation at the specimen, and normally FEM analyses are required. That is the reason why the strains were controlled in the experiments. It was further decided to apply a same strain ε_y into the main loading direction in all tests of one test series and to apply strains of the same magnitude at the subaxis perpendicular to the main direction, once at a phase angle between both axes of 0° ($\lambda_\varepsilon = +1$), once at a constant phase angle between the two axes of 180° ($\lambda_\varepsilon = -1$), and once where no external load was applied at the subaxis ($\lambda_\varepsilon = -\nu$). Four different test series with strain amplitudes ε_y of 1.0%, 0.7%, 0.5% and 0.4% were run, except in the $\lambda_\varepsilon = +1$ case, where the test section of the hourglass specimens buckled at strain amplitudes larger than 0.5%. As can be seen from Figure 5.36, for $\lambda_\varepsilon = +1$ considerably smaller crack initiation life values were observed. The results were compared to the equivalent strain hypothesis, the maximum shear strain hypothesis, a hypothesis by Brown and Miller [946] (hypotheses which all require proportional loading situations) and the incremental plastic work hypothesis of Ott [940].

Figure 5.37 shows a comparison of predicted and the actual behaviour. The individual test data should coincide as closely as possible with the curves in Figure 5.37 which represent the predictions by the individual prediction methods. From the figure it can be seen that the predictions and the actual results differed considerably, except for the $\lambda_\varepsilon = -\nu$ condition. In this case the stress/strain conditions in the biaxial tests are close to those under uniaxial conditions. For the $\lambda_\varepsilon = -1$ case the predictions turned out to be mainly conservative, which would be acceptable from an engineering viewpoint. However, for the $\lambda_\varepsilon = +1$ condition all predictions were unconservative up to factors of 3 to 4. This is *not* acceptable from an engineering viewpoint. Although the incremental plastic work hypothesis of Ott showed the smallest overall scatter band between the most conservative and the most unconservative predictions, all hypotheses still have to be further improved, before they can be recommended for more general engineering application.

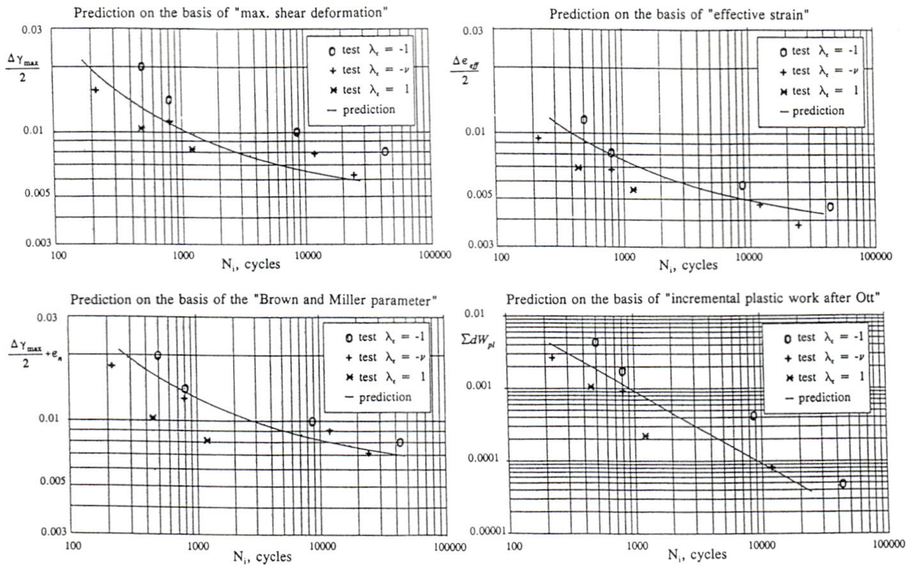

Figure 5.37 Comparison of crack initiation life predictions and of actual test results

Up to now only some macrocrack propagation studies have been performed. These showed that the multiaxial effect on crack propagation was not so strong. However, further systematic studies will have to be performed in the future.

It is further highly important to consider also the new generation of aircraft materials, such as hybrid or full composites. Systematic investigations of these materials have just been started within a European research and development project.

Determination of Crack Resistance Curves

The load bearing capacity of structural parts with cracks or crack-like defects is usually calculated on the basis of critical material parameters like K_{Ic}, J_{Ic}, or $J = f(\Delta a)$ curves (crack resistance curves), which are derived from tests with standard specimens under uniaxial loading. This procedure has to be questioned, especially if the loading characteristics of the laboratory specimens and of the structural parts are significantly different. This applies, for instance, to cracks in notches and to cracks in a multiaxial load field.

A theoretical analysis shows that the load bearing strength of thin-walled parts increases with increasing biaxial tension load ratio, and decreases if compression

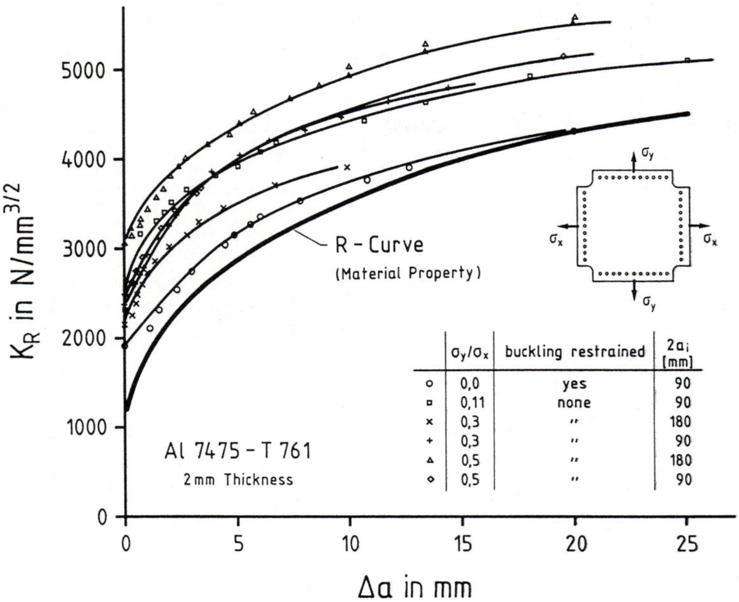

Figure 5.38 Crack resistance curves: Influence of biaxial loading conditions on the failure behaviour of thin Al 7475-T761 sheet material

loads are applied parallel to the crack. These conclusions were verified for 2 mm thick sheets of Al 7475-T761 alloy. Crack resistance curves of cruciform specimens were measured for $0 \leq \sigma_y/\sigma_x \leq 0.5$. The initial crack length was 90 mm and 180 mm in an 800 mm x 800 mm test section.

Uniaxial tension loading of precracked panels is usually accompanied by buckling in the crack tip region. Therefore, for determining crack resistance curves in uniaxial tests, mechanical buckling prevention devices were applied. The biaxial experiments show that a tension stress of 10% of the main tension stress applied parallel to the crack plain is sufficient to prevent this local buckling. This results in an increase of the crack resistance under monotonic increasing loads. The results of all tests are in good agreement to the theoretical calculations. Crack resistance increases with increasing λ. This means that the load bearing capacity of cracked thin sheet structures is higher under biaxial tension loads compared to uniaxial load. The lower broadly drawn curve in Figure 5.38 is the lower limiting curve of the crack growth resistance of Al 7475-T761 measured under uniaxial load with buckling prevention devices applied.

5.3.3 Chevron Notched Specimen Testing

G. Marci, M. Spiegelberg and J. Eschweiler

Fracture is the final stage of degradation processes to which every structure or component is subject by the forces of nature. This final stage of degradation can be characterized by the fracture mechanics parameter "fracture toughness", which is experimentally determined. Plane strain fracture toughness testing in accordance with the "ASTM E 399-90" standard [947] is a test method requiring skilled and experienced personnel in addition to time consuming evaluation procedures. A simplified "Plane Strain Fracture Toughness" test designed especially for evaluating thick-section Al alloy products is ASTM E 1304-89 [948]. The latter test method utilizes a chevron notched specimen. With these specimens, there is no need for precracking and crack length determination. The present paper concerns itself with a test method for estimating fracture toughness by use of chevron notched specimens. Unlike ASTM E 1304-89, the question of plane strain versus plane stress will be of minor importance in this report, and ASTM E 1304-89 will not be discussed further. The present report is to show that chevron notched specimen testing is an economical way, and in many cases the only way,to estimate material's resistance to fracture.

Many industrial semi-finished products - either metallic or nonmetallic - do not lend themselves to plane strain fracture toughness testing due to geometrical size limitations. Yet, it is often desirable to test the materials fracture resistance. With chevron notched specimens, fracture toughness testing is possible down to 12 by 12 mm cross section of semi-finished products in the ease of metallic materials. Even smaller cross sections are sufficient for ceramet and ceramic materials. The fracture resistance values may be used to insure a certain level of fracture resistance as a criterion of quality control, or may be used for comparing batches of different materials. In research and product development it is quite common that only very small quantities of materials are available for testing purposes. Fracture resistance is often one of the important material parameters which have to meet a set requirement. Here too, chevron notched specimen testing offers an efficient way to aid the decision making process in the early stages of development.

Introduction of a crack with a predetermined geometry at a certain position in a specimen is nearly impossible to achieve in ceramic materials and very difficult in ceramet materials. While not easy, chevron notches of high precision can be ground in specimens made of these materials. Thus, chevron notched specimen testing of ceramic and ceramet materials is the common way to obtain fracture resistance data. Many materials producer today are required to furnish fracture resistance data for their products. Complete automation - no manual operation from raw stock to the finished test report - cannot be achieved with the ASTM E 399 standard procedure due to required crack length measurements. chevron notched specimen testing would allow complete automation; in addition, machining of the specimens does not require as many machining steps.

In the following, specimens and loading fixtures are presented which can be used for many different applications. The specimens shown are by far not all available for chevron notched specimen testing. References to further information will be given. Subsequently, the theory behind chevron notched specimen testing is elaborated. Finally, comparative results and results from special applications are presented.

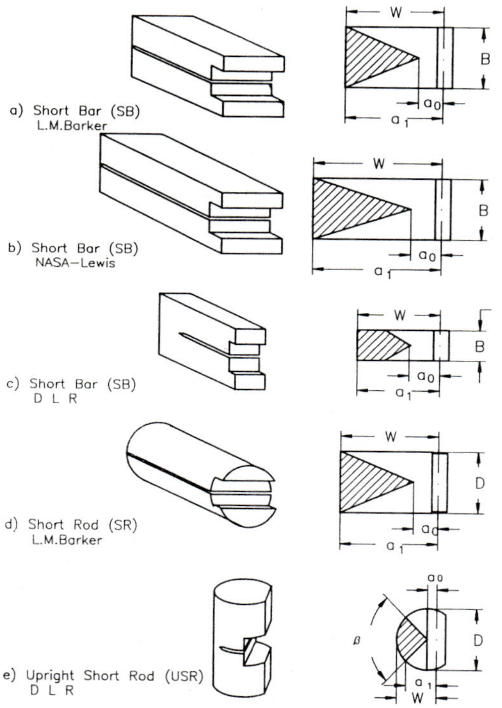

Figure 5.39 Examples of chevron notched specimens for fracture toughness testing

Testing of Chevron Notched Specimens

The successful application of chevron notched specimen testing depends largely on the available specimen geometries and the necessary mathematical fracture mechanics characterization of these specimens. Figure 5.39 shows some chevron notched specimens [949-953] which are representative of what is available presently as far as specimen geometries are involved. There are many more specimens available which vary in certain aspects of the specimen or chevron geometry [954]. Figure 5.40 shows schematically how these specimens are loaded in a test machine. It should be obvious that the specimen grips have to be adjusted to the size of the specimen.

	Type	W	2H	a_0	a_1
a	SB	1.45B	B	0.2W	W
b	SB	2.0B	B	0.2W	W
c	SB	2.5B	2B	0.35W	0.7W
d	SR	1.45D	D	0.2W	W
e	USR	2/3D	1.5D	0.25W	0.53W

Table 5.2 Geometrical definitions for chevron notched specimens shown in Figure 5.39

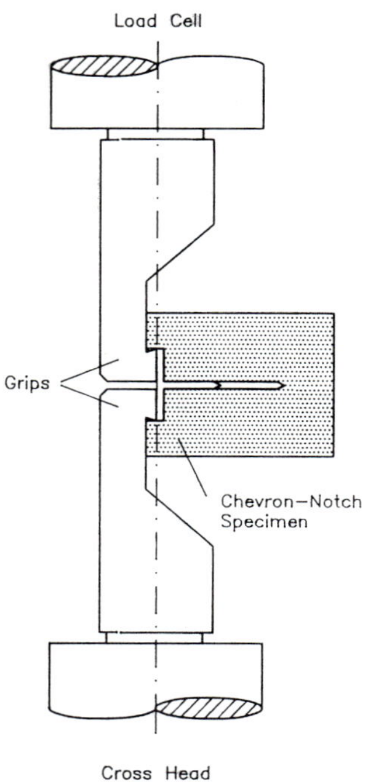

Figure 5.40 Gripping arrangement for chevron notched specimen testing

Theory of Chevron Notched Specimen Testing

It was the perceptive ideas of Barker [949] which led to fracture toughness testing via chevron notched specimens. A crack initiates at the tip of a chevron under low loading conditions because of the stress concentration there. Under increasing load, the crack propagates into the chevron in a stable fashion due to the increase in width "b" of the chevron. At the point of instability, the increase in energy needed for incremental crack advance is more than compensated by energy set free by the specimen; when this occurs, the maximum load is reached. The assumption is made here that no R-curve behaviour occurs in the material, i.e., K_{Ic} is independent of the amount of crack extension, or equivalently K_{Ic} is a constant material property at a given temperature. A further assumption is that the chevron machined by cutting thin slits into the material produces a strong constraint against plastic deformations such that plane strain conditions are established in comparatively small specimens.

In order to simplify the analytical derivations, the crack length and the chevron notch geometry are normalized as follows (see Figure 5.39):

$$\alpha = \frac{a}{W} \qquad \alpha_0 = \frac{a_0}{W} \qquad \alpha_1 = \frac{a_1}{W} \tag{5.44}$$

The energy made available for crack extension $\Delta\alpha$ by the change in potential energy is:

$$\Delta U = \frac{P^2}{2W} \cdot \frac{dC}{d\alpha} \cdot \Delta\alpha \qquad \text{C=compliance. P=load} \tag{5.45}$$

The energy needed for crack extension is

$$\Delta W = G_{Ic} \cdot b \cdot \Delta\alpha = \frac{K_{Ic}^2}{E'} \cdot b \cdot \Delta\alpha \tag{5.46}$$

where E' is Young's modulus under plane strain.
In case of crack extension, there has to be $\Delta U = \Delta W$

$$\frac{P^2}{2W} \cdot \frac{dC}{d\alpha} = \frac{K_{Ic}^2}{E'} \cdot b \tag{5.47}$$

$$K_{Ic} = \frac{P}{\sqrt{2W}} \cdot \sqrt{\frac{dC}{d\alpha} \cdot \frac{E'}{b}} \tag{5.48}$$

$$b = B \cdot \frac{\alpha - \alpha_0}{\alpha_1 - \alpha_0} \tag{5.49}$$

$$C^* = C \cdot E' \cdot B \tag{5.50}$$

$$K_{Ic} = \frac{P}{B \cdot \sqrt{W}} \cdot \sqrt{\frac{1}{2} \cdot \frac{dC^*}{d\alpha} \cdot \frac{\alpha_1 - \alpha_0}{\alpha - \alpha_0}} \tag{5.51}$$

$$K_{Ic} = \frac{P}{B \cdot \sqrt{W}} \cdot Y^* \tag{5.52}$$

$$Y^* = \sqrt{\frac{1}{2} \cdot \frac{dC^*}{d\alpha} \cdot \frac{\alpha_1 - \alpha_0}{\alpha - \alpha_0}} \tag{5.53}$$

For the experimental determination of K_{Ic} of a material from a chevron notched specimen it is necessary to know the correction factor Y* as a function of the normalized crack length $\alpha = a/w$.

The assumption was made that K_{Ic} is a constant material property. It is obvious from Eq. (5.52) and the chevron geometry that at the point of unstable crack propagation the load reaches the maximum value and Y* the minimum value. Since the maximum load P_{max} is furnished by the experimental result, only the minimum correction factor Y^*_{min} needs to be known to determine K_{Ic}.

$$K_{Ic} = \frac{P_{max}}{B \cdot \sqrt{W}} \cdot Y^*_{min} \tag{5.54}$$

The compliance C of a chevron notched specimen with crack length a is determined either experimentally, analytically or by FEM. C is the slope of a load versus displacement curve of the respective chevron notched specimen. The compliance C of a certain chevron notched specimen geometry in which all dimensions of one specimen are scaled relative to the dimensions of another specimen can be normalized by multiplication with Young's modulus and the width of the chevron (B or D).

$$C^* = CE'B \quad \textit{for SB specimen} \tag{5.55}$$

$$C^* = CE'D \quad \textit{for USR, SR specimen} \tag{5.56}$$

The normalized compliance values are put in a - if possible, simple - functional relation to the normalized crack length α, which can be differentiated with respect to the normalized crack length.

Equations (5.57) through (5.61) are the numerical expressions of the normalized compliance C* as a function of α for some of the specimens shown in Figure 5.39.

$$C^* = -276.6 + 2129.2 \cdot \alpha - 4754.2 \cdot \alpha^2 + 3754.5 \cdot \alpha^3 \quad (SB) \tag{5.57}$$

$$C^* = e^{4.1 - 6.2 \cdot \alpha + 33.2 \cdot \alpha^2 - 47.8 \cdot \alpha^3 + 25.5 \cdot \alpha^4} \quad (SR, 0.2) \tag{5.58}$$

$$C^* = e^{29.9 - 1159.2 \cdot \alpha + 377.4 \cdot \alpha^2 - 394.4 \cdot \alpha^3 + 156.5 \cdot \alpha^4} \quad (SR, 0.5) \tag{5.59}$$

$$C^* = 0.053 \cdot e^{11.2 \cdot \alpha} + 27.54 \qquad (USR, 60°) \tag{5.60}$$

$$C^* = 0.04 \cdot e^{11.7 \cdot \alpha} + 32.15 \qquad (USR, 90°) \tag{5.61}$$

Knowing C*, Y* is calculated from Eqs. (5.57) through (5.61) by differentiating C* with respect to α and taking the square root of 1/2 times the differential $\sqrt{1/2 \cdot dC^*/d\alpha}$. This expression is multiplied with the square root of the geometrical term $\sqrt{(\alpha_1 - \alpha_0)/(\alpha - \alpha_0)}$. Finally, the minimum of Y* is calculated and put into Eq. (5.54) to get the expression for fracture toughness. Y* and the respective minima Y^*_{min} and α_{min} were obtained for Eqs. (5.57), (5.60), and (5.61) this way and are given in Eqs. (5.62), (5.63), and (5.64). For Eqs. (5.58) and (5.59). Y^*_{min} and α_{min} are given in Eqs. (5.65) and (5.66), respectively.

$$Y^* = 48.9 - 19.3 \cdot \frac{1}{\alpha} + 0.0116 \cdot e^{\frac{3}{\alpha}}, \; Y^*_{min} = 14.69, \; \alpha_{min} = 0.475 \tag{5.62}$$

$$Y^* = 0.439 \cdot e^{5.6 \cdot \alpha} \cdot (\alpha - 0.25)^{-\frac{1}{2}}, \; Y^*_{min} = 9.83, \; \alpha_{min} = 0.339 \tag{5.63}$$

$$Y^* = 0.352 \cdot e^{5.85 \cdot \alpha} \cdot (\alpha - 0.25)^{-\frac{1}{2}}, \; Y^*_{min} = 8.56, \; \alpha_{min} = 0.335 \tag{5.64}$$

$$Y^*_{min} = 25.73. \; \alpha_{min} = 0.49 \tag{5.65}$$

$$Y^*_{min} = 28.27, \; \alpha_{min} = 0.65 \tag{5.66}$$

Experimental results show that intolerable scatter can occur for small specimens and with certain materials, if the correction factor Y^*_{min} is obtained from one equation equivalent to Eqs. (5.62) through (5.66). This scatter is primarily because such equations presume a certain nominal chevron notch geometry (α_0. α_1), while actual chevron notched specimens can deviate from this nominal geometry due to machining tolerances. The relative scatter becomes larger as the specimen gets smaller. These manufacturing tolerances are of minor importance for the C* versus α relation and the $dC^*/d\alpha$ term. But, Eq. (5.53)

$$Y^* = \sqrt{\frac{1}{2} \cdot \frac{dC^*}{d\alpha} \cdot \frac{\alpha_1 - \alpha_0}{\alpha - \alpha_0}} = \sqrt{\frac{1}{2} \cdot \frac{dC^*}{d\alpha}} \cdot \sqrt{\frac{\alpha_1 - \alpha_0}{\alpha - \alpha_0}} \tag{5.67}$$

contains a geometry-sensitive term, namely

$$\sqrt{\frac{\alpha_1 - \alpha_0}{\alpha - \alpha_0}} \tag{5.68}$$

Therefore, Y* has to be divided into a geometry-sensitive term and a geometry-insensitive term.

$$Y = \sqrt{\frac{1}{2} \cdot \frac{dC^*}{d\alpha}} \tag{5.69}$$

$$Y^* = Y \cdot \sqrt{\frac{\alpha_1 - \alpha_0}{\alpha - \alpha_0}} \tag{5.70}$$

$$Y^*_{min} = Y(\alpha_{min}) \cdot \sqrt{\frac{\alpha_1 - \alpha_0}{\alpha_{min} - \alpha_0}} \tag{5.71}$$

$Y(\alpha_{min})$ is obtained by differentiating equations like Eqs. (5.57) through (5.61) with respect to α and determining $dC^*/d\alpha$ for α_{min}, where α_{min} is obtained from equations like Eqs. (5.62) through (5.66); this value is inserted into Eq. (5.69). Similarly α_{min} is inserted for α in Eq. (5.70). Combining the two equations gives Eq. (5.71). $Y(\alpha_{min})$ for Eqs. (5.62) through (5.64) is given in Eqs. (5.72) through (5.74).

$$\begin{aligned} Y(\alpha_{min}) = & -3.851 + 73.63 \cdot \alpha - 65.85 \cdot \alpha^2 \\ & - 94.43 \cdot \alpha^3 + 179.8 \cdot \alpha^4 \end{aligned} \quad (SB) \tag{5.72}$$

$$Y(\alpha_{min}) = 0.545 \cdot e^{5.6 \cdot \alpha}, \quad (USR, 60°) \tag{5.73}$$

$$Y(\alpha_{min}) = 0.483 \cdot e^{5.85 \cdot \alpha}, \quad (USR, 90°) \tag{5.74}$$

Some Results of Fracture Toughness Determinations

In Table 5.3 fracture toughness values K_{Ic} determined in accordance with the ASTM E 399 standard for the Al alloy 7475-T7351 are compared to the data obtained with DLR short bar specimens. As can be seen from Table 5.3, there is good agreement between the two test methods. This is true in general for all short bar or short rod specimens if their thickness (diameter) is in the range of the thickness requirement of standard ASTM E 399. As was pointed out before, the assumption in chevron notched specimen testing is the absence of R-curve behaviour in the material to be tested. Materials which exhibit R-curve behaviour tend to furnish higher fracture toughness data when tested with chevron notched specimens. In such a case, a chevron notch geometry α_0 and α_1 should be selected to reduce crack propagation between α_0 and α_{min}.

Figure 5.41 shows some fracture toughness data obtained on 12 mm dia. extruded rods by means of 11.5 mm dia. short rod specimens. The material was an experimental alloy Al6Fe made from melt spun ribbons after different metallurgical treatment [115]. Such fracture toughness values can only be obtained from short rod specimens.

Table 5.4 gives the fracture toughness values obtained with DLR short bar specimens in 60 mm dia. extruded rods of a mechanically alloyed Al alloy. The alloying element was carbon. The individual specimens had different positions relative to the axis of the extruded rod and different orientations. Such an analysis of the variation of fracture toughness in a plane perpendicular to the axis of the rod can be made only with chevron notched specimens.
Table 5.3 and Table 5.4 are examples of applications for fracture toughness testing by means of chevron notched specimens which are not possible by any other experimental method.

Material	Orien-tation	Width B mm	CT Specimen K_{Ic} $MPa\sqrt{m}$	Width B mm	chevron Specimen K_{SB} $MPa\sqrt{m}$
7475-T7351	LT	25	43.5	20	47.4
7475-T7351	LT	25	43.5	20	47.1
7475-T7351	LT	25	41.5	20	44.7
7475-T7351	LT	25	41.5	20	44.5
7475-T7351	LT	25	41.5	20	43.7
7475-T7351	LT	25	41.5	20	44.8
7475-T7351	TL	38	40.5	30	42.6
7475-T7351	TL	25	38.7	20	36.8
7475-T7351	TL	25	38.7	20	36.8
7475-T7351	TL	25	38.7	20	37.7
7475-T7351	TL	25	38.7	20	36.3
7475-T7351	TL	25	38.7	20	38.2
7475-T7351	TL	25	38.7	20	37.7
7475-T7351	TL	25	38.7	20	38.5
7475-T7351	TL	25	38.7	20	38.0
7475-T7351	TL	25	38.7	20	38.2
7475-T7351	TL	25	38.7	20	37.5
7475-T7351	TL	25	38.7	20	38.5
7475-T7351	TL	25	37.0	20	37.5
7475-T7351	TL	25	37.0	20	39.3
7475-T7351	TL	25	38.7	20	36.9
7475-T7351	TL	25	38.7	10	39.1
7475-T7351	TL	25	38.7	10	39.6
7475-T7351	TL	25	38.7	10	40.7
7475-T7351	TL	25	38.7	10	46.3
7475-T7351	TL	25	38.7	10	43.3
7475-T7351	TL	25	38.7	10	39.3
7475-T7351	TL	25	38.7	10	39.3
7475-T7351	TL	25	38.7	10	43.9

Table 5.3 Comparison between K_{SB} and K_{Ic} values for the Al alloy 7475-T7351 based on DLR SR specimens (see Figure 5.39c)

Finally it may be stated that chevron notched specimen testing allows fracture toughness determinations on many materials and product forms which are not amenable with standardized testing methods. These test results can be used in a comparative manner or as quality control requirements.

Orientation, ASTM	K_{SB} in $MPa\sqrt{m}$
C - L	22.2, 22.2, 22.1 mean value = 22.2
C - L	21.9, 22.0, 21.9 mean value = 21.9
C - L	24.4, 23.0, 23.0 mean value = 23.4
R - L	19.1, 19.4, 19.3 mean value = 19.3
L - C	34.4, 34.8, 35.2 mean value = 34.8

Table 5.4 Measured fracture toughness with DLR SB specimens at different positions and orientations in a 60 mm dia. extruded PM aluminium rod

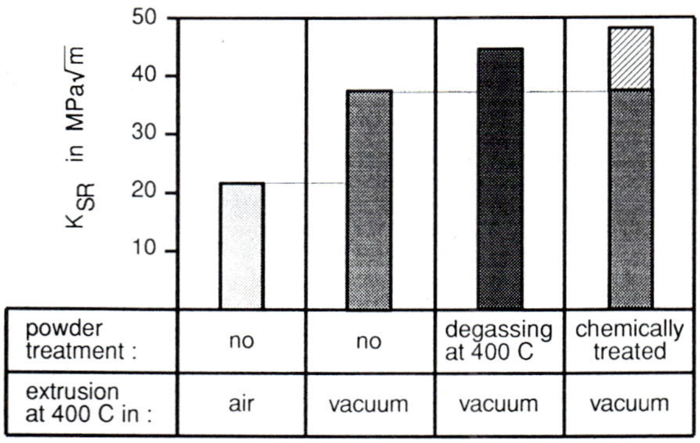

Fracture toughness of Al6Fe extrusions (short rod specimens)

Figure 5.41 Mechanical properties of Al-6Fe extrusions from melt spun ribbons

Bibliography

[1] Bunk, W.; Esslinger, P.; Kellerer, H.; Proc. SAMPE 1989, S. Benson, E. Trewin, R.M. Turner, Eds., Birmingham, England, Materials and Processing - Move into the 90s, 327-341.

[2] *Aluminium-Lithium Alloys,* M. Peters, P.-J. Winkler, Eds., DGM, Oberursel, 1992.

[3] Peters, M.; Bunk, W.; J. Aircraft, Vol. 27 (1990) 456-458.

[4] Fournier, P.; Barbaux, Y.; Rendigs, K.H.; in [2], (to be published).

[5] Arendts, F.J.; Proc. BDLI Werkstofftag 1991, Hamburg, (to be published).

[6] Neite, G.; Matucha, K.H.; Dudek, H.J.; Bunk, W.; Proc. EUROMAT, Aachen, DGM, 1989, 1325-1330.

[7] Gilman, P.S.; Zedalis, M.S.; Peltier, J.M.; Das, S.K.; Industrial Heating, 56, No. 2 (1989) 30-34.

[8] Leatham, A.G., Brooks, R.G.; Coombs, J.S., Ogilvy, A.J.; Proc. ICSF, Swansea, England, 1990, 1-12.

[9] Schulte, K.; Bunk, W.; AGARD Conf. Proc. 444, 26-1.

[10] Hayashi, T.; Sadhana 11 (1987) 299-325.

[11] private communication; ESA/ESTEC, 1990.

[12] Minoru Taya; Materials Trans, JIM, 1991, 1.

[13] Kumpfert, J.; Ward, C.H.; Grundhoff, K.J.; Schurmann, H.; Lee, Y.T.; Peters, M.; in [5], (to be published).

[14] Marissen, R.; in : Advances in Fatigue Science and Technology, C.M. Branco, L.G. Rosa, Eds., Kluwer Acad. Pub., 1989, p. 697.

[15] Bunk, W.; Esslinger, P.; Kellerer, H.; in [1], p. 327.

[16] Krüger, W.; Bunk, W.; Proc. Ceramic Materials for Engines 1991, Göteborg, (to be published).

[17] Krammer, P.; Schwab, R.R.; MTU FOCUS, Germany, 1992, 5-14.

[18] Aerospacial Comp., 1990.

[19] Dietrich, G.; Meistring, R.; Proc. 2. Symp. Materialforschung 1991, Dresden, 1561-1606.

[20] J.E. Hatch (Ed.); *Aluminium, Properties and Physical Metallurgy.* American Society for Metals, Metals Park, Ohio, 1984.

[21] Polmear, I.J.; *Light Alloys, Metallurgy of the Light Metals.* 2nd Edition, E. Arnold, London, 1989.

[22] Vasudevan, A.K.; Doherty, R.D.; *Aluminum Alloys - Contemporary Research and Applications.* Academic Press Inc., San Diego, CA, 1989.

[23] Peters, M.; Welpmann, K.; Sanders, Jr., T.H.; Proc. Conf. on Advanced Materials Research and Developments for Transport, Strasbourg, Nov. 1985, R.J.H. Wanhill, W.J.G. Bunk and J.G. Wurm, Eds., Les Editions de Physique, Les Ulis Cedex, France, 1985, p.63.

[24] McDarmaid D.S.; Peel, C.J.; in: [52], p. 993.

[25] Schelb, W.; Haszler, A.; Jäger, H.; Welpmann, K.; Peters, M.; in: [2], in press.

[26] Summerson, T.J.; Sprowls, D.O.; in: Aluminum Alloys, their Physical and Mechanical Properties, EMAS, Cradley Heath, U.K., 1986, p. 1576.

[27] Metals Handbook, 9th Edition, Vol. 2: Properties and Selection: Nonferrous Alloys. ASM, Metals Park, Ohio, 1979.

[28] anonym; Tempers for Aluminum and Aluminum Products. The Aluminium Association Inc., Washington, 1989.

[29] Tack, W.T.; Heubaum, F.H.; Pickens, J.R.; Scripta Metall. 24 (1990) 1685.

[30] Cho, A.; Greene, R.E.; Edwards, H.M.; Skillingberg, M.H.; in: [2], in press.

[31] Cieslak, S.J.; Alcoa Aluminum Alloy 6013. Alcoa Green Letter No. 225, 1987.

[32] Hahn, G.T.; Rosenfield, A.R.; Met. Trans., 6A(1975)653.

[33] Suresh, S.; Vasudevan, A.K.; Tosten, M.; Howell, P.R.; Acta Met., 35(1987)25.

[34] Bunk, W.G.J.; Mat. Sci. Eng., A134 (1991) 1087.

[35] Grimes, R.; Baker, C.; Stowell, M.F.; Watts, B.M.; Aluminium 51(1975)720.

[36] Wert, J.A.; Paton, N.E.; Hamilton, C.H.; Mahoney, M.W.; Metall. Transactions 12A(1981)1267.

[37] Hamilton, C.H.; Bampton, C.C.; Paton, N.E.; *Superplasticity in High Strength Aluminum Alloys,* in: Superplastic Forming of Structural Alloys, N.E. Paton, C.H. Hamilton (Eds.), 173-189, TMS, Warrendale, PA, USA 1982.

[38] Friedrich, H.E.; Winkler, P.-J.; Advanced Materials & Processes, August, 1991, 16.

[39] Liesner, C.; Aluminium 65(1989)249.

[40] Grimes, R.; Davis, T.; Saxty, H.J.; Fearon, J.E.; in: [51], p. C3-11.

[41] Fridlyander, I.N.; Bratukhin, A.G.; Davydov, V.G.; in: [2], in press.

[42] Miller, W.S.; White, J.; Lloyd, D.J.; in: [51], p.C3-139.

[43] Meyer, P.; Dubost, B.; in: [50], p.37.

[44] Bretz, P.E.; Sawtell, R.R.; in: [50], p. 47.

[45] Wakiyama, T.; Tsuji, Y.; Ohzono, T.; in: [2], in press.

[46] Pickens, J.R.; Heubaum, F.H.; Langan, T.J.; Kramer, L.S.; in: [52], p. 1397.

[47] Shrimpton, G.R.D.; Angus, H.C.; in: [52], p. 1565.

[48] Aluminum-Lithium Alloys, Eds. T.H. Sanders, E.A. Starke, TMS-AIME, Warrendale, 1981.

[49] Aluminum-Lithium Alloys II, Eds. T.H. Sanders, E.A. Starke, TMS-AIME, Warrendale, 1984.

[50] Proceedings of the Third International Aluminium-Lithium Conference, Eds. C. Baker, P.J. Gregson, S.J. Harris, C.J. Peel, The Institute of Metals, London, 1986.

[51] 4th International Aluminium Lithium Conference, Paris, Eds. G. Champier, B. Dubost, D. Miannay, L. Sabetay, Les éditions de physique, Les Ulis, 1987.

[52] Proceedings of the Fifth International Aluminum-Lithium Conference, Williamsburg, Eds. T.H. Sanders and E.A. Starke, Materials and Component Engineering Publications, Birmingham, 1989.

[53] Sanders, T.H.; Starke, E.A.; in: [52], p. 1.

[54] Starke, E.A.; Quist, W.E.; in: AGARD Conference Proceedings No.444, New Light Alloys, Neuilly sur Seine, 1989, paper 4.

[55] Polmear, I.J.; Proceedings of the 2nd. Risø International Symposium, Eds. Hansen et al., Risø National Laboratory, Roskilde, 1981, p.99.

[56] Gatenby, K.M.; Reynolds, M.A.; White, J.; Palmer, I.G.; in: [52], p. 909.

[57] Doudeau, M.; Meyer, P.; Constant, D.; in: AGARD Conference Proceedings No.444, New Light Alloys, Neuilly sur Seine, 1989, paper 2.

[58] Lee, E.W.; Waldman, J.A.; Witters, J.J.; Rioja, R.J.; Divecha, A.P.; in: [2], in press.

[59] Welpmann, K.; Buhl, H.; Braun, R.; Peters, M; in: [51], p. C3-677.

[60] 't Hart, W.G.J.; Schra, L.; McDarmaid, D.S.; Peters, M.; in: AGARD Conference Proceedings No.444, New Light Alloys, Neuilly sur Seine, 1989, paper 5.

[61] Miller, W.S.; Thomas, M.P.; Lloyd, D.J.; Creber, D.; Materials Science and Technoloy 2(1986)1210.

[62] Miller, W.S.; White, J.; Reynolds; M.A.; in: 1st International SAMPE Metals Conference, Cherry Hill, New Jersey, 1987, p.308.

[63] Venkateswara Rao, K.T.; Yu, W.; Ritchie, R.O.; Metall.Trans.A 19A(1988)549.

[64] Kaufman, J.G.; Santer, J.S.; in: Application of Fracture Mechanics for Selection of Metallic Structural Materials, Eds. J.E. Champbell, W.W. Gerberich, J.H. Underwood, American Society for Metals, Metals Park, Ohio, 1982, p.169.

[65] Peel, C.J.; in: AGARD Lecture Series No.174, New Light Alloys, September 1990, paper 1.

[66] Terlinde, G.; Sauer, D.; Fischer, G.; Smith, A.; in: [2], in press.

[67] Peters, M.; Welpmann, K.; McDarmaid, D.S.; t' Hart, W.G.J.; in: AGARD Conference Proceedings No.444, New Light Alloys, Neuilly sur Seine, 1989, paper 6.

[68] Gregson,P.J.; Newman,J.; Gray,A.; Materials Science and Technoloy 5(1989)65.

[69] Fox, S.; Flower, H.M.; McDarmaid, D.S.; in: [50], p. 263.

[70] Venkateswara Rao, K.T.; Yu, W.; Ritchie, R.O.; Metall.Trans.A 19A(1988)563.

[71] Rinker, J.G.: Marek, M.; Sanders, T.H.; Materials Science and Engineering 64(1984)203.

[72] Balmuth, E.S.; Schmidt, R.; in: [48], p. 69.

[73] Lane, P.L.; Gray, J.A.; Smith, C.J.E.; in: [50], p. 273.

[74] Peters, M; Bachmann, V.; Welpmann, K.; in: [51], p.C3-785.

[75] Holbein, R.; Dembeck, A.; Werkst.Korros. 41(1990)609.

[76] Hack, T.; Koch, U.; in: [2], in press.

[77] Braun, R.; Werkst.Korros. 40(1989)661.

[78] Smith, C.J.E.; Gray, J.A.; Schra, L.; Boogers, J.A.M.; Braun, R.; Buhl, H.; Vaessen, G.; in: AGARD Conference Proceedings No.444, New Light Alloys, Neuilly sur Seine, 1989, paper 7.

[79] Braun, R.; in: [2], in press.

[80] Braun, R.; Buhl, H.; in: [2], in press.

[81] Colvin, E.L.; Murtha, S.J.; in: [52], p. 1251.

[82] Schra, L.; t' Hart, W.G.J.; Boogers, J.A.M.; National Aerospace Laboratory NLR, NLR TP 90272 L, Amsterdam, 1990.

[83] Reboul, M; Meyer, P.; in: [51], p. C3-881.

[84] Dorward, R.C.; Hasse, K.R.; Corrosion 44(1988)932.

[85] Gray, A.; in: [51], p. C3-891.

[86] Gray, A.; Holroyd, N.J.H.; White, J.; in: [52], p. 1175.

[87] Kolkman, H.J.; Schra, L.; National Aerospace Laboratory NLR, NLR TP 91399 L, Amsterdam, 1991.

[88] Venkateswara Rao, K.T.; Piascik, R.S.; Gangloff, R.P.; Ritchie, R.O.; in: [52], p. 955.

[89] Chellman, D.J.; Rainen, R.A.; Witters, J.J.; James, R.S.; in: [2], in press.

[90] Reynolds, M.A.; Creed, E.; Court, S.A.; Gregson, P.J.; in: [2], in press.

[91] Smith, A.F.; in: [2], in press.

[92] Rioja, R.J.; Newman, J.M.; Govada, A.K.; in: [2], in press.

[93] Witters, J.J.; Lee, E.W.; Lisagor, W.B.; Herner, S.B.; Kilmer, R.J.; Talia, J.E.; in: [2], in press.

[94] Kehl, W; Fischmeister, H.F.; Powder Metallurgy, 3(1980)113.

[95] Benjamin, J.S.; DGM Conference "New Materials by Mechanical Alloying Techniques", Hirsau, Germany, 1988.

[96] Jones, H.; Aluminium, 54(1978)274.

[97] Voss, D.; Final Report, AFOSR 77-3440, EOARD-TR-80-1, (1979).

[98] Sawtell, R.R.; International Conference of PM Aerospace Materials, Lausanne, Switzerland, 1991.

[99] Gilman, P.S.; Das, S.K.; in: Proc. Int. Conf. on Aereospace Materials, Luzern, No. 2-4, 1987, Metal Powder Rep., 1987, p.27.

[100] Maringer, R.E.; J. Vac. Sci. Techn., 11(1974)1067.

[101] Gurney. F.J.; Abson, D.J.; Depierre, V.; Technical Report AFML-TR-73-252 (1973).

[102] Champier, G.; Agard Lecture Series No.174 on New Light Alloys, Madrid, Spain, 1990.

[103] Suryanarayana, C.; Froes, F.H.; Quist, W.E.; *Advances in Powder Metallurgy-1991,* MPIF and APMI, Princeton, USA, (1991)15.

[104] Frazier, W.E.; Thompson, J.J.; *Light-Weight Alloys for Aerospace Applications,* E.W. Lee, E.H. Chia and N.J. Kim, Eds., The Minerals, Metals & Materials Society, 1989.

[105] Bunk, W.G.J.; Materials Science and Engineering, A134(1991)1087.

[106] Irmann, R.; Technische Rundschau, Bern, 36(1949)19.

[107] Arnhold, V.; Müller-Schwelling, D.; International Congress and Exposition, SAE, Warrendale, PA, USA,1991.

[108] Kumpfert, J.; Thumann, M.; Staniek, G.; Kleinekathöfer, W.; in: Structural Applications of Mechanical Alloying, ASM, 1990, p. 176.

[109] Schelleng, R.D.; Kemppinen, A.I.; Weber, J.H.; in: [52], p. 731.

[110] Gilman, P.S.; Journal of Metals, 8(1991)7.

[111] Ravichandra, K.S.; Dwarakadasa, E.S.; Journal of Metals, 5(1987)28.

[112] Hunt, W.H.; International Conference of PM Aerospace Materials, Lausanne, Switzerland, 1991.

[113] K. Wefers, C. Misra; *Oxides and Hydroxides of Aluminium,* Alcoa Technical Paper No. 19, Pittsburgh, 1987.

[114] Kim, Y.-W.; Froes, F.H.; 8th ILMT, Leoben-Vienna, 1987, p. 756.

[115] Staniek, G.; Wefers, K.; in: Advances in Powder Metallurgy-1991, MPIF and APMI, Princeton, USA, (1991)31.

[116] Poulsen, E.R.; Hall, J.A.; J. of Metals 35(6/1983)60.

[117] Bomberger, H.B.; Froes, F.H.; Morton, P.H.; in: Titanium Technology, 3-17, F.H. Froes, D. Eylon, H.B. Bomberger, Eds., TDA, Dayton, OH, USA, 1985.

[118] Donachie, M.J., Jr.; Ed., Titanium and Titanium Alloys, ASM, 1982.

[119] Jaffee, R.I.; Promisel, N.E.; Eds., The Science, Technology and Application of Titanium, Pergamon Press, New York, 1970.

[120] R.I. Jaffee, H.M. Burte; Eds., Titanium Science and Technology, Plenum Press, New York, 1973.

[121] J.C. Williams, A.F. Belov; Eds., Titanium and Titanium Alloys, Plenum Press, New York, 1982.

[122] H. Kimura, O. Izumi; Eds., Titanium '80, TMS, Warrendale, PA, USA, 1980.

[123] G. Lütjering, U. Zwicker, W. Bunk, Eds., Titanium Science and Technology, DGM, Oberursel, Germany, 1985.

[124] P. Lacombe, B. Tricot, G. Béranger, Eds., Sixth World Confernce on Titanium, Les Editions de Physique, Les Ulis Cedex, France, 1989.

[125] F.H. Froes, I.L. Caplan, Eds., Seventh World Conference on Titanium, TMS, Warrendale, PA, USA, 1992.

[126] Flower, H.M.; Materials Science and Technology 6(1990)1082.

[127] Rosenberg, H.W.; in: [119], p. 851.

[128] Hall, J.A.; in: [124], p. 1.

[129] Peters, M.; Lee, Y.T.; Grundhoff, K.-J.; Schurmann, H.; Welsch, G.; in: Microstructure/Property Relationships in Titanium Aluminides and Alloys, TMS, Warrendale, PA, USA, (1991), 533.

[130] Peters, M.; Bachmann, V.; Trautmann, K.-H.; Lee, Y.T.; Ward, C.H.; Schurmann, H.; in: [125], .

[131] Prandi, B.; Alheritiere, E.; Schwartz, F.; in: [124], Vol. 2, p. 811.

[132] Bania, P.J.; Parris, W.M.; in: Titanium 1990, Vol 2, p.784, TDS, Dayton, OH, 1990.

[133] Margolin, H.; Chesnutt, J.C.; Lütjering, G.; Williams, J.C.; in: [122], Vol. 1, p.169.

[134] Peters, M.; Lütjering, G.; Ziegler, G.; Z. Metallkde. 74(1983)274.

[135] Peters, M.; Gysler, A.; Lütjering, G.; in: [122], Vol. 3, p. 1777.

[136] Lütjering, G.; Gysler, A.; in: [123], Vol. 4, p. 2065

[137] Peters, M.; Williams, J.C. Metallurgical Transactions, 15A(1984)1589.

[138] Döker, H.; Peters, M.; in: Fatigue '84, Vol. I, p.275, EMAS, Birmingham, England, 1984.

[139] Marissen, R.; Peters, M.; Schulte, K.; Trautmann, K.-H.; Nowack, H.; in: [123], Vol. 4, p. 2289.

[140] Peters, M.; Welpmann, K.; Döker, H.; in: [123], Vol. 4, p. 2267.

[141] Peters, M.; Williams, J.C.; in: [123], Vol. 3, p. 1843.

[142] Lee, Y.T.; Peters, M.; Wirth, G.; Materials Science and Engineering, 93(1988)105.

[143] Peters, M.; Lütjering, G.; in: [122] Vol.2, p.929.

[144] Peters, M.; Gysler, A.; Lütjering, G.; Metallurgical Transactions, 15A(1984)1597.

[145] Rätzer-Scheibe, H.-J.; Buhl, H.; in: [123], Vol.4, p. 2641.

[146] Kelly, E.J.; Modern Aspects of Electrochemistry, Ed.: L.O'M. Bockris, B. E. Conway, R. E. White; Plenum Press, New York, London, Vol. 14, 1982, S. 319.

[147] Griess, Jr., J.C.; Corrosion, 24(1968)96.

[148] McKay, P.; Her Majesty's Stationery Office, London 1987, Paper 7, p.107.

[149] Schutz, R.W.; Grauman, J.S.; STP 917, ASTM, 1986, 130.

[150] Kobayashi, M.; in: [124], Vol. 4, p. 1695.

[151] Koizumi, T.; Furuya, S.; in: [120], Vol.4, p.2383.

[152] Szklarska-Smialowska, Z.; *Pitting Corrosion of Metals,* NACE, Houston, 1986, p.236.

[153] Bomberger, H.B., Jr.; Meyn, D.A.; Fraker, A.C.; in: [123], p.2435.

[154] Brown, B.F.; Lennox, T.J., Jr.; Newbegin, R.L.; Peterson, M.H.; Smith, J.A.; Waldron, L.J.; NRL Memorandum Report 1574, Second Interim Report of Progress, U.S. Naval Research Laboratory, Washington, D.C., Nov. 1964.

[155] Blackburn, M.J.; Feeney, J.A.; Beck, T.R.; in: Advances in Corrosion Science and Technology, Eds., M.G.Fontana, R.W.Staehle, Plenum Press New York-London, 1973, Vol.3, S.67

[156] Koch, G.H.; Bursle, A.J.; Liu, R.; Pugh, E.N.; Metallurgical Trans., 12A(1981)1833.

[157] Scully, J.C.; *Stress Corrosion Research,* Sijthoff & Noordhoff, Alphen aan den Rijn, Holland, 1979, p.209.

[158] Padawer, G.M.; Adler, P.N.; Stauber, M.C.; Grumman Aerospace Corp., Report RE-522, 1976.

[159] Xiao, H.Z.; Gao, S.J.; Wan, X.J.; Scripta Met., 21(1987)265.

[160] Vassel, A.; Lapasset, G.; Aucouturier, M.; Comptes Rendues, 274(1972)494.

[161] Katz, J.; Gerberich, W.W.; Int. J. Fract. Mech., 6(1970)219.

[162] Papazoglov, T.P.; Hepworth, M.T.; Trans. TSM-AIME, 242(1968)682.

[163] Wasilewski, R.J.; Kehl, G.L.: Metallurgica, 50(1954)225.

[164] Holman, W.R.; Crawford, R.W.; Parades, F.; Trans. TMS-AIME, 233(1965)1836.

[165] Adler, P.N.; Schulte, R.L.; Scripta Met., 10(1976)669.

[166] Vassel, A.; in: Environment-Sensitive Fracture of Engineering Materials, Ed., Z.A. Foroulis, AIME, Warrendale, PA, USA, 1979, p.277.

[167] Koch, G.H.; Bursle, A.J.; Pugh, E.N.; Met. Trans.,9A(1978)129.

[168] Buhl, H.; ASTM STP 665, Stress Corrosion Cracking - The Slow Strain Rate Technique, Eds., G.M.Ugiansky and J.H.Payer, ASTM, 1979, p.333.

[169] Wanhill, R.J.H.; NLR-MP-78039-U, Nationaal Lucht- en Ruimtevaartlaboratorium, Amsterdam, The Netherlands, 1978.

[170] Speidel, M.O.; in: Advances in Fracture Research (Fracture 84), Vol. 1, Pergamon Press, Oxford and New York, 1986, S. 379-398.

[171] Ryuichiro, E.; Yoshikazu, Y.; Akira, G.; in: Corrosion Fatigue, Mechanics, Metallurgy, Electrochemistry, and Engineering, ASTM STP 801, Eds., T.W. Crooker and B.N. Leis, ASTM, 1983, p.135.

[172] Wei, R.P.; Gao, S.J.; Scripta Met., 17(1983)959.

[173] Kofstad, P.; J. Less-Com. Met., 12(1967)449.

[174] Bertrand, G.; Jarraya, K.; Chaix, K.; Oxid. of Met., 21(1983)1.

[175] Kofstad, P.; Elsevier Applied Science, London and New York, 1988.

[176] Winkler, P.-J.; Däubler, M.A.; Peters, M.; in: [125].

[177] Eylon, D.; Froes, F.H.; Gardiner, R.W.; J. of Metals 35(1983)35.

[178] Winkler, P.-J.; in: [124], Vol. 3, p. 1135

[179] Peters, M.; Winkler, P.-J.; Bunk, W.; in: [123], Vol. 2, p. 681.

[180] Lee, Y.T.; Schurmann, H.; Grundhoff, K.-J.; Peters, M.; Powder Metallurgy International, 22(1990)11.

[181] Froes, F.H.; Smugeresky, J.E., Eds., Powder Metallurgy of Titanium Alloys, TMS, Warrendale, PA, USA, 1980.

[182] Lipsitt, H.A.; in: High-Temperature Ordered Intermetallic Alloys, C.C. Koch, C.T. Lin, N.S. Stoloff, Eds., 1985, Pittsburgh, PA, Materials Research Society, p.351.

[183] Murray, J.L.; in: Binary Alloy Phase Diagrams, T.B. Massalski, Ed., 1986, Metals Park, ASM International, p. 173.

[184] McCollough, C.; Valencia, J.J.; Levi, C.G.; Mehrabian, R.; Acta metall., 37(1989)1321.

[185] McAndrew, J.B.; Simcoe, C.R.; WADD Technical Report 60-99, Wright-Patterson AFB, 1960.

[186] Blackburn, M.J.; Ruckle, D.L.; Bevan, C.E.; AFWAL Technical Report 78-18, Wright-Patterson AFB, 1978.

[187] Blackburn, M.J.; Smith, M.P.; AFWAL Technical Report 82-4086, Wright-Patterson AFB, 1982.

[188] Marquardt, B.J.; unpublished research, General Electric Company, 1989.

[189] Blackburn, M.J.; Smith, M.P.; WRDC Technical Report 89-4095, Wright-Patterson AFB, 1989.

[190] Rowe, R.G.; in: High Temperature Aluminides and Intermetallics, S.H. Whang, C.T. Lin, D.P. Pope, J.O. Stiegler, Eds., 1989, Warrendale, TMS, p.375.

[191] Larsen, J.M.; Williams, K.A.; Balsone S.J.; Stucke, M.A.; in: [190], p.521.

[192] Koss, D.A.; Banerjee, D.; Lukasak, D.A.; Gogia, A.K.; in: [190], p.175.

[193] Dimiduk, D.M.; Miracle D.B.; Ward, C.H.; Mat. Sci. and Tech., 1992, in press.

[194] Ward, C.H.; Int. Mat. Rev., 1992, in press.

[195] Court, S.A.; Löfvander, J.P.A.; Loretto M.H.; Fraser, H.L.; Phil. Mag. A, 61(1990)109.

[196] Marquardt, B.J.; Scarr, G.K.; Chesnutt, J.C.; Rhodes C.G.; Fraser, H.L.; in: [124], p. 955

[197] Baeslack III, W.A.; Broderick, T.F.; Scripta Met., 24(1990)319.

[198] Marquardt, B.J.; Scarr, G.K.; Chesnutt, J.C.; Rhodes C.G.; Fraser, H.L.; WRDC Technical Report 89-4133, Wright-Patterson AFB, 1989.

[199] Banerjee, D.;Gogia, A.K.; Nandy T.K.; Joshi, V.A.; Acta Met., 36(1988)871.

[200] Mozer, B.; Bendersky, L.A.; Boettinger W.J.; Rowe, R.G.; Sripta Met., 24(1990)2363.

[201] Banerjee, D.; private communication, Wright Laboratory, Dec. 1990.

[202] Kestner-Weykamp, H.T.; Ward, C.H.; Broderick T.F.; Kaufman, M.J.; Scripta Met., 23(1989)1679.

[203] Muraleedharan, K.; Gogia, A.K.; Nandy, T.K.; Banerjee, D.; Lele, S.; (Part I), submitted to Met. Trans. A, 1991.

[204] Muraleedharan, K.; Gogia, A.K.; Nandy, T.K.; Banerjee, D.; Lele, S.; (Part II), submitted to Met. Trans. A, 1991.

[205] Kestner-Weykamp, H.T.; Baker, D.R.; Paxton D.M.; Kaufman, M.J.; Scripta Met., 24(1990)445.

[206] Ward, C.H.; Williams, J.C.; Thompson, A.W.; Rosenthal, D.G.; Froes, F.H.; in: [124], p.1103.

[207] Kumpfert, J.; Grundhoff, K.J.; Schurmann, H.; Lee, Y.T.; Ward, C.H.; Peters, M.; in: Proc. Conf. Euromat 91; 1991, London, The Institute for Metals.

[208] Blackburn, M.J.; Smith, M.P.; AFWAL Technical Report 80-4175, Wright-Patterson AFB, 1980.

[209] Lukasak, D.A.; Koss, D.A.; Met. Trans., 21A(1990)135.

[210] DeLuca, D.P.; Cowles, B.A.; WRDC Technical Report 90-4075, Wright-Patterson AFB, 1990.

[211] Chan, K.S.; Met. Trans., 21A(1990)2687.

[212] Chan, K.S.; Met. Trans., 23A(1991)183.

[213] Soboyejo, W.O.; Met. Trans. A, 1992, in press.

[214] Marquardt, B.J. in: AeroMat 90, Long Beach, CA, USA, May 1990, ASM International.

[215] Cho, W.; Thompson A.W.; Williams, J.C.; Met. Trans., 21A(1990)641.

[216] Hayes, R.W.; Acta Met., 39(1991)569.

[217] Balsone, S.J.; Maxwell, D.C.; Khobaib, M.; Nicholas, T.; in: Fatigue 90, H. Kitagawa and T. Tanaka, Eds., 1990, Warley, UK, Materials and Components Engineering Publications, Ltd, p.1173.

[218] Nicholas, T.; Mall, S.; in: [217], p. 1905.

[219] Aswath, P.B.; Suresh, S.; Met. Trans., 21A(1991)817.

[220] Balsone, S.J.; in: Oxidation of High Temperature Intermetallics, T. Grobstein and J. Doychak, Eds., 1988, Warrendale, TMS, p. 219.

[221] Ward, C.H.; Balsone, S.J.; in: Microstructure/Property Relationships in Titanium Aluminides and Alloys, Y-W. Kim and R.R. Boyer, Eds., 1991, Warrendale, TMS, p. 373.

[222] Kim, Y-W.; Dimiduk, D.M.; JOM, 43(Aug.1991)40.

[223] Huang, S.C.; Hall, E.L.; Gigliotti, M.F.X.; in: N.S. Stoloff, C.C. Koch, C.T. Liu, O. Izumi, Eds., High Temperature Intermetallic Alloys II, MRS Symp. Proc. 81(1987)481.

[224] Kim, Y-W.; JOM, 41(July 1989)24.

[225] Tsujimoto, T.; Hashimoto, K.; Structures and in: C.T. Liu, A.I. Taub, N.S. Stoloff, C.C. Koch, Eds., High Temperature Intermetallic Alloys III, MRS Symp. Proc. 133(1989)391.

[226] Kawabata, T.; Tamura, T.; Izumi, O.; in: [225], p. 329.

[227] Kawabata, T.; Tadano, M.; Izumi, O.; Scripta Met., 22(1988)1725.

[228] Huang, S.C.; Hall, E.L.; Structures and Properties of Gamma-TiAl Alloys Containing Interstitial Elements. in: L.A. Johnson, D.P. Pope, J.O. Stiegler, Eds., High Temperature Intermetallic Alloys IV, MRS Symp. Proc. 213(1991)827.

[229] Meyn, D.A.; Alloys. In: Y-W. Kim, R.R. Boyer (eds.), in: [221], p. 275.

[230] Dimiduk, D.M.; Miracle, D.B.; Kim, Y-W.; Mendiratta, M.G.; ISIJ International, 31(1991)1223.

[231] Huang, S.C.; Hall, E.L.; Met. Trans., 22A(1991)2619.

[232] Loiseau, A.; Lasalmonie, A.: Mater. Sci. Eng., 67(1984)163.

[233] Nishiiyama, Y.: Miyashita, T.; Isobe, S.; Noda, T.; in: S.H. Whang, C.T. Liu, D.P. Pope, J.O. Stiegler, Eds., High Temperature Aluminides and Intermetallics, TMS Warrendale, 1990, p. 557.

[234] Fukutomi, H.; Hartig, C.; Mecking, H.; Z. Metallkde., 81(1990)272.

[235] Simon Shong, D.;Kim, Y.W.; Scripta Met., 23(1989)257.

[236] Seeger, J.; Hartig, C.; Bartels, A.; Mecking, H.; in: [228], p. 157.

[237] Welsch, G.; Kahveci, A.I.; in: [220],p. 207.

[238] Huang, S.C.; Hall, E.L.; Gigliotti, M.F.X.; in : [124], p. 1109.

[239] Inui, H.; Nakamura, A.; Yamaguchi, M.; in: [228], p. 569.

[240] Beaven, P.A.; Appel, F.; Dogan, B.; Wagner, R.; in: Ordered Intermetallics-Physical Metallurgy and Mechanical Behaviour, NATO Advanced Research Workschop, Irsee, Germany, 1991.

[241] Huang. S.C.; Hall, E.L.; Scripta Met. 25(1991)1805.

[242] Lipsitt, H.A.; Shechtman, D.; Schafrik, R.E.; Met. Trans., 6A(1975)1991.

[243] Shih, D.S.; Huang, S.C.; Scarr, K.G.; Jang, H.J.; Chesnut, J.C.; in: [221], p. 135.

[244] Huang, J.S.; Kim, Y.W.; Scripta Met., 25(1991)1901.

[245] Takashi, T.; Oikawa, H.; in: [221], p. 227.

[246] Mitao, S.; Tsuyama, S.; Minikawa, K.; in: [221], p. 297.

[247] Tsuyama, S.; Mitao, S.; Minakawa, K.; in: [221], p. 213.

[248] Dowling, Jr., W.E.; Donlon, W.T.; Allison, J.E.; in: [228], p. 561.

[249] Bradley, E.F., Superalloys, ASM, Metals Park, OH, USA, 1988

[250] Ilschner, B., Hochtemperaturplastizität, Springer, Berlin, Heidelberg, New York, 1973

[251] Singer, R.F., Mats. Sci. Techn. 3(1987)726-732

[252] Harris, K., Erickson, G.L., Schwer, R.E., *Directionally Solidified and Single-Crystal Superalloys*. In: Metals Handbook, Vol. 1, 10th Edition, 995-1006, ASM, Metals Park, OH, USA, 1989

[253] Gibbons, T.B., *The Superalloys - Present Status and Future Prospects*. In: Advanced Materials and Processes, Vol 1, 315-325, DGM, Oberursel, 1990

[254] Fleetwood, M.J. Mats. Sci. Techn., 2(1986)1176-1182

[255] Liu, C.T., Sikka, V.K., J. of Metals 38(5/1986)19-21

[256] Darolia, R., J. of Metals 43(3/1991)44-49

[257] Aoki, K., Izumi, O., Nippon Kinzoku Gakkaishi, 43(1979)1190

[258] R.A. Rapp, Ed.; *High Temperature Corrosion,* NACE-6, Houston, 1983.

[259] *'High-temperature corrosion of superalloys',* Mat. Sci. Techn., 2(1987)481-583.

[260] Rätzer-Scheibe, H.J.; ESA-TT-887, 1985, see also DLR, Köln, DFVLR-Mitt. 84-04, 1984.

[261] Kohl, F.J.; Stearns, C.A.; Fryburg, G.C.; *Metal-Slag-Gas Reactions and Processes,* Z.A. Foroulis, W.W. Smeltzer, Eds., Electrochem. Soc., New York, 1975, p.649.

[262] Bianco, R.; J. Electrochem. Soc., 136(1989)265C.

[263] Goebel, J.A.; Pettit, F.S.; Met. Trans., 1(1970)1943.

[264] Bornstein, N.S.; DeCrescente, M.A.; Met. Trans., 2(1971)2875.

[265] Shores, D.A.; in: [258], p. 493.

[266] Stern, K.H.; NRL-MR-4772, Naval Research Lab., Washington D.C., 1982.

[267] Luthra, K.L.; in: [258], p. 507.

[268] Rätzer-Scheibe, H.J.; AGARD-CP-461, 1989, p.17-1.

[269] Fritscher, K.; Rätzer-Scheibe, H.J.; Wirth, G.; Winstone, M.R.; *High Temperature Materials for Power Engineering,* in: E. Bachelet et.al., Eds., Part II, Kluwer Academic Publishers, Dordrecht, 1990, p.1238.

[270] Rätzer-Scheibe, H.J.; Winstone, M.R.; in: Proc. 3rd Int. Symp. on High Temperature Corrosion, Les Embiez, France, May 25-29, 1992.

[271] Peichl, L.; Wortmann, J.; Rätzer-Scheibe, H.J.; *High Temperature Alloys: Their Exploitable Potential,* in: J.B. Marriott et.al., Eds., Elsevier Applied Science, London and New York, 1988, p. 363.

[272] Sims, Ch.T.; *in: High Temperature Alloys for Gas Turbines", D.Coutsouradis et al.,Eds.,* Applied Science Publishers, London (1978) p. 13.

[273] Goward, G.W.; Mater. Sci. Technol., 2(1986)194.

[274] Meetham,G.W.; J. Mater. Sci., 26(1991)853.

[275] Haefer, R.A.; *Oberflächen- und Dünnschicht-Technologie* WFT Bd.5, Springer Verlag (1987).

[276] Hocking, M.G.; Vasantasree, V.; Sidky, P.S.; *Metallic and Ceramic Coatings: Production, HT Properties and Application,* Longman Scientific & Technical (1989).

[277] Restall, J.E.; Wood, M.I.; Mater. Sci. Technol., 2(1986)225.

[278] Parzuchowski, R.S.; in: Metallurgical Coatings, R.F. Bunshah, Ed., Am. Vac. Soc., 1977, p.349.

[279] Lehnert, G.; Meinhardt, H.W.; DEW Techn. Rev., 11(1971)236.

[280] Grossklaus, W.D.,Jr.; Katz, G.B.; Wortman, D.J.; *in: High Temperature Coatings, M. Khobaib et al., Eds.,* Met. Soc., 1987, p.67.

[281] Toriz, F.C.; Thakker, A.B.; Gupta, S.K.; Surf. Coat. Technol., 39/40(1989)161.

[282] Grünling, H.W.; Schneider, K.; Singheiser, L.; *Application of Advanced Materials for Turbomachinery and Rocket Propulsion,* AGARD-CP-449, 1989, p. 14-1.

[283] Gilgengorn, I.S.; Rogelberg, I.L.; Phys. Met. Metallography, 20(1965)60.

[284] Fryburg, G.C.; Stearns, C.A.; Kohl, F.J.; J. Electrochem. Soc., 124(1977)1147.

[285] Fritscher, K.; Patent DE 3737361 C2 (15.11.90), and U.S. 5, 080, 981 (Jan. 14, 1992).

[286] Duret, C.; Pichoir, R.; in: Coatings for High Temperature Applications, E. Lang, Ed., Applied Science Publ. Barking, England (1983) p.33.

[287] Boone, D.H.; Mat. Sci. Technol., 2(1986)220.

[288] Fritscher, K.; Bunk,W. *in: Proceedings FGM'90, M.Yamanouchi et al., Eds.*, The Society of Non-Traditional Technology, Tokyo, Japan, 1990, p.91.

[289] Halnan, W.K.; Lee, D.; in: High Temperature Protective Coatings, S.C. Singhal, Ed., Met. Soc. AIME, Atlanta, Georgia, 1983, p.3.

[290] Schulz, U.; Kröder, C.; Fritscher, K.; in: Proceedings SURTEC 91, H. Vesper, Ed., DFO, Düsseldorf, Germany, 1991, p.97.

[291] Bauer, V.; Feuerstein, A.; Hartig, K.; Kienel, G.; Lämmermann, H.; Patent DE 3627151 A1, 1986.

[292] Gill, B.J.; Tucker, R.C., Jr.; Mat. Sci. Technol., 2(1986)207.

[293] Nicoll, A.R.; Gunner, H.; Wuest, G.; Keller, S.; Mat. Sci. Technol., 2(1986)214.

[294] Miller, R.A.; J. Amer. Ceram. Soc., 67(1984)517.

[295] Ulion, N.E.; Ruckle, D.L.; European Patent EP 0044329B1.

[296] Smialek, J.L.; Gibala, R.; Met. Trans. A, 14(1983)2143

[297] Saunders, S.R.J.; Nicholls, J.R.; Mat. Sci. Technol., 5(1989)780.

[298] Gupta, T.K.; Bechtold,J.H.; Kuznicki, R.C.; Cadoff, L.H.; J. Mater. Sci., 12(1977)2421.

[299] Wu,B.Ch.; Chang, E.; Chang, Sh.F.; Tu, D.; J. Amer. Ceram. Soc., 72(1989)212;

[300] Rhys-Jones, T.N.; Toriz, F.C.; High Temp. Technol., 7(1989)73.

[301] Lee, E.Y.; Biederman, R.R.; Sisson, Jr. R.D.; *in: Plasma Surface Engineering, E. Broszeit et al., Eds.*, DGM Oberursel, Germany, 1989, p.365.

[302] Lelait, L.; Alpérine, S.; Diot, C.; Mévrel, M.; Mater. Sci. Eng., A121(1989)475.

[303] Alpérine, S.; *in: High Temperature Surface Interactions,* AGARD-CP-461, 1989, p. 6-1.

[304] Sato, T.; Shimada, M.; J. Am. Ceram. Soc., 68(1985)356.

[305] Fritscher, K.; Schulz, U.; unpublished results

[306] Chou, T.W.; Kelly, A.; Okura, A.; Composites 16(1985)187.

[307] Stöckel, D.; *Faserverbundwerkstoffe,* in: Metallische Verbundwerkstoffe, 1977, p. 132.

[308] McElman, J.A.; Mittnick, M.A.; Proc. Advancing with Composites, Int. Conf. on Comp.Mat., Crivelli-Visconti, Ed., May 10-12, 1988, Milano, Italy, p. 177.

[309] Bayoumi, M.A., Suéry, M.; Proc. ICCM-VI and ECCM-2, Metthews et al., Eds., Elsevier Sci. Publisher, London, 1987, Vol. 2, p. 2481.

[310] Schulte, K.; Minoshima, K.; Proc. 12th Risø Nat. Symp. on Materials Science: Metal Matrix Composites, Hansen et al., Eds., Risoslash. Nat. Lab., Roskilde, DK, 1991, p. 123.

[311] Warren, R.; Proc. 9th Risø Int. Symp. on Metallurgy and Materials Science, Andersen et al., Eds., Sept. 5-9, 1988, p. 233.

[312] Mortensen, A.; in [311], p. 141.

[313] Masson, J.J.; Weber, K.; Miketta, M.; Schulte, K.; in [310], p. 509.

[314] Masson, J.J.; Kleine, A.; Proc. IPCM '91, J. Verpoest, Ed., Butterworth-Heinemann, Oxford, 1991, p. 232.

[315] Schulte, K.; Trautmann, K.-H.; Girot, F.; Z. Materialprüfung, 32(1990)343.

[316] Schulte, K.; Proc. DVM-Tagung Bauteil '90, Deutscher Verband für Materialforschung und -prüfung, May 16-18, 1990, Berlin, p. 93.

[317] Minoshima, K.; Schulte, K.; Proc. ECCM-5, T. Bunsel et al., Eds., April 5-10, 1992, Bordeaux, France, p. 569.

[318] Bijlmer, P.F.A.; Proc. ICAS, Lisbon, 1978.

[319] Koch, G.A.; Proc. ICAF, Brussels, May 1979.

[320] Koch, G.A.; Machielse, L.A.H.; Fokker Report R-2430 (1979).

[321] Marissen, R.; Vogelesang, L.B.; Proc. SAMPE, Cannes, Jan. 1981.

[322] Vogelesang, L.B.; Marissen, R.; Schijve, J.; Proc. ICAF, 1981, p. 3.4/1.

[323] Schijve, J.; Vogelesang, L.B.; Marissen, R.; US patent numbers 4,500589 and 4,489,123; European patent number 0055289.

[324] Lin, C.T.; Kao, P.W.; Yang, F.S.; Composites, 22(1991)135.

[325] Marissen, R.; Master Thesis, Dept. of Aerospace Eng., Delft University (1980).

[326] Marissen, R.; Vogelesang, L.B.; Schijve, J.; Proc. DVM-Ak Bruchvorgänge und Betriebsfestigkeit 1982.

[327] Marissen, R.; MRS-Europe, Nov. 1985, p. 179.

[328] Marissen, R.; DLR, Köln, DFVLR-FB 88-56 (1988).

[329] Marissen, R.; Trautmann, K.-H.; Foth, J.; Nowack, H.; Proc. Fatigue 84 Conf., 1984, p. 1081.

[330] Nowack, H.; Marissen, R.; Trautmann, K.-H.; Peters, M.; Schulte, K.;, Sontani, D.; Proc. SAMPE 1991, p. 495.

[331] Martineau, P., Pailler, R., Lahaye, M., Naslain, R.; J. Mat. Sci., 19(1984)2749.

[332] Smith, P.R., Froes, F.H.; J. Metals, 36(1984)19.

[333] Chou, T.W., Kelly, A. Okura, A.; Composites, 16(1985)187.

[334] Stephens, J.R.; NASA Techn. Mem. 100212, ASM Int. Comp. Ses., Cincinnati, Ohio, 1987.

[335] Gabb, T.P., Gayda, J., Lerch, B.A., Halford, G.R.; Scripta Met. Mat., 25(1991)2879.

[336] MacKay, R.A., Brindlay, P.K., Froes, F.H.; JOM 1991, May, p.23.

[337] Dudek, H.J., Leucht, R., G. Ziegler, G.; in : [123], Vol. 3, p.1773.

[338] Leucht, R., Dudek, H.J., Ziegler, G.; in: Haftung als Basis für Stoffverbunde und Verbundwerkstoffe, Ed. W. Brockmann, DGM, Oberursel 1986, p.101.

[339] Leucht, R., Dudek,H.J., Ziegler, G.; Z. Werkstofftech., 18(1987)27.

[340] Leucht, R., Dudek, H.J., Ziegler, G.; Proc. EUROMAT, Aachen 1989, p.

[341] Leucht, R., Dudek, H.J., Ziegler, G.; Proc. 4th Int. Conf. Fibre Reinforced Composites, Liverpool, UK, 1990, p.279.

[342] Dudek, H.J., Leucht, R., Ziegler, G.; Proc. 4th Eur. Conf. Comp. Mat., Stuttgart, 1990, p.339.

[343] Ward-Close, C.M., Partridge, P.G.; J. Mat. Sci., 1990.

[344] Leucht, R., Dudek, H.J.; unpublished results.

[345] Ning, X.J., Pirouz, P., Lagerlof, K.P.D., DiCarlo, J.; J. Mater. Res., 5(1990)2865.

[346] Ning, X.J., Pirouz, P.; J. Mater. Res., 6(1991)2234.

[347] Ning, X.J., Pirouz, P., Bhatt, R.T.; Mat. Res. Soc. Symp. Proc., 250(1992), in press.

[348] Le Petitcorps, Y.; Ph.D. Thesis, University Bordeaux I, 1985.

[349] Pailler, R., Martineau, P., Lahaye, M., Naslain, R.; Rev. de Chim. Minerale, 18(1981)520.

[350] Wright, P.K., Nimmer, R., Smith, G., Sensmeier, M., Brun, M.; in: Interfaces in Metal-Matrix Composites, Eds.: Lin, R.Y., Arsenault, R.J., Martins, G.P., Fishman, S.G., Min., Met. & Mat. Soc., Warrendale, 1990, p.559.

[351] Hartley, M.V.; Proc. Riso Conf., 1988, p.383.

[352] Naik, R.A., Pollock, W.D., Johnson, W.S.; J. Mat. Sci., 26(1991)2913.

[353] Hemptenmacher, J., Leucht, R., Dudek, H.J.; Tagungsband DGM-Metallographietagung 1991, Friedrichshafen, Germany.

[354] Leucht, R., Peters, M., Dudek, H.J.; DGM-Conf., Verbundwerkstoffe und Werkstoffverbunde, Chemnitz, 1992.

[355] Dudek, H.J., Leucht, R. Borath, R., Ziegler, G.; Microchim. Acta, 1990-II, p.137.

[356] Nutt, S.R., Wawner, F.E.; J. Mat. Sci., 20(1985)1953.

[357] Lancin, M., Thibault-Desseaux, J., Bour, J.S.; J. Microsc. Spectrosc. Electron., 13(1988)503.

[358] Rhodes, C.G., Spurling, R.A.; ASTM STP 864, 1983, p.585.

[359] Park, Y.H., Narayen, D., Marcus, H.L.; Mat. Sci. Eng., 67(1984)79.

[360] Yang, J.-M., Jeng, S.M.; JOM, Nov., 1989, p.56.

[361] Gundel, D.B., Wawner, F.E.; Scripta Met. Mat., 25(1991)437.

[362] Yang, J.-M., Jeng, S.M., Yang, C.J.; Mat. Sci. Engn. A, 138(1991)169.

[363] Le Petitcorps, Y., Pailler, R. Naslain, R.; Comp. Sci. Technol., 35(1989)207.

[364] Jansson, S., Deve, H.E., Evans, A.G.; Met. Trans. A, 22(1991)2975.

[365] Eldridge, J.I., Brindley, P.K.; J. Mat. Sci. Lett., 8(1989)1451.

[366] Drzal, L.T., Herrera-Franco, P.J.; in: Engineering Materials Handbook, Vol.3, ASM International, 1991, p. 391.

[367] Herrera-Franco, P., Wu, W.-L., Madhukar, M., Drzal, L.T.; Proc. 46th Ann. Conf. Comp. Inst., 1991, 1/Session p.14B1.

[368] Vassel, A., Merienne, M.C., Pautonnier, F., Molliex, L., Favre, J.P.; in : [124], p. 919.

[369] Molliex, L., Favre, J.P., Vassel, A.; Proc. Int. Conf. Titanium Prod. Appl., Orlando, 1990, ONERA, Tire a part No 1990-133.

[370] Curtin, W.A.; J. Mat. Sci., 26(1991)5239.

[371] Yang, J.-M., Jeng, S.M., Yang, C.J.; Mat. Sci. Engn. A, 138(1991)155.

[372] Hemptenmacher, J., Leucht, R., Dudek, H.J.; DGM-Conf., Verbundwerkstoffe und Werkstoffverbunde, Chemnitz, 1992.

[373] Marshall, D.B.; J. Amer. Ceram. Soc., 67(1984)C259.

[374] Marshall, D.B., Oliver, W.C.; J. Amer. Ceram. Soc., 70(1987)542.

[375] Le Petitcorps, Y., Lahaye, M., Pailler, R. Naslain, R.; Comp. Sci. Technol., 32(1988)31.

[376] Martineau, Ph., Le Petitcorps, Y., Pailler, R. Naslain, R.; Actes ECCM-1, Eds.: A.R. Bunsell et.al., Bordeaux, 1985, p.725.

[377] Chang, D.J., Kao, W.H.; SAMPE Journal, March/April 1988, p.13.

[378] Smith, P.R., Froes, F.H., Cammett, J.T.; in: Mechanical Behavior of Metal-Matrix Composites, Eds.: Hack, J.E., Amateau, M.F., Met. Soc. AIME, p. 143.

[379] Park, Y.H., Narayen, D., Schmerling, M., Marcus, H.L.; J. Mat. Sci., 19(1984)2239.

[380] Bhatt, R.T., Grimes, H.H.; NASA Tech. Mem. 79223, 1979.

[381] Soumelidis, P., Quenisset, J.M., Naslain, R., Stoloff, N.S.; J. Mat. Sci., 21(1986)895.

[382] Yang, J.-M., Jeng, S.M., Yang, C.J.; Mat. Sci. Engn. A, 138(1991)181.

[383] Ibbotson, A,R., Beevers, C.J. Bowen, P.; Scripta Metall. Mater., 25(1991)1781.

[384] Bowen, P., Cotterill, P.J., Ibbotson, A.R.; IOP Short Meeting Series No 28, Inst. Phys. 1991, p. 82.

[385] Le Petitcorps, Y., Pailler, R. Naslain, R.; MRS-Europe, Nov. 1985, p.147.

[386] Huang, S.C., Hall, E.L.; Scripta Met. Mat., 25(1991)1805.

[387] Driver, D.; IOP Short Meeting Series No 28, Inst. Phys., 1991, p.1.

[388] Leucht, R., Dudek, H.J., Ziegler, G.; Proc. 4th Eur. Conf. Comp. Mat., Stuttgart, Elsevier, London, 1990, p. 393.

[389] Dudek, H.J., Leucht, R.; Deutsches Patent DE 40 18340 A1.

[390] Dudek, H.J., Leucht, R.; Deutsches Patent DE 40 21547 A1.

[391] Kleine, A., Dudek, H.J., Ziegler, G.; Proc. 4th Eur. Conf. Comp. Mat., Stuttgart, 25-28.Sept.1990, Elsevier, London, p. 267.

[392] Neite, G., Matucha, K.H., Dudek, H.J., Bunk, W.; Advanced Materials and Processes (EUROMAT'89), Vol.2., Eds., H.E.Exner, V.Schumacher, DGM-Verlag, Stuttgart, 1990, p. 1325.

[393] Ziegler, G., Dudek, H.J.; Mikrochim. Acta Suppl., 8(1979)421.

[394] Levi, C.G., Abbaschian, G.J., Mehrabian, R.; Met. Trans., A9(1978)697.

[395] Dudek, H.J., Kleine, A., Borath, R., Neite, G.; unpublished results.

[396] Cappleman, G.R., Watts, J.F., Clyne, T.W.; J. Mat. Sci., 20(1985)2159.

[397] Dudek, H.J., Kleine, A., Borath, R., Leucht, R., Mucha, H.; in: [466], p. 145.

[398] Leucht, R., Dudek, H.J., Ziegler, G.; Z. Werkstofftech., 18(1987)27.

[399] Le Petitcorps, Y.; Ph.D. Thesis, University Bordeaux I, 1985.

[400] Naidich, V., Ju., in: Progr. Surf. Membr. Sci., Vol.14, Eds.: D.A.Candehead, J.F. Danielli, Academic, New York, 1981, p. 353.

[401] Farid, B., Godby, R.W.; Phys. Rev. B, 43(1991-I)14248.

[402] Chatain, D., Coudurier, L., Eustathopoulos, N.; Revue Phys. Appl., 23(1988)1055.

[403] Rivollet, I., Chatain, D., Eustathopoulos, N.; Acta Met., 35(1987)835.

[404] Li, J.G., Hausner, H.; J. Mat. Sci. Lett., 10(1991)1275.

[405] Laurent, V., Chatain, D., Chatillon, C., Eustathopoulos, N.; Acta Metall., 36(1988)1797.

[406] Mcevoy, A.J., Williams, R.H., Higginbotham, I.G.; J. Mat. Sci., 11(1976)297.

[407] Shinoda, T., Hua Liu, Mishima, Y., Suzuki, T.; Mat. Sci. Engn. A, 146(1991)91.

[408] Laurent, V., Chatain, D., Eustathopoulos, N.; Mat. Sci. Engn. A, 135(1991)89.

[409] Eustathopoulos, N., Joud, J.C., Desre, P., Hicter, J.M.; J. Mat. Sci., 9(1974)1233.

[410] Yang, M., Scott, V.D.; Carbon, 29(1991)877.

[411] Ramquist, L.; Intern. J. Powder Met., 1(1965)2.

[412] Li, J.G.; Rare Metals, 10(1991)255.

[413] Elsner, G., Petzow, G.; ISIJ International, 30(1990)1011.

[414] Burger, K., Mader, W., Rühle. M.; Ultramicroscopy, 22(1987)1.

[415] Burger, K., Rühle, M.; Ultramicroscopy, 29(1989)88.

[416] Taylor, P.A., Nelson, J.S., Dodson, B.W.; Phys. Rev. B, 44(1991-I) 5834.

[417] Eustathopoulos, N., Passerone, A.; 1978, in: [459].

[418] Peden, C.H.F., Kidd, K.B., Shinn, N.D.; J. Vac. Sci. Technol., A9(1991)1518.

[419] Lang, G.; Aluminium, 49(1973)231.

[420] Crispin, R.M., Nicholas, M.; J. Mat. Sci., 11(1976)17.

[421] John, H., Hausner, H.; J. Mat. Sci., 5(1986)549.

[422] Lijun, Z., Jinbo, W., Jiting, Q., Qiu, N.; TMS, 1990, p. 213.

[423] Madeleno, U., Liu, H., Shinoda, T., Mishima, Y., Suzuki, T; J. Mat. Sci., 25(1990)3273.

[424] Ownby, P.D., Ke Wen K. Li, Weirauch, D.A. Jr.; J. Am. Ceram. Soc., 74(1991)1275.

[425] Mehrotra, S.P., Chaklader, A.C.D.; Met. Trans., B16(1985)567.

[426] Mellul, S., Chevalier, J.-P.; Phil. Mag. A, 64(1991)561.

[427] Ohuchi, F.S., Kohyama, M.; J. Am. Ceram. Soc., 74(1991)1163.

[428] Ramquist, L.; Jernkout. Ann., 153(1969)159.

[429] Samsonow, G.V., Panasyuk, A.D., Kozina, G.K.; Sov. Powder Metall Mat. Ceram., 1968, p. 874.

[430] Eustathopoulos, N., Chatain, D., Couduier, L.; Mat. Sci. Engn. A, 135(1991)83.

[431] Standing, R., Nicholas, M.; J. Mat. Sci., 13(1978)1509.

[432] Weaver, J.H.; Physics Today, January 1986, 2.

[433] Hasegawa, S., Nakamura, S., Kawamoto, N., Kishibe, H., Mizokawa, Y.; Surf. Sci., 206(1988)L851.

[434] Schmeisser, D., Jacobi, K.; Surf. Sci., 88(1979)138.

[435] Reichel, R., Gaukler, K.H.; Surf. Interf. Anal., 15(1990)211.

[436] Reichel, R., Gaukler, K.H.; Surf. Interf. Anal., 17(1991)751.

[437] Taylor, T.N., Martin, J.A.; J. Vac. Sci. Technol. A, 9(1991)1840.

[438] Rühle, M.; Fresenius J. Anal. Chem., 341(1991)369.

[439] Jaycock, M.J., Parfitt, G.D.; *Chemistry of Interfaces,* Ellis Horwood Ltd., Chichester, 1981.

[440] Weyl, W.A.; in: Structure and Properties of Solid Surfaces, Eds., R.Gomer, C.S. Smith, University of Chicago Press, 1953, p. 147.

[441] Chang, C.C.; J. Appl. Phys., 39(1968)5570.

[442] French, T.M., Somorjai, G.A.; J. Phys. Chem., 74(1970)2489.

[443] Arbab, M., Chottiner, G.S., Hoffman, R.W.; Mat. Res. Soc. Symp. Proc.,153(1989)66.

[444] Timsit, R.S., Waddington, W.G. Humphreys, C.J., Hutchison, J.L.; Ultramicroscopy, 18(1985)387.

[445] Merkle, K.L.; Ultramicroscopy, 37(1991)130.

[446] Tanaka, N., Mihima, K.; Appl. Surf. Sci., 33/34(1988)472.

[447] Ponce, F.A., Aranovich, J.; Appl. Phys. Lett., 38(1981)439.

[448] Knauss, D., Mader, W.; Ultramicroscopy, 37(1991)247.

[449] Elßner, G., Petzow, G.; Z. Metallkde., 64(1973)280.

[450] Rühle, M., Evans, A.G.; Mat. Sci. Eng. A, 107(1989)187.

[451] Necker, G., Mader, W.; Phil. Mag. Lett., 58(1988)205.

[452] Ernst, F., Pirouz, P., Heuer, A.H.; Mat. Res. Soc. Symp. Proc., 138(1989)557.

[453] Landman, U., Luedtke, W.D., Burnham, N.A., Colton, R.J.; Science, 248(1990)454.

[454] Burnham, N.A., Dominguez, D.D., Movery, R.L., Colton, R.J.; Phys. Rev. Lett., 64(1990)1931.

[455] Johnson, K.H., Pepper, S.V.; J. Appl. Phys., 53(1982)6634.

[456] Nath, K., Anderson, A.B.; Phys. Rev. B, 36(1989-I)1013.

[457] Bordier, G., Noguera, C.; Phys. Rev. B44(1991-II)6361.

[458] Baglin, J.E.E.; Mat. Sci. Eng. B, 1(1988)1.

[459] Delanny, F., Froyen, F., Deruyttere, A.; J. Mat. Sci., 22(1987)1.

[460] Neite, G., Dudek, H.J., Kleine, A., Borath, B.; Proc. ECCM4, Elsevier, London, 1990, 237.

[461] Hallstedt, B., Liu, Z.-K., Agren, J.; J. Mat.Sci., A25(1990)135.

[462] Pfeifer, M., Rigsbee, J.M., Chawla, K.K.; J.Mat.Sci., 25(1990)1563.

[463] Mucha, H., Overath, H., Dudek, H.J.; unpublished results.

[464] Dudek, H.J., Leucht, R., Ziegler, G.; in: [123], p. 1773.

[465] Naslain, R.; Pailler, R.; Martineau, P.; in: Physical Chemistry of the Solid State: Applications to Metals and their Compounds, Elsevier, Amsterdam, 1984, p. 481.

[466] Leucht, R., Dudek, H.J., Ziegler, G.; in: Haftung als Basis für Stoffverbunde und Verbundwerkstoffe", Ed. W. Brockmann, DGM-Verlag, Oberursel, 1986, p. 101.

[467] Leucht, R., Dudek, H.J., Ziegler, G.; in: Verbundwerkstoffe, Stoffverbunde in Technik und Medizin, Ed. G. Ondracek, DGM-Verlag, 1988, p. 123.

[468] Dudek, H.J., Larson, L.A., Browning, R.; Surf. Interf. Anal., 6(1984)274.

[469] Ning, X.J., Pirouz, P., Bhatt, R.T.; Mat. Res. Symp. Soc., 250(1992); in press

[470] Dudek, H.J., Leucht, R., Borath, R., Ziegler, G.; Mikrochim. Acta, 1990-II, p. 137.

[471] Knoch, H.; Ziegler, G.; Science of Ceramics, 9(1977)494.

[472] Gazza, G.E.; Katz, R.N.; Priest, H.F.; J. Am. Ceram. Soc., 64(1981)C-161.

[473] Lange, F.F.; Amer. Ceram. Soc. Bull., 62(1983)1369.

[474] Wötting, G.; Ziegler, G.; Ceramics Int., 10(1984)18.

[475] Deeley, G.G.; Herbert, J.M.; Moore, N.C.; Powd. Met., 8(1961)145.

[476] Wild, S.; Grieveson, P.; Jack, K.H.; Latimer, M.J.; in: Special Ceramics 5, Ed. P. Popper, Brit. Cer. Res. Assn., Stoke-on-Trent, 1972, p. 377.

[477] Weston, R.J.; Carruthers, T.G.; Proc. Brit. Ceram. Soc., 22(1973)197.

[478] Terwilliger, G.R.; Lange, F.F.; J. Am. Ceram. Soc., 57(1974)25.

[479] Kingery, W.D.; J. Appl. Phys., 30(1959)301.

[480] Drew, P.; Lewis, M.H.; J. Mater. Sci., 9(1974)261.

[481] Iskoe, J.L.; Lange, F.F.; in: Ceramic Microstructures, Eds. R.M. Fulrath, J.A. Pask, Westview Press, Boulder, CO, 1977, p. 669.

[482] Hampshire, S.; Jack, K.H.; in: Special Ceramics 7, Eds. D.E. Taylor, P. Popper, Proc. Brit. Ceram. Soc., 31(1981)37.

[483] Jack, K.H.; in: Processing of Crystalline Ceramics, Materials Science Research, Eds. H. Palmour, R.F. David, T.M. Hare, Plenum Press, New York, 11(1978)561.

[484] Tien, T.Y.; Petzow, G.; Gauckler, L.J.; Weiss, J.; in: Progress in Nitrogen Ceramics, Ed. F.L. Riley, Proc. Nato Adv. Study Inst., Martinus Nijhoff, The Hague, 1983, p.89.

[485] Müller, R.; Ph.D.Thesis, University of Stuttgart, 1981.

[486] Gazza, G.E.; Am. Ceram. Soc. Bull., 54(1975)778.

[487] Rae, A.W.J.M.; Thompson, D.P.; Pipkin, N.J.; Jack, K.H.; in: Special Ceramics 6, Ed. P. Popper, Brit. Ceram. Res. Assn., Stoke-on-Trent, 1975, p. 347.

[488] Jack, K.H.; J. Mater. Sci., 11(1976)1135.

[489] Tsuge, A.; Kudo, H.; Komeyo, K.; J. Am. Ceram. Soc., 57(1974)269.

[490] Rae, A.W.J.M.; Thompson, D.P.; Jack, K.H.; in: Ceramics for High Performance Application II, Eds. J.J. Burke, E.N. Lenoe, R.N. Katz, Brook Hill Publishing Company, Chestnut Hill, Mass., 1978, p. 1039.

[491] Gauckler, L.J.; Hohnke, H.; Tien, T.Y.; J. Am. Ceram. Soc., 63(1980)35.

[492] Thompson, D.P.; paper read by S. Hampshire for the author at 6th Journées d'Etudes sur les Nitrures (JENI 6), St. Etienne, France, 1984.

[493] Lange, F.F.; Singhal, S.C.; Kuznicke, R.C.; J. Am. Ceram. Soc., 60(1977)249.

[494] Bowen, L.J.; Carruthers, T.; Brook, R.J.; J. Am. Ceram. Soc., 61(1978)335.

[495] Claussen, N.; Wagner, R.; Gauckler, L.J.; Petzow, G.; J. Am. Ceram. Soc., 61(1978)369.

[496] Rice, R.W.; McDonough, W.J.; J. Am. Ceram. Soc., 58(1975)264.

[497] Weiss, J.; Gauckler, L.J.; Tien, T.Y.; J. Am. Ceram. Soc., 62(1979)632.

[498] Bellosi, A.; Vincenzini, P.; Babini, G.N.; J. Mater. Sci., 23(1988)2348.

[499] Luxem, W.; unpublished data, DLR, 1991.

[500] Collongues, R.; Gilles, J.C.; Lejuc, A.M.; Perez y Jorba, M.; Michel, D.; Mat. Res. Bull., 2(1967)837.

[501] McDonough, W.M.; Wu, C.C.; Morgan, P.E.D.; J. Am. Ceram. Soc. 64(1981)C-45.

[502] Weiss, J.; Gauckler, L.J.; Lukas, H.L.; Petzow, G.; Tien, T.Y.; J. Mater. Sci., 16(1981)2997.

[503] Larsen, D.C.; Adams, J.W.; Semiannual Interim Tech. Rep. 16, IIT Research Inst., Chicago, IL, 1984, p.87.

[504] Dutta, S.; Buzek, B.; J. Am. Ceram. Soc., 67(1984)89.

[505] Lange, F.F.; Falk, L.K.L.; Davis, B.T.; J. Mater. Res., 2(1987)66.

[506] Scott, H.G.; J. Mater. Sci., 10(1975)1527.

[507] Falk, L.K.L., Holmström, M.; in: Euroceramics, Vol. 1, Eds. G. de With, R.A. Terpstra, R. Metselaar, Elsevier Appl. Sci., London, 1989, p.1.373.

[508] Cheng, Y.; Thompson, D.P.; Brit. Ceram. Trans. J., 87(1988)107.

[509] Kim, J.R.; Kim, D.-H.; Kim, C.H.; J. Am. Ceram. Soc., 73(1990)2567.

[510] Kim, J.R.; Kim, C.H.; J. Mater. Sci., 25(1990)493.

[511] Ekström, T.; Falk, L.K.L.; Knutson-Wedel, E.M.; J. of Mater. Sci. Letters, 9(1990)823.

[512] Ekström, T.; Falk, L.K.L.; Knutson-Wedel, E.M.; J. Mater. Sci., 26(1991)4331.

[513] Takahashi, T.; Ceram. Trans., 7(1990)674.

[514] Palm, J.A.; Greskovich, C.D.; Am. Ceram. Soc. Bull., 59(1980)447.

[515] Tabata, H.; Abe, O.; Machida, M.; Ohashi, M.; Kanzaki, S.; in: Proc. 2nd Int. Symp. Ceramic Components and Materials for Engines, Eds. W. Bunk, H. Hausner, DKG, Bad Honnef, 1986, p. 713.

[516] Tsuge, A.; Nishida, K.; Komatsu, M.; J. Am. Ceram. Soc., 58(1975)323.

[517] Smith, J.T.; Quackenbusch, C.L.; Am. Ceram. Soc. Bull., 59(1980)516.

[518] Wötting, G.; Kanka, B.; Ziegler, G.; in: Proc. Int. Conf. Non-Oxide Technical and Engineering Ceramics, Ed. S. Hampshire, Elsevier Appl. Sci., Barking, 1986, p. 83.

[519] Knickerbocker, S.H.; Zangvil, A.; Brown, S.D.; J. Am. Ceram. Soc., 68(1985)C-99.

[520] Giachello, A.; Martinengo, P.C.; Tommosini, G.; Popper, P.; Am. Ceram. Soc. Bull., 59(1980)1212.

[521] Pomeroy, M.J.; Saruhan, B.; McArdle, T.; Hampshire, S.; Revue de Chimie Minerale, 22(1985)555.

[522] Mahoney, F.M.; Hoffmann, M.J.; Petzow, G.; Proc. 4th Symp. on Ceramic Components for Engines, 1991, in press.

[523] Saruhan, B.; Pomeroy, M.J.; Hampshire, S.; Proc. Int. Conf. Non-Oxide Technical and Engineering Ceramics, Ed. S. Hampshire, Elsevier Appl. Sci., London, 1986, p. 69.

[524] Hampshire, S.; Pomeroy, M.J.; Saruhan, B.; in: High Tech Ceramics, Ed. P. Vincenzini, Elsevier Sci. Pub. B.V., Amsterdam, 1987, p. 941.

[525] Faber, K.T.; Evans, A.G.; Acta Metall., 31(1983)565.

[526] Rouby, D.; cfi/Ber. DKG 66(1989)208.

[527] Faber, K.T.; Ph.D.Thesis, University of Berkeley, 1982.

[528] Milewski, J.V.; Gac, F.D.; Petrovic, J.J.; Skaggs, S.R.; J. Mater. Sci., 20(1985)1160.

[529] Liu, H.; Weisskopf, K.L.; Petzow, G.; J. Am. Ceram. Soc., 72(1989)559.

[530] Nickel, K.G.; Hoffmann, M.J.; Greil, P.; Petzow, G.; Adv. Ceram. Mater., 3(1989)557.

[531] Karasek, K.R.; Bradley, S.A.; Donner, J.T.; Martin, M.R.; Haynes, K.L.; Yeh, H.C.; J. Mater. Sci., 24(1989)1617.

[532] Braue, W.; Hölscher, A.; Saruhan, B.; Ziegler, G.; in: Euro-Ceramics, Vol. 3, Engineering Ceramics, Elsevier Appl. Sci., London and New York, 1989, p. 3263.

[533] Birchall, J.-D.; Stanley, D.R.; Mockford, M.; Pigott, G.H.; Pinto, P.J.; J. Mater. Sci. Letters, 7(1988)350.

[534] Marshall, D.B.; Evans, A.G.; J. Am. Ceram. Soc., 68(1985)225.

[535] Lundberg, R.; Kahlman, L.; Pompe, R.; Carlsson, R.; Am. Ceram. Soc. Bull., 66(1987)330.

[536] Shalek, P.D.; Petrovic, J.J.; Hurley, G.F.; Gac, F.D.; Am. Ceram. Soc. Bull., 65(1986)351.

[537] Bellosi, A.; De Portu, G.; Mater. Sci. and Eng., A109(1989)357.

[538] Porter, J.R.; Lange, F.F.; Chokshi, A.H.; Mat. Res. Soc. Symp. Proc., 78(1987)289.

[539] Wada, H.; Wang, M.J.; Tien, T.Y.; J. Am. Ceram. Soc., 71(1988)837.

[540] Hayami, R.; Ueno, K.; Kondou, I.; Tamari, N.; Toibana, Y.; in: Tailoring Multiphase and Composite Ceramics, Ed. R.E. Tressler, G.L. Messing, G.G. Pantaro, R.E. Newnham, Plenum Press, New York, 1986, p. 663.

[541] Saruhan, B.; Ziegler, G.; in: Proc. of 7th Cimtec World Congress, Montecatini, Italy, 1990.

[542] Hoffmann, M.J.; Nagel, A.; Greil, P.; Petzow, G.; J. Am. Ceram. Soc., 72(1989)765.

[543] Marshall, D.B.; Cox, B.N.; Evans, A.G.; Acta Metall., 33(1985)2013.

[544] Guo, J.-K.; Mao, Z.-Q.; Bao, C.-D.; Wang, R.-H.; Yan, D.-S.; J. Mater. Sci., 17(1982)3611.

[545] Warren, R.; Andersson, C.H.; Composites, 5(1984)101.

[546] Jaskowiak, M.H.; DiCarlo, J.A.; J. Am. Ceram. Soc. 72(1989)192.

[547] Saruhan, B.; Ziegler, G.; Silicates Industries, [1-2](1991)29.

[548] Corbin, N.D.; Rossetti, G.A.; Hartline, S.D.; Ceram. Eng. Sci. Proc., 7(1986)958.

[549] Bhatt, R.T.; *Effects of fabrication conditions on the properties of SiC fibre reinforced RBSN matrix composites;* NASA Tech. Rep. No. 86C2, January 1986.

[550] Sakamoto, H.; Kodama, H.; Miyoshi, T.; J. Ceram. Soc. Jpn. Inter. Ed., 95(1987)817.

[551] Kanka, B.; Ziegler, G.; poster presented at 7th Cimtec World Congress, Montecatini, Italy, 1990.

[552] Foulds, W.; Lecostaouec, I.F.; Landry, C.; DiPietro, S.; Vasilos, T.; Ceram. Eng. Sci. Proc. 10(1989)1083.

[553] Kodama, H.; Sakamoto, H.; Miyoshi, T.; J. Am. Ceram. Soc., 72(1989)551.

[554] Mah, T.; Hecht, N.L.; McCullum, D.E.; Hoenigman, J.R.; Kim, H.M.; Katz, A.P.; Lipsitt, H.A.; J. Mater. Sci., 19(1984)1191.

[555] Clark, T.J.; Arons, R.M.; Stamatoff, J.B.; Rabe, J.; Ceram. Eng. Sci. Proc., 6(1985)576.

[556] Luthra, K.L.; J. Am. Ceram. Soc., 69(1986)C-231.

[557] Lindley, M.W.; Godfrey, D.J.; Nature (London), 229(1971)192.

[558] Caputo, A.J.; Lackey, W.J., Stinton, D.P.; Ceram. Eng. Sci. Proc., 6(1985)644.

[559] Lange, F.F.; J. Am. Ceram. Soc., 56(1973)445.

[560] Buljan, S.T.; Pasto, A.E.; Kim, H.J.; A. Ceram. Soc. Bull., 68(1989)387.

[561] Buljan, S.T.; Zilberstein, G.; Mat. Res. Soc. Sym. Proc., 78(1987)273.

[562] Greil, P.; Petzow, G.; Tanaka, H.; Ceram. Int. 13(1987)19.

[563] Duclos, R.; Crampon, J.; Cales, B.; in: Euro-Ceramics, Vol. 3, Engineering Ceramics, Elsevier Appl. Sci., London and New York, 1989, p. 3.273.

[564] Saruhan, B.; Braue, W.; Götting, J.; unpublished results.

[565] Niihara, K.; Hirano, T.; Nakahira, A.; Suganuma, K.; Izaki, K.; Kawakami, T.; J. Jpn. Soc. Powder and Powder Metall., 36(1989)243.

[566] Janssen, R.; Heussner, K.H.; Powd. Metall. Int., 23(1991)241.

[567] Niihara, K.; Hirai, T.; Seramikkusu, 26(1986)598.

[568] Niihara, K.; Seramikkusu, 99(1991)974.

[569] *Prüfung von keramischen Hochleistungswerkstoffen, 4-Punkt Biegeprüfung,* Normenausschuß für Materialprüfung, NMP 291 Nr., 20-28.

[570] *Flexural Strength of High Performance Ceramics at Ambient Temperatures,* U.S. Military Standard, MIL-STD-1942(MR), 1983.

[571] *Testing Methode for Flexural Strength (Modulus of Rupture) of High Performance Ceramics,* Japanese Industrial Standard, (JIS) R 1601, 1981.

[572] Weibull, W.; *A Statistical Theory of the Strength of Materials,* Ingvetenskaps Handl. Nr.151, Stockholm, 1939.

[573] Nadler, P.; Hermsdorfer Technische Mitteilungen, Teil 1: 33(1972)1031, Teil 2: 40(1974)1262.

[574] Papst, R.; Z. Werkstofftech., 6(1975)17.

[575] Maier, H.R.; Nick, H.; Krauth, A.; Ber. Dt. Keram. Ges., 54(1979)413.

[576] Lindner, H.A.; Wiedenhorn, G.; Thoma, H.J.; Ber. Dt. Keram. Ges., 53(1976)309.

[577] Ikeda, K.; Igaki, H.; Kruoda, T.; Am. Ceram. Soc. Bull., 65(1986)683.

[578] Schweiger, G.; Erben, W.; Heckel, K.; Materialprüf., 26(1984)340.

[579] Munz, D.; Frac. Mech. Ceram., 6(1983)1.

[580] Mendiratta, M.G.; Petrovic, J.J.; J. Am. Ceram. Soc., 61(1978)226.

[581] Evans, A.G.; Wiederhorn, S.M.; J. Mater. Sci., 9(1974)270.

[582] Tighe, N.J.; J. Mater. Sci., 13(1978)1455.

[583] Evans, A.G.; Int. J. Fract., 10(1974)251.

[584] Evans, A.G.; Johnson, H.; J. Mater. Sci., 10(1975)214.

[585] Fett, T.; *Lebensdauervorhersage an keramischen Werkstoffen mit den Methoden der Bruchmechanik bei elastischem und viskoelastischem Materialverhalten,* Ph.D. Thesis, Techn. University of Karlsruhe, (TH), 1983.

[586] Gugel, E.; cfi/Ber. DKG, 2/85, 89-91.

[587] Prochazka, S.; Special Ceramics, 6(1975)171.

[588] Braue, W.; Göring, J.; Ziegler, G.; Proceedings. Third Int. Symp., Ceramic Materials for Heat Engines, V.J. Tennery, Ed., Las Vegas, Nov. 28-30, 1988, p. 817.

[589] Schnitzeler, F.; Göring, J.; Ziegler, G.; Fortschrittsberichte der DKG, Vol. 6, 1991, p. 95.

[590] Jack, K.H.; *Sialons: A Study in Materials Development,* in: Non-Oxide Technical and Engeneering Ceramics, Elsevier, London, New York, S. Hampshire, Ed., 1986 p. 1.

[591] R.J.Lumby, P.Hodgson, N.E.Cother, A.Szweda, Soc. of Automotive Engineers, SAE Paper 850521, 1985.

[592] Lumby, R.J.; in Ceramics Engeneering and Science Proceedings, Jan. - Feb., 1982, Vol. 3, No. 1-2, p. 50.

[593] Kleebe, H.J.; *Einflußfaktoren auf das Nachsinterverhalten von reaktionsgebundenem Siliciumnitrid,* Ph.D. Thesis, University of Köln, 1988.

[594] Li, Z.; Brandt, R.C.; J. Am. Ceram. Soc., 72(1981)459.

[595] Popp, G.; *Untersuchungen zum Rißausbreitunsverhalten von Siliziumkarbid und Aluminiumoxid an Luft und im Hochvakuum bei hohen Temperaturen,* Ph.D. Thesis, University of Stuttgart, 1981.

[596] K.Kriz, Forschungsbericht BMFT-FB-T-84-285, 1984.

[597] Cohrt, H.; *Mechanik und Mechanismen des Biegekriechens von reaktionsgebundenem Siliciumcarbid,* Ph.D. Thesis, Techn. University of Karlsruhe, (TH), 1984.

[598] Evans, A.G.; Russell, L.R.; Richerson, D.W.; Met. Trans. A, 6(1975)707.

[599] Danzer, R.; Keramische Zeitschrift, 42(1990)313.

[600] Grathwohl, G.; Materialwiss. und Werkstofftechnik, 19(1988)113.

[601] Kawakubo, T.; Okabe, N.; Mori, T.; *Static and cyclic fatigue behaviour in ceramics,* Fatigue 90, Vol.II, Ed., H.Kitagawa, T.Tanaka, July, 1990.

[602] Fett, T.; Himsolt, G.; Munz, D.; Advanced Ceramic Materials, Vol.1, No.2, April 1986, p. 179.

[603] Heigl, H.; Heckel, K.; *Bending Strength of Silicon Nitride at 1200°C under Static and Cyclic Bending Load,* cfi-Ber. DKG 2/3-83, 69.

[604] Masuda, M.; Soma, T.; Matsui, M.; Oda, I.; J. Ceram. Soc. Jpn. Ed., 96(1988)275.

[605] Kawakubo, T.; Komeya, K.; J. Am. Ceram. Soc., 70(1987)400.

[606] Winkler, H.G.F.; *Petrogenesis of metamorphic rocks.* Springer, New York, Heidelberg, Berlin, 1974.

[607] Bowen, N.L.; Greig, J.W.; Zies, E.G.; J. Wash. Acad. Sci., 14(1924)183.

[608] Davis, R.F.; Pask, J.A.; in: A.M. Alper; Eds.: Refractory Materials. High Temperature Oxides, Part IV, Refractory Glasses, Glass-Ceramics, and Ceramics, Academic Press, New York and London, 1971, p.37.

[609] Pask, J.A.; Tomsia, A.P.; J. Amer. Ceram. Soc., 74(1991)2367.

[610] Somiya, S.; Hirate, Y.; Ceram. Bull., 70(1991)1624.

[611] Aksay, I.; Dabbs, D.M.; Sarikaya, M.; J. Am. Ceram. Soc., 74(1991)2343.

[612] Schneider, H.; *Formation, properties and high-temperature behaviour of mullite.* Habilitation Thesis, Faculty of Chemistry, University of Münster, Germany, (1985).

[613] Somiya, S.; Davis, R.F.; Pask, J.A.; in : Ceramic Trans., Vol. 6, The Am. Cer. Soc., 1990.

[614] Butt, D.P.; Mecholsky, Jr., J.J.; Topical Report for Glass Research Institute, CAM-8806, GRI-88/0134, NTIS PB-194088 (1988)132.

[615] Butt, D.P.; Mecholsky, J.J.; J. Amer. Ceram. Soc., 73(1990)2690.

[616] Braue, W.; Schneider, H.; unpublished results.

[617] Schneider, H.; Henne, R.; unpublished results.

[618] Schneider, H.; Ceramic Trans., 6(1990)135.

[619] Schneider, H.; Eberhard, E.; J. Amer. Ceram. Soc., 73(1990)2073.

[620] Kanzaki, S.; Tabata, H.; Kumazawa, T.; Ceramic Transactions, 6(1990)339.

[621] Kanka, B.; Schneider, H.; unpublished results.

[622] Okamoto, Y.; Fukudome, H.; J. Eur. Ceram. Soc., 6(1990)161.

[623] Kumazawa, T.; Ohta, S.; Kanzaki, S.; Ceramic Transactions, 6(1990)401.

[624] Dokko, P.C.; Pask, J.A.; Mazdiyasni, K.S.; J. Amer. Ceram. Soc., 60(1977)150.

[625] Mah, T.-I.; Mazdiyasni, K.S.; J. Amer. Ceram. Soc., 66(1983)699.

[626] Ohnishi, H.; Maeda, K.; Nakamura, T.; Kawanami, T.; Ceramic Transactions, 6(1990)605.

[627] Kanzaki, S.; Kumazawa, T.; Asaumi, J.; Abe, O.; Tabata, H.; J. Ceram. Soc. Japan, 93(1985)407.

[628] Kollenberg, W.; Schneider, H.; J. Amer. Ceram. Soc., 72(1989)1739.

[629] Nixon, R.D.; Chevacharoenkul, S.; Davis, R.F.; Tiegs, T.N.; Ceramic Transactions, 6(1990)579.

[630] Heynes, A.P.; Doremus, R.H.; J. Amer. Ceram. Soc., 74(1991)2469.

[631] Tiegs, T.; Becher, P.; Angelini, P.; Ceramic Transactions, 6(1990)463.

[632] Becher, P.F.; J. Amer. Ceram. Soc., 74(1991)255.

[633] Wei, G.C.; Becher, P.F.; Bull. Amer. Ceram. Soc., 56(1977)559.

[634] Wei, G.C.; Becher, P.F.; Ceram. Bull., 64(1985)298.

[635] Claussen, N.; Petzow, G.; in: Tailoring Multiphase and Composite Ceramics, Vol. 20, Plenum Press, New York, 1986, p.649.

[636] Wagner, R.; *Elastische Eigenschaften von keramischen Verbundwerkstoffen bei hohen Temperaturen;* Ph.D.Thesis, University Cologne (1989).

[637] Borom, M.P.; Brun, M.K.; Szala, L.E.; Adv. Ceram. Mater., 3(1988)491.

[638] Baudin, C.; Moya, J.S.; J. Amer. Ceram. Soc., 73(1990)1417.

[639] Hirata, Y.; Matsushita, S.; Ishihara, Y.; J. Amer. Ceram. Soc. 74(1991)2438.

[640] Boch, P.; Chartier, T.; Rodrigo, P.D.D.; Ceramic Transactions, 6(1990)353.

[641] Leriche, A.; Ceramic Transactions, 6(1990)541.

[642] Nischik, C.; Seibold, M.M.; Travitzky, N.A.; Claussen, N.; J. Amer. Ceram. Soc., 74(1991)2464.

[643] Claussen, N.; In: Structural Ceramics-Processing, Microstructure, and Properties, Risø Int.Symp.on Metallurgy and Materials Science, Risø National Laboratory, Roskilde, Denmark, 1990.

[644] Wallace, J.S.; Petzow, G.; Claussen, N.; In: Advances in Ceramics, 12: Science and Technology of Zirconia II, The American Ceramic Society (1984) 436.

[645] Ishitsuka, M.; Sato, T.; Endo, T.; Shimada, M.; J. Amer. Ceram. Soc., 70(1987)C342.

[646] Rundgren, K.; Elfing, P.; Tabata, H.; Ceramic Transactions, 6(1990)553.

[647] Ruh, R.; Mazdiyasni, K.S.; Mendiratta, M.G.; J. Amer. Ceram. Soc., 71(1988)503.

[648] Kamiaki, H.; Yamagishi, C.; Asaumi, J.; Ceramic Transactions, 6(1990)509.

[649] Okada, K.; Otsuka, N.; J. Amer. Ceram. Soc., 74(1991)2414.

[650] Ismail, M.G.M.U.; Tsunatori, H.; Nakai, Z.; J. Amer. Ceram. Soc., 73(1990)537.

[651] Kim, E.; Zografou, C.; Krönert, W.; Proc. UNITECR 91, Aachen (1991) p.135.

[652] Dechamps, P.; Sakaguchi, S.; Poorteman, M.; Cambier, F.; J. Amer. Ceram. Soc., 74(1991)2476.

[653] Dreyer, W.; *Materialverhalten anisotroper Festkörper.* Springer, Wien-New York (1974).

[654] Kingery, W.D.; Bowen, H.K.; Uhlmann, D.R.; *Introduction to ceramics.* 2nd Ed., John Wiley and Sons, New York-London-Sidney-Toronto (1976).

[655] Hilpert, K.; Schneider, H.; unpublished results.

[656] Winter, J.K.; Ghose, S.; Amer. Min., 64(1979)573.

[657] Mantell, C.L.; *Carbon and Graphite Handbook,* Interscience Publishers, NY, 1968.

[658] Gordon, J.E.; *The New Science of Strong Materials,* Penguin Books Ltd., Harmondsworth, Middlesex, England, 1984.

[659] Böder,H.; Gölden, D.; Rose, P.; Wurmseher, H.; Z. Werkstofftechnik, 11(1980)275.

[660] Buckley, J.D.; Am. Ceram. Soc. Bull., 67(1988)364.

[661] Oberlin, A.; *Chemistry and Physics of Carbon,* Vol. 22, Marcel Dekker, NY, 1989.

[662] Pleger, R.; Braue, W.; *Interfacial Structures in 2D C/C Composites: an Assessment by TEM,* 20th. Int. Carbon Conf., St. Barbara, USA, 1991, p.398.

[663] Kelly, B.T.; *Physics of Graphite,* Applied Science Publishers, New York, 1981, p.35.

[664] Fitzer, E.; Hüttner, W.; J. Phys., D 14(1981)347.

[665] Mantell, Ch.L.; *Carbon and Graphite Handbook,* Applied Science Publishers, New York, 1968, p.9.

[666] Wege, E.; *Process Mineralogy of Ceramic Materials,* W. Baumgart, A.C. Dunham, G.C. Amstutz, Eds., Ferdinand Enke Publ., Stuttgart, 1984, p.125.

[667] Strife, J.R.; Sheehan, J.E.; Am. Ceram. Soc. Bull., 67(1988)369.

[668] Dresselhaus, M.S.; *Graphite Fibers and Filaments,* Springer Verlag, Berlin, Heidelberg, New York, Springer Series in Material Science Vol.5, 1988.

[669] Ergun, S.; Carbon, 6(1968)7.

[670] Rousseaux, F.; Tchoubar, D.; Carbon, 15(1977)55.

[671] Fitzer, E.; Schäfer, W.; Yamada, S.; Carbon, 7(1969)613.

[672] Lersmacher, B.; Lydtin, H.; Chemie Ing. Tech., 39(1967)833.

[673] Huttepain, M.; Oberlin, A.; Carbon, 28(1990)103.

[674] Franklin, R.S.; Proc. R. Soc., London, A 209(1951)196.

[675] Jenkins, G.M.; Kawamura, K.; *Polymeric Carbons,* Cambridge Univ. Press, London, 1976.

[676] Rothwell, W.S.; J. Appl. Pys., 39(1968)1840.

[677] Guigon, M.; Oberlin, A.; Desarmot, G.; Fib. Sci. Techn., 20(1984)55.

[678] Yoshida, A.;Kaburagi, Y.; Hishiyama, Y.; Carbon, 29(1991)1107.

[679] Marshall, D.B.; J. Am. Ceram. Soc., 68(1985)225.

[680] Manocha, L.M.; Bahl, O.P.; Singh, Y.K.; *Effect of Fiber/Matrix Interactions on Mechanical Properties of Carbon/Carbon Composites,* Proc. Interfacial Phenomena in Composite Materials, Sheffield, UK, 1989, p.310.

[681] Lee, S.M.; *International Encyclopedia of Composites,* VCH, New York, 1989, p.158.

[682] Bennet, S.C.; Johnson, D.J.; Carbon, 17(1979)25.

[683] Kamiya, K.; Inagaki, M.; Carbon, 11(1975)429.

[684] Hishiyama, Y.; Inagaki, M.; Kimura, S.; Yamada, S.; Carbon, 12(1974)249.

[685] according to Aetzbauer, cited in: Menges, G.; *Werkstoffkunde der Kunststoffe;* Carl Hanser Verlag, München, Wien, 1979.

[686] Rosen, B.W.; Dow, N.F.; Hashin, Z.; *Mechanical Properties of Fibrous Composites;* NASA-CR 31, 1964.

[687] Harlow, D.G.; Phoenix, S.L.; Int. Journ. of Fracture, 15(1979)321.

[688] Manders, P.W.; Bader, M.G.; Chou, T.W.; Fibre Science and Techn., 17(1982)183.

[689] Bader, M.G.; Priest, A.M.; in: Progress in Sci. and Eng. of Comp., Eds. T. Hayashi, K. Kawata, S. Umekawa, Tokyo, 1982, p. 1129.

[690] Street, K.N.; Ferte, J.P.; Proc. ICCM-I, Geneva, 1976, p. 137.

[691] Chi, Z.; Chou, T.W.; Shen, G.; J. Mat. Sc., 19(1984)3319.

[692] Helmer, T.; Wanner, A.; Kromp, K.; *Zugversuche an Faserbündeln und deren statistische Auswertung*. in: DKG: Werkstoffe - Verfahren - Anwendung, Vol. 5, 1990.

[693] Peters, P.W.M.; Schmauch, J.; Weiß, R.; Bader, M.G.; Proc. ECCM-V, Bordeaux, 1992, p. 157.

[694] Chamis, C.C.; J. of Comp. Techn. and Review, 11(1989)3.

[695] Hashin, Z.; *Theory of Fiber Reinforced Materials;* NASA CR-1974, 1972.

[696] Kies, J.A.; Composites, NLR-Rep. 5752, AD-274560, US Naval Research Laboratory, Washington, DC, 1962.

[697] Chamis, C.C.; DMSMD Rep. No. 9, Case Western Reserve Univ., Cleveland, Ohio, 1967.

[698] Sørensen, B.F.; Talreja, R.; Sø, O.T.; Proc. 4th. Int. Symp. on Ceramic Materials and Components for Engines, Göteborg/Sweden, 1991.

[699] Peters, P.W.M.; J. of Comp. Mats., 18(1984)545.

[700] Peters, P.W.M.; ASTM STP 907 (1986)84.

[701] Highsmith, A.L.; Stinchcomb, W.W.; Reifsnider, K.K.; *Stiffness Reduction Resulting from Transverse Cracking in Fibre-Reinforced Composite Laminates;* VPI-E-81.33, Dept. of Engineering, Science and Mechanics, Virginia Polytechnic Institute and State University, Blacksburg, VA, 1981.

[702] Peters, P.W.M.; poster presented at Conf. on Testing Evaluation and Quality Control of Composites, University of Surrey, Guildford, UK, 1983.

[703] Peters, P.W.M.; Chou, T.W.; Composites, 15(1987)40.

[704] Peters, P.W.M.; Andersen, S.I.; J. of Comp. Mats., 23(1989)944.

[705] Peters, P.W.M.; *Interface Dominated Mechanical Properties of Aramid/Epoxy;* Report to AKZO, 1989.

[706] Peters, P.W.M.; Meusemann, H.; Proc. ICCM-VI, ECCM-II, London, 1987, p. 3.508.

[707] Peters, P.W.M.; ASTM-STP 1012, 1989, p. 103.

[708] Peters, P.W.M.; Proc. Advances with Composites, Milan, Italy, 1988.

[709] Peters, P.W.M.; Proc. Interfacial Phenomena in Comp. Mats., Sheffield, 1989, p. 59.

[710] Flaggs, D.L.; Kural, M.H.; J. of Comp. Mats., 16(1982)103.

[711] Parvizi, A.; Garret, K.W.; Bailey, J.E.; J. Mat. Sci., 13(1978)195.

[712] Nuismer, R.J.; Tan, S.C.; in: Mechanics of Composite Materials, Proc., Int. Union of Theoretical and Applied Mechanics, Symposium on Mechanics of Composite Materials, Virginia Polytechnic Institute and State University, Blacksburg, VA, 1982.

[713] Peters, P.W.M.; unpublished data.

[714] Peters, P.W.M.; *Influence of Cure Temperature and Fibre Surface Treatment on Fibre/Matrix Bond Strength;* to be published in J. Comp. Mat.

[715] Schulte, K.; Proc. Europ. Symposium on Damage Development and Failure Processes in Comp. Materials, Leuven/Belgium, 1987, p. 39.

[716] Schulte, K.; Reese, E.; Chou, T.W.; Proc. ICCM-VI/ECCM-2, London/UK, 1987, p. 4.89.

[717] Baron, Ch.; Schulte, K.; Harig, H.; Composite Science and Technology 29(1987)257.

[718] Schulte, K.; Masson, J.J.; Proc. ECCM-3, Bordeaux, 1989, p. 501.

[719] Masson, J.J.; Baron, Ch.; Schulte, K.; Proc. JNC-6, Paris, 1988, p. 661.

[720] Schulte, K.; Stinchcomb, W.W.; In: Application of Fracture Mechanics to Composite Materials, K. Friedrich, Ed., Elsevier Science Publisher, Amsterdam, 1989, p. 273.

[721] Schulte, K.; Marissen, R.; Composite Science and Technology, 45(1992), in press.

[722] Schulte, K.; Proc. Int. Conf. on Testing, Evaluation and Quality Control of Composites, TEQC-83, 1983, p. 233.

[723] Jamison, R.D.; Schulte, K.; Reifsnider, K.L.; Stinchcomb, W.W.; ASTM-STP 836, 1984, p.21.

[724] Reifsnider, K.L.; Schulte, K.; Duke, I.C.; ASTM STP 813, 1983, p.136.

[725] Baron, Ch.; *Mechanische Eigenschaften kohlenstoffaserverstärkter Kunststoffe (CFK) bei Variation der Matrixduktilität und der Bruchdehnung der Fasern.* Ph.D. Thesis, University Bremen, 1990.

[726] Salkind, M.J.; Fatigue of Composites, ASTM STP 497, 1972, p.143.

[727] Reimer, L; *Transmission Electron Microscopy,* Springer Verlag, Berlin, 1984.

[728] Hirsch, P., A. Howie, R.B. Nicholson, D.W. Pashley, M.J. Whelan; *Electron Microscopy of Thin Crystals,* R.E. Krieger Comp., Malabar, Florida, 1977.

[729] Bradley, S., W. King; *Fronties of Electron Microscopy in Materials Science,* North-Holland, Amsterdam, 1987.

[730] Siegel, B.M., Beaman, D.R.; *Physical Aspects of Electron Microscopy and Microbeam Analysis,* Wiley, New York, 1975.

[731] Hren, J.J., J.I. Goldstein, D.C. Joy; *Introduction to Analytical Electron Microscopy,* Plenum, New York, 1979.

[732] Schattschneider, P.; *Fundamentals of Inelastic Electron Scattering,* Springer, Wien, 1986.

[733] Egerton, R.F.;Electron Energy-Loss Spectroscopy in the *Electron Microscope,* Plenum, New York, 1986.

[734] Benninghoven, A., F.G. Rüdenauer, H.W. Werner; *Secondary Ion Mass Spectrometry, Basic Concepts, Instrumental Aspects, Applications and Trends,* Wiley, New York, 1987.

[735] Czanderna, A.W., D.M. Hercules, Eds.; *Ion Spectroscopies for Surface Analysis,* Plenum Press, New York, 1991.

[736] Russell, T.P.; *Small-Angle Scattering,* in: Handbook on Synchrotron Radiation, Vol.3, Eds., G.S. Brown, D.E. Moncton, North-Holland, Amsterdam, 1991, Chap. 11, p.379.

[737] Schülke, W.; *Inelastic Scattering by Electronic Excitations,* in: [736], p. 565.

[738] Fabian, D.J., Ed.; *Soft X-Ray Band Spectra,* Academic Press, London, 1968.

[739] Spence, J.C.H.; Ultramicroscopy, 18(1985)165.

[740] Grasserbauer, M., H.J. Dudek, M.F. Ebel; *Angewandte Oberflächenanalyse,* Springer, Berlin, 1986.

[741] Ramarker, D.E.; Crit. Rev. Solid State Mat. Sci., 17(1991)211.

[742] Skiff, W.M., R.W. Carpenter, S.H. Lin, A. Higgs; Ultramicroscopy, 25(1988)47.

[743] Dudek, H.J.; Z. Angew. Physik, 35(1970)331.

[744] Rosencwaig, A.; Ann. Rev. Mat. Sci., 15(1985)103.

[745] Faruqi, A.R.; Nucl. Instr. Meth. Phys. Res., A273(1988)754, and A310(1991)14.

[746] Culhane, J.L.; Nucl. Inst. Meth. Phys. Res., A310(1991)1.

[747] Michette, A.G.; Rep. Prog. Phys., 51(1988)1525.

[748] Born, M., E. Wolf; *Principles of Optics,* Pergamon Press, Oxford, 1980.

[749] Valdre, U., Ed.; *Electron Microscopy in Material Science,* Academic, New York, 1971.

[750] Hawkes, P.W.; *Electron Optics and Electron Microscopy,* Taylor & Francis LtD, 1972.

[751] Kay, D.H., Ed.; *Techniques for Electron Microscopy,* Blackwell, Oxford, 1965.

[752] Joy, D.C., A.D. Romig Jr., J.I. Goldstein; *Principles of Analytical Electron Microscopy,* Plenum, New York, 1986.

[753] Weiss, J.K., R.W. Carpenter, A.A. Higgs; Ultramicroscopy, 36(1991)319.

[754] Hall, E.L.; J. Physique, C6(1982)239.

[755] Doig, P., P.E.J. Flewitt; Met. Trans. A13(1982)1397.

[756] Spence, J.C.H.; *Experimental High-Resolution Electron Microscopy,* Clarendon Press, Oxford, 1981.

[757] Buseck, P.R., Ed.; *High-Resolution Electron Microscopy,* North-Holland, Amsterdam, 1985.

[758] Kihlborg, L., Ed.; *Direct Imaging of Atoms in Crystals and Molecules,* Nobel Symposium 47, Royal Swedish Academy Science, Chemica Scripta, 1978.

[759] Hawkes, P.W.; *Computer Processing of Electron Microscope Images,* Springer, Berlin, 1980.

[760] Barer, R., V.E. Cosslett, Eds.; *Advances in Optical and Electron Microscopy,* Vol.9, Academic Press, London, 1984.

[761] Krivanek, O.L., Ed.; Ultramicroscopy, 14(1984)1.

[762] Crewe, A.V.; Chimia Scripta, 14(1978/79)17.

[763] Cowley, J.M.; Chimia Scripta, 14(1978/79)33.

[764] Goldstein, J.I., D.E. Newbury, P. Echlin, D.C. Joy, Ch. Fiori, E. Lifshin; *Scanning Electron Microscopy and X-Ray Microanalysis,* Plenum, New York, 1981

[765] Wells, O.C., A. Boyde, E. Lifshin, A. Rezanowich; *Scanning Electron Microscopy,* McGraw-Hill, New York, 1974

[766] Holt, D.B., M.D. Muir, P.R. Grant, I.M. Boswarva: *Quantitative Scanning Electron Microscopy,* Academic, London, 1974

[767] Chiang S., R.J. Wilson; *Scanning tunneling microscopy,* in Examinating the Submicron World, Eds., F. Feder, et.al., Plenum, New York, 1986.

[768] Kaiser, W.J.; Jaklevic, R.C.; IBM J. Res. Dev. 30(1986)411.

[769] Burnham, N.A., R.J. Colton; *Force Microscopy,* in: Scanning Tunneling Microscopy: Theory and Application, Ed., D.Bonnell, VCH Publ. 1992.

[770] Frommer, J., E. Meyer; J. Phys.: Condens. Matter, 3(1991)S1.

[771] Hoffman, W.P., W.C. Hurley, P.M. Liu, T.W. Owens; J. Mat. Res., 6(1991)1685.

[772] Bauer, E., W. Telieps; J. Vac. Sci. Technol., A6(1988)573.

[773] Ai, R., M.I. Buckett, D. Dunn, T.S. Savage, J.P. Zhang, L.D. Marks; Ultramicroscopy, 39(1991)387.

[774] Lehmpfuhl, G., C.E. Warble; Ultramicroscopy, 19(1986)135.

[775] Krakow, W., V. Castano; Mat. Res. Soc. Symp. Proc., 83(1987)219.

[776] Takayanagi, K., Y. Tanishiro, K. Kobayashi; Jap. J. Appl. Phys., 26(1987)L957.

[777] Lehmpfuhl, G., Y. Uchida; Ultramicroscopy, 26(1988)177.

[778] Uchida, Y., G. Lehmpfuhl; Surface Science, 188(1987)364.

[779] Ogawa, S., Y. Tanishiro, K. Takayanagi, K. Yagi; J.Vac. Sci. Technol., A5(1987)1735.

[780] Drucker, J., M. Krishnamurthy, G. Hembree; Ultramicroscopy, 35(1991)323.

[781] Miller, M.K., G.D.W. Smith; *Atom Probe Microanalysis: Principles and Applications to Materials Science,* Mat. Res. Soc., Pittsburgh, 1989.

[782] Chabala, J.M., R. Levi-Setti, Y.L., Wang Appl. Surf. Sci., 32(1988)10.

[783] Barofsky, D.F. ACS Symp., Ser. 291(1985)113.

[784] Cowley, J.M.; *Diffraction Physics,* North-Holland, Amsterdam, 1975.

[785] Pinsker, Z.G.; *Electron Diffraction,* Butterworths, London, 1953.

[786] Tanaka, M., M. Terauchi; *Convergent-Beam Elektron Diffraction,* Jeol Ltd, 1418 Nakagami, Akishima, Tokyo, 1985.

[787] Liu, P., J. Skogsmo; Acta Cryst., B47(1991)425.

[788] Reimer, L., G. Pfefferkorn; *Raster-Elektronenmikroskopie,* Springer Verlag, Berlin, 1973.

[789] Pendry, J.B.; *Low Energy Electron Diffraction,* Academic Press, London, 1974.

[790] Petroff, P.M., C.H. Chen, D.J. Werder; Ultramicroscopy, 17(1985)185.

[791] Lehmpfuhl, G., W.C.T. Dowell; Acta Cryst., A42(1986)569.

[792] Shannon, M.D., A.J. Eades, M.E. Meichle, P.S. Tutner, B.F. Buxton; Phys. Rev. Lett., 53(1984)2125.

[793] Yamamoto, Y., S. Ino, T. Ichikawa; Jap. J. Appl. Phys., 25(1986)L331.

[794] Sakamoto, T., N.J. Kawai, T. Nakagawa, K. Ohta, T. Kojima; Appl. Phys. Lett., 47(1985)617.

[795] Fadley, C.S.; *Progr. Surf. Sci., 16(1984)275.*

[796] Margoninski, Y.; Contemp. Phys., 27(1986)203.

[797] Siegbahn, K.; *Photoelectron Spectroscopy: Retrospects and Prospects,* Uppsala University, Institute of Physics, UUIP-1136, 1985.

[798] Frank, D.G.; Batina, N.; Golden, T.; Lu, F.; Hubbard, A.T.; Science, 247(1990)182.

[799] Frank, D.G.; Chyan, O.M.R.; Golden, T.; Hubbard, A.T.; Vac. Sci. Technol., A10(1992)158.

[800] Duke, P.J.; *Synchrotron Radiation and the Submicron World: Selected Activities at the Daresbury Laboratory, UK,* in [833].

[801] Sparks, C.J. Jr.; *Synchrotron Radiation: Research with x-rays,* Physics Today, Mai, 1981, p.40.

[802] Harp, G.R., B. Tonner; Proc. Mat. Res. Soc., 1988,

[803] Norman, D.; *X-Ray Absorption Spectroscopy (EXAFS and XANES) at Surfaces,* J. Phys. C: Solid State Phys., 1986

[804] Wong, J.; Mat. Sci. Eng., 80(1986)107.

[805] Woodruff, D.P.; Surf. Interf. Anal., 11(1988)25.

[806] Krivanek, O.L., Ed.; Microsc. Microanal. Microstruct. 2(1991)143.

[807] Ahn, C.C., O.L. Krivanek; *EELS Atlas,* Gatan, Inc., 780 Commonwealth Drive, Warrendale, PA 15086, USA, 1983.

[808] Bauer, R., U. Hezel, D. Kurz; Optik, 77(1987)171.

[809] Mory, C., H. Kohl, M. Tence, C. Colliex; Ultramicroscopy, 37(1991)191.

[810] Thomas, J.M., R.F.S., T.G. Sparrow, M.K. Uppal, B.G. Williams; Phil. Trans. R. Soc. Lond., A318(1986)259.

[811] Batson, P.E., K.L. Kavanagh, C.Y. Wong, J.M. Woodall; Ultramicroscopy, 22(1987)89.

[812] Erskine, J.L.; Solid State Mat. Sci., 13(1987)311.

[813] Briggs, D., M.P. Seah; *Practical Surface Analysis,* Wiley, Chichester, 1983.

[814] Thomson, M., M.D. Baker, A.Christie, J.F. Thyson; *Auger Electron Spectroscopy,* Wiley, New York, 1985.

[815] Hofmann, S. Progr. Surf. Sci., 36(1991)35.

[816] Feuerbacher, B., B. Fitton, R.F. Willis; *Photoemission and Electronic Properties of Surfaces,* Wiley, Chichester, 1978.

[817] Taylor, T.N.; J. Mat. Res., 4(1989)189.

[818] Marcus, P., I. Olefjord; Surf. Interf. Anal., 9(1986)377.

[819] Castle, J.E.; Surf. Interf. Anal., 9(1986)345.

[820] Asami, K., K. Hashimoto; Langmuir, 3(1987)897.

[821] Olefjord, I., B. Brox, U. Jelvestam; J. Electrochem. Soc., 132(1985)2854.

[822] Lieng-Huang Lee, Ed.; *Characterization of Metal and Polymer Surfaces,* Vol.1, Metal Surfaces, Vol.2, Polymer Surfaces, Academic press, New York, 1977.

[823] Drummond, I.W., F.J. Street, L.P. Ogden, D.J. Surman; Scanning 13(1991)149.

[824] Drummond, I.W., L.P. Ogden, F.J. Street, D.J. Surman; J. Vac. Sci. Technol., A9(1991)1434.

[825] Ninomiya, K., Y. Hirai, A. Momose, S. Aoki, K. Suzuki; zJ. Vac. Sci. Technol., A9(1991)1244.

[826] Westhof, J., G. Meister, F. Lodders, R. Matzdorf, R. Hennig, E. Janssen, A. Goldmann; Appl. Phys., A53(1991)410.

[827] Niehus, H., R. Spitzl; Surf. Interf. Anal., 17(1991)287.

[828] Müller, K.-H., K. Seifert, M. Wilmers; J. Vac. Sci. Technol., A3(1985)1367.

[829] Otten, M.T.; Philips Electron Optics Bull., 126(1989)21.

[830] Krishnan, K.M.; Ultramicroscopy, 24(1988)125.

[831] Dudek, H.J., W. Bunk; Fresenius Z. Anal. Chem., 329(1987)422.

[832] Dudek, H.J.; Materialwiss. und Werkstofftechnik, 21(1990)48.

[833] Feder, R., J.Wm. McGowan, D.M., Shinozaki; *Examining the Submicron World,* Plenum Press, New York, 1986.

[834] Nowack, H.; *Schwingfestigkeit bei einstufigen und variablen Beanspruchungen: Werkstoffphysikalische und mechanische Gesichtspunkte - Vorhersagemethoden.* Habilitation Thesis, Univ.-GH Essen, 1988.

[835] Jacoby, G.H.; Nowack, H.; *Auslegungskonzepte für Betriebs- und Gestaltfestigkeit.* VDI-Berichte No. 385, VDI-Verlag, Düsseldorf, 1980.

[836] Jacoby, G.H.; in: ASTM-STP 462 (1968)184.

[837] Foth, J.; *Einfluß von einstufigen und nicht-einstufigen Schwingbelastungen auf Rißbildung und Mikrorißausbreitung in Al 2024-T3,* Ph.D. Thesis, TH Hamburg-Harburg, 1984.

[838] Palmgren, A.; VDI-Z., 69(1929)337.

[839] Miner, M.; Journ. of Appl. Mechanics (1945)A159.

[840] Hanschmann, D.; Maldfeld, E.; Nowack, H.; Proc. 5th Int. Spring Meeting, Société de Métallurgie et des Matériaux, Paris, France, 1992.

[841] Gaßner, E.; Materialprüfung 22(1980)155.

[842] Gaßner, E.; Kreutz, P.; Fortschr.-Ber. VDI-Z. Series 5, No. 80, 1984.

[843] Hanschmann, D.; *Ein Beitrag zur rechnergestützten Anrißlebensdauervorhersage schwingbeanspruchter Kraftfahrzeugbauteile aus Aluminiumwerkstoffen.* Ph.D. Thesis, RWTH Aachen, 1981.

[844] Crews, J.H.; Hardrath, H.F.; Experimental Mechanics, 6(1966)313.

[845] Ott, W.; *Neues Verfahren zur Betriebsfestigkeitsanalyse komplexer Bauteile auf der Grundlage der FEM.* Ph.D. Thesis, RWTH Aachen, 1987.

[846] Viswanathan, R.; *Damage Mechanisms and Life Assessment of High-Temperature Components.* ASM International, Metals Park, Ohio, USA, 1989.

[847] Agatonovic, P.; Taylor, N.; Proc. 4th Int. Conf. on Creep and Fracture of Engineering Materials and Structures, Swansea, 1990, p. 803.

[848] El Haddad, M.H.; Smith, K.N.; Topper, T.H.; Journ. Engng. Mater. Technol. 101(1979)42.

[849] El Haddad, M.H.; Dowling, N.E.; Topper, T.H.; Smith, K.N.; Int. Journ. of Fracture, 16(1980)15.

[850] Vormwald, M.; *Anrißlebensdauervorhersage auf der Basis der Schwingbruchmechanik für kurze Risse.* Ph.D. Thesis, TH Darmstadt, 1989.

[851] Nowack, H.; Wu, Y.; Trautmann, K.-H.; Marissen, R.; *Ermüdungsrißforschritt an einer hochfesten Aluminiumlegierung bei durchplastiziertem Probenquerschnitt.* Werkstoffkunde - Beiträge zu den Grundlagen und zur interdisziplinären Anwendung, DGM-Verlag, Oberursel, 1991.

[852] Foth, J.; Marissen, R.; Trautmann, K.-H.; Nowack, H.; The Behaviour of Short Fatigue Cracks, EGF Pub.-1, K.J. Miller, E.R. de los Rios, Eds., Mechanical Engineering Publications, London, 1986, p. 353.

[853] Elber, W.; In: ASTM STP 486, 1971, p. 230.

[854] Castro, D.E.; Marci, G.; Munz, D.; in : ASTM STP 969, 1988, p. 818.

[855] Schijve, J.; Broek, D.; Aircraft Engineering, Nov. 1962, p. 314.

[856] Hertzberg, R.W.; Newton, C.H.; and Jaccard, R.; In: ASTM STP 982, Eds. J.C. Newman, Jr. and W. Elber, 1988, p. 139.

[857] Marci, G.; Castro, D.E.; Bachmann, V.; in : ASTM STP 1074, J.P. Gudas, J.A. Joyce and E.M. Hackett, Eds., 1990, p. 563.

[858] Marci, G.; Hartmann, K.; Bachmann, V.; Mat.-wiss. u. Werkstofftechn., 21 (1990) 174-184.

[859] Schijve, J.; J.C. McMillan and R.M.N. Pelloux, in: ASTM STP 415, 1967, p. 533.

[860] Hudson, C.M.; Seward, S.K.; Engng. Fract. Mechanics, 8(1976)315.

[861] Sugano, M.; Kanno, S.; and Satake, T.; Acta Metall., 37(1989)1811.

[862] Wei, R.P.; Shih, T.T.; Int. Journ. Fracture, 10(1974)77.

[863] Ranganathan, N.; Petit, J.; Nadeau, A.; 2nd European Colloquium on Fracture (ECF2) Darmstadt, 9-11 Oct. 1978, Fortschr.-Ber. VDI-Z. 18, No. 6, p. 337.

[864] Ranganathan, N.; Petit, J.; Bouchet, B.; Engng. Fract. Mechanics, 11(1979)775.

[865] Mill, W.J.; Hertzberg, R.W.; Engng. Fract. Mechanics, 7(1975)705.

[866] Matsuoka, S.; Tanaka, K.; Engng. Fract. Mechanics, 13(1980)293.

[867] Schijve, J.; Delft Univ. of Technology, Dept. of Aerospace Engineering, Memorandum M-336, 1979.

[868] De Koning, A.U.; ASTM STP 743, Richard Roberts, Ed., 1981, p. 63.

[869] Katcher, M.; Kaplan, M.; ASTM STP 559, 1974, p. 264.

[870] Clerivet, A.; Bathias, C.; Engng. Fract. Mechanics, 12(1979)599.

[871] J. Schijve; Engng. Fract. Mechanics, 14(1981)461.

[872] Zhang, C.; Marissen, R.; Schulte, K.; Trautmann, K.-H.; Nowack, H.; Schijve, J.; Fatigue Fract. Engng. Mater. Struct., 10(1987)315.

[873] Baudin, G.; Robert, M.; ONERA T.P. No. 1981-44, 1981.

[874] Hudson, C.M.; Hardrath, H.F.; NASA Technical Note D-1803.

[875] Bathias, C.; Vancon, M.; Engng. Fract. Mechanics, 10(1978)409.

[876] Castro, D.E.; DLR, Köln, DLR-Report DLR-FB 89-33 (1989).

[877] Schijve, J.; In:Advances in Aerospace Sciences, 1961.

[878] Hertzberg, R.W.; In: ASTM STP 415, 1967, p. 205.

[879] Castro, D.E.; Marci, G.; In: Mikrostruktur und Gebrauchseigenschaften von Werkstoffen, DVM, Deutscher Verband für Materialprüfung (1987) 363-373.

[880] Hudson, C.M.; Raju, K.N.; NASA Technical Note D-5702.

[881] Rice, R.C.; Stephens, R.I.; In: ASTM STP 536, 1973, p. 95.

[882] Wei, R.P.; Shih, T.T.; NASA Contract Report-2239.

[883] Himmelein, M.K.; Hillberry, B.M.; In: ASTM STP 590, 1976, p. 321.

[884] Trebules, Jr.; V.W., Roberts, R.; Hertzberg, R.W.; In: ASTM STP 536, 1973, p. 115.

[885] Brown, R.D.; Weertman, J.; Engng. Fract. Mechanics, 10(1978)867.

[886] Matsuoka, S.; Tanaka, K.; Kawahara, M.; Engng. Fract. Mechanics, 8(1976)507.

[887] Bernard, P.J.; Lindley, T.C.; Richards, C.E.; In: ASTM STP 595, 1976, p. 78.

[888] Hudson, C.M.; Hardrath, H.F.; NASA Technical Note D-960.

[889] Adefifa, O.A.; Gowda, C.V.B.; Topper, T.H.; In: ASTM STP 595, 1976, p. 142.

[890] Marci, G.; in : ASTM STP 677, C.W. Smith, Ed., 1979, p. 168.

[891] Marci, G.; Brändle, M.; Bachmann, V.; Mat.-wiss. u. Werkstofftechn. 21 (1990) 174-184.

[892] Marci, G.; Bachmann, V.; In: ASTM STP 791, Eds. J.C. Lewis and G. Sines, 1983, II-47-II-66.

[893] Dao-Lun Chen; Weiss, B.; Stickler, R.; Int. Journ. Fatigue, 13(1991)327.

[894] Paris, P.C.; Gomez, M.P.; Anderson W.E.; The Trend in Engineering, Vol. 13, No. 7, 1961, p. 9.

[895] Forsyth, P.J.E.; in : Proc. of the Crack Propagation Symposium, Cranfield/England, Vol. 1, 1962, p. 76.

[896] Foreman, R.G.; Kearney, V.E.; Engle, R.M.; J. of Basic Engineering, 89(1967)459.

[897] Liu, H.W.; Discussion in: Fatigue, An Interdisciplinary Approach, Proc. 10th Sagamore Army Materials Conference, Syracuse University Press, 1964, 127.

[898] Elber, W.; Damage Tolerance in Aircraft Structures, ASTM STP 486, 1971, p. 230.

[899] Marci, G.; Fracture Mechanics, ASTM STP 677, C.W. Smith, Ed., 1979 p. 168.

[900] Marci, G.; Castro, D.E.; Bachmann, V.; J. Testing and Evaluation, 19(1989)28.

[901] Marci, G.; Engin. Frac. Mech., 41(1992)367.

[902] GARTEUR TP 54, produced by GARTEUR ACTION GROUP SM (AG) 04, Annex A3, (DLR) issued by Onera, 1991.

[903] Newman, J.C., Jr.; Methods and Models for Predicting Fatigue Crack Growth under Random Loading, ASTM STP 748, J.B. Chang and C.M. Hudson, Eds., 1981, p. 53.

[904] Newman, J.C., Jr.; Design of Fatigue and Fracture Resistance Structures, ASTM STP 761, P.R. Abelkis and C.M. Hudson, Eds., 1982, p. 255.

[905] Baudin, G.; Robert, M.; Onera T.P. No. 1981-44 presented at: Int. Committee on Aeronautical Fatigue (ICAF), Noordwijkerhout/Netherlands, 1981, p. 18.

[906] Marci, G.; Bazant, E.; Materialprüfung, 19(1977)455.

[907] Döker, H.; Bachmann, V.; In: Mechanics of Fatigue Crack Closure, J.C. Newman, Jr. and W. Elber, Eds., ASTM STP 982, 1987, p. 247.

[908] Marci, G.; Hartmann, K.; Bachmann, V.; Materialwiss. u. Werkstofftechn., 21(1990)174. 1981.

[909] De Koning, A.U.; in: Fracture Mechanics, 13th Conf., ASTM STP 743, Richard Roberts, Ed., 1981, p. 63.

[910] Marci, G.; Conf. Proc. DVM-Arbeitskreise Betriebsfestigkeit und Bruchvorgänge, DVM, Berlin, 1982, p. 203.

[911] Hollingsworth, E.H.; Hunsicker, H.Y., Metals Handbook, Ninth Edition, Vol. 13: Corrosion, ASM International, Metals Park, Ohio, 1987, p.583.

[912] Sprowls, D.O., Metals Handbook, Ninth Edition, Vol. 13: Corrosion, ASM International, Metals Park, Ohio, 1987, p.245.

[913] Ailor, W.H.; *Handbook on Corrosion Testing and Evaluation,* John Wiley and Sons, New York, London, Sydney, Toronto, 1971.

[914] Annual Book of ASTM Standards, Part 10, Metals - Physical, Mechanical, Corrosion Testing, American Society for Testing and Materials.

[915] *Deutsche Luftfahrt-Norm,* Beuth Verlag, W-1000 Berlin, Burggrafenstr. 6.

[916] Speidel, M.O., Metall.Trans. A, 6(1975)631.

[917] Speidel, M.O., The Theory of Stress Corrosion Cracking in Alloys, Ed. J.C.Scully, NATO Scientific Affairs Division, Brussels, 1971, p.289.

[918] Sprowls, D.O.; Bucci, R.J.; Ponchel, B.M.; Brazill, R.L.; Bretz, P.E.; NASA Contractor Report 172387, August 31, 1984.

[919] Bucci, R.J.; Brazill, R.L.; Sprowls, D.O.; Ponchel, B.M.; Bretz, P.E.; Corrosion Cracking, Conference Proceedings, Ed. V.S.Goel, American Society for Metals, 1986, p. 267.

[920] Parkins, R.N.; in: ASTM STP 665, G.M. Ugiansky and J.H. Payer, Eds., 1979, p. 5.

[921] Parkins, R.N.; Proc. 5th Symp. Pipe Line Res., AGA Cat. No. L 30174, 1974.

[922] Parkins, R.N.; Mazza, F.; Royuela, J.J.; Scully, J.C.; Br. Corros. J. 7(1972)154.

[923] Buhl, H.; in: [920], p. 333.

[924] Buhl, H.; Mierke, G.; Aluminium, 60(1984)E704.

[925] Holroyd, N.J.H.; Scamans, G.M.; in: ASTM STP 821, S.W. Dean, E.N. Pugh and G.M. Ugiansky, Eds., 1984, p.202.

[926] Tien, J.K.; Thompson, A.W.; Bernstein, I.M.; Richards, R.J.; Metall. Trans. A, 7(1976)821.

[927] Ratke, L.; Gruhl, W.; Z. Metallkunde, 71(1980)568.

[928] Holroyd, N.J.H.; Hardie, D.; Corr. Science, 21(1981)129.

[929] Buhl, H.; Braun, R.; Yuan, C.; Helin, J.; Deutsche Forschungsanstalt für Luft- und Raumfahrt, 5000 Köln 90, DLR-FB 91-35.

[930] Craig, J.G.; Newman, R.C.; Jarrett, M.R.; Holroyd, N.J.H.; in: [51], p. C3-825.

[931] Moran, J.P.; Stoner, G.E.; in: [52], p. 1187.

[932] Braun, R.; Werkst. Korros., 41(1990)547

[933] Braun, R.; Proceedings of the 9th European Congress on Corrosion, Utrecht, Royal Netherlands Industries Fair, 1989, Vol.I, FU-093.

[934] Braun, R.; in: Conf. Proc. Corrosion-Deformation Interactions, Fontainebleau, France, Oct. 5-7, 1992, to be published.

[935] Braun, R.; Werkst. Korros., to be published.

[936] Braun, R.; Werkst. Korros., 43(1992)106.

[937] Ricker, R.E.; Fink, J.L., Vasudevan, A.K.; Metall. Trans. A, 22(1991)264.

[938] Mathiak, F.; Krawietz, A.; Nowack, H.; Trautmann, K.-H.; German patent No. 39 14 966 (1989).

[939] Mroz, Z.; Acta Mechanica, 7(1969)199.

[940] Ott, W.; *Neues Verfahren zur Betriebsfestigkeitsanalyse komplexer Bauteile auf der Grundlage der FEM.* Ph.D.Thesis, TH Aachen, 1987.

[941] Krawietz, A.; *Materialtheorie - Mathematische Beschreibung des phänomenologischen thermomechanischen Verhaltens.* Springer-Verlag, Heidelberg, Berlin, 1986.

[942] Trautmann, K.-H.; Internal DLR report (1990).

[943] Nowack, H.; *Schwingfestigkeit bei einstufigen und variablen Beanspruchungen: Werkstoffphysikalische und mechanische Gesichtspunkte - Vorhersagemethoden.* Habilitation Thesis, Universität-GH Essen, 1988.

[944] Nurtjahjo, A.; Ott, W.; Marissen, R.; Trautmann, K.-H.; Nowack, H.; 3rd Int. Conf. on Low Cycle Fatigue and Elasto-Plastic Behaviour of Mats., Berlin, 1992, to be published in Conf. Proc.

[945] Hanschmann, D.; Maldfeld, E.; Nowack, H.; 5th Int. Spring Meeting, Société Francaise de Métallurgie et de Matériaux, Paris, June 15-17, 1992, to be published in Conf.Proc.

[946] Brown, M.W.; Miller, K.J.; Proc. Institute of Mechanical Engineers, Vol. 187, No. 65, 1973.

[947] *Standard Test Method for Plane Strain Fracture Toughness of Metallic Materials.* Designation: ASTM E 399-90.

[948] *Standard Test Method for Plane-Strain (Chevron-Notch) Fracture Toughness of Metallic Materials.* Designation: ASTM E 1304-89.

[949] Barker, L.M.; Engin.Fract. Mech., 9(1977)361.

[950] Barker, L.M.; ASTM STP 678, 1979, p. 73.

[951] Munz, D.; Bubsey, R.T.; Srawley, J.E.; Int. Journ. Fract., 16(1980)359.

[952] Bubsey, R.T.; Munz, D.; Pierce, W.S.; Shannon, Jr., J.L.; Int. Journ. Fracture, 18(1982)125.

[953] Marci, G.; Eschweiler, J.; Mat.-wiss. u. Werkstofftech., 19(1988)317.

[954] Underwood, J.H.; Freiman, S.W.; Baratta, F.I.; Eds., ASTM STP 855, Philadelphia, 1984.

Abbreviations

ADAM	angular distribution Auger microscopy
AES	Auger electron spectroscopy
AC	American Matrix Company
AFM	atomic force microscopy
AFRP	aramid fibre reinforced plastic
APS	air plasma spraying
AR	as received
ARALL	aramid reinforced aluminium laminates
ARXPS	angle resolved X-ray photoelectron spectroscopy
ASPS	argon shrouded plasma spraying
ATEM	analytical transmission electron microscopy
BLM	breaking load method for SCC testing
CBED	convergent beam electron diffraction
CCA	composite cylinder assemblage
CF	corrosion fatigue
CFRP	carbon fibre reinforced plastic
COD	crack opening displacement
CTEM	conventional transmission electron microscopy
CVD	chemical vapor deposition
CVI	chemical vapor infiltration
DCB	double cantilever beam
DB	diffusion bonding
DLR	German aerospace research establishment
DS	directional solidification
EB-PVD	electron-beam physical vapor deposition
EDX	energy dispersive X-ray analysis
EELS	electron energy loss spectroscopy
ESK	Göring?
EPMA	electron probe microbeam analysis
EXAFS	extended X-ray absorption fine structures
FALSTAFF	fighter aircraft loading standard for fatigue
FCP	fatigue crack propagation
FGM	functional gradient materials
FIM	field ion microscopy
FPF	first ply failure
FWHM	full width at half maximum
GFRP	glass fibre reinforced plastic
GLARE	glass fiber reinforced aluminium laminates
GP	Guinier-Preston
GPB	precursor of S'
HIP	hot isostatic pressing
HREELS	high resolution electron energy loss spectroscopy
HREM	high resolution electron microscopy

ISS	ion scattering spectroscopy
LAMMA	laser ablation microprobe mass spectrometry
LCF	low cycle fatigue
LEED	low energy electron diffraction
LEEM	low energy electron microscopy
LM	light microscopy
LN	Luftfahrt-Norm, German aviation standard
LPPS	low pressure plasma spraying
MA	mechanical alloying
M(T)	specimen configurations in accordance with ASTM
MMC	metal matrix composite
MP	melting piont
ODS	oxide-dispersion strengthening
PAM	polyacrylnitril
PFZ	precipitate free zone
PIXE	particle induced X-ray emmission spectroscopy
PM	powder metallurgy
PMC	polymer matrix composites
PMMC	polymethyl methacrylate
PS	plasma speaying
RBS	backscattering spectroscopy
RBSN	reaction bonded silicon nitride
REM	reflection electron microscopy
RFE	relative fracture energy
RHEED	reflection high energy electron diffraction
RRA	retrogression and re-aging
RS	rapid solidification
RT	room temperature
SAD	selected area diffrctometry
SAP	sintered aluminium powder
SB	short bar
SC	single crystal alloys
SCC	stress corrosion cracking
SECP	selected area channelling pattern
SEM	scanning electron microscopy
SEXAFS	surface-sensitive EXAFS
SHT	solution heat treatment
SIMS	Secondary ion mass spectroscopy
SMF	strain magnification factor
SNMS	Secondary neutrals mass spectroscopy
SPF	superplastic forming
SR	short rod
SSR	slow strain rate
SSRHE	slow strain rate hydrogen embrittlement
SSSS	supersaturated solid solution
STEM	scanning transmission electron microscopy

STM	Scanning tunnelling microscopy
TA	Tateho Company
TBC	thermal barrier coating
TBO	time between overhauls
TCP	topologically close packed
TEC	thermal expansion coefficient
TEM	transmission electron microscopy
TIT	turbine gas inlet temperature
TK	Tokai Carbon Company
TMT	thermomechanical treatment
TOF	time of flight spectroscopy
TWIST	transport aircraft wing structures
UHV	ultra high vacuum
UTS	ultimate tensile strength
WDX	wave length dispersive X-ray analysis
XPD	X-ray photoelectron diffraction
XPS	X-ray photoelectron spectroscopy
XRD	X-ray diffraction
YPSZ	yttria partially stabilized zirconia
YS	yield stress

Index

F

G

H

N

O

P

Q

R

S

T

V

W

X

Y

Laudatio

A Tribute to Distinguished Service

DLR, Germany's Aerospace Research Center, rests on a distinct scientific base made up of individually organized institutes. They operate in consort with DLR's engineering and facility departments, and provide the scientific rationale without which the organization's overall accomplishments would not be conceivable. Each institute is headed by one director.

Wolfgang Bunk, longtime director of DLR's Institute of Materials Research and full professor of the renowned Technical University of Aachen, has certainly been one of DLR's eminent institute leaders whose reputation stands out markedly. It is with pride and pleasure that I pay tribute to his service and achievements.

When Professor Dr. Wolfgang G.J. Bunk joined DLR in 1970, the institution had just been formed through a merger of several laboratories dating back to pre-war times. One of them, the Deutsche Versuchsanstalt für Luftfahrt (DVL), comprised an institute for materials research set up in 1967 and directed by the late Professor Friedrich-Carl Althof. It was still a rather modest enterprise when Professor Bunk took over. In the ensuing, formative years of DLR Dr. Bunk's leadership skills, developed during his industrial career, proved highly beneficial. The Institute of Materials Research soon attracted special attention and support from inside and outside the DLR. Its staff grew in quality and quantity; modern test equipment was installed and the institute building expanded. Today, after more than twenty years of continuous growth and recognition, the institute ranks high in its field, not only nationally but also internationally.

This achievement was more than the result of adroit research management. As the head of a devoted group of collaborators, Dr. Bunk soon brought to fruition a second talent, namely that of a science educator. Whoever visits his institute will notice an impressive listing on display in the seminar room of all those post-graduate students and scientists who obtained a doctorate or a professoral habilitation whilst involved in the institute's work. They totalled thirty-six by the beginning of 1992.

In his effort to put unrelenting stress on excellence, Dr. Bunk soon extended his educational coaching beyond the individual member of the institute and implemented, beginning as early as 1973, the institute's "Materials Science Colloquia". Each year in December one or two of the institute's divisions present the status and results of their research activities to a critical auditorium composed of colleagues and outside experts. Thus, in a rather informal but at the same time highly demanding manner scientific competence can be demonstrated and excellence put on record. When DLR some years later established the practice of formally evaluating, at regular intervals, all its institutes, Dr. Bunk gave valuable advice for drawing up the rules.

There is a third notable trait in Dr. Bunk's approach to science management: He is highly cooperation minded. Being responsible for a materials science institute, he was placed in a service position with regard to DLR's primary research goals related to aeronautics, spaceflight, and renewable energy technology. In such a setting interdisciplinary cooperation is mandatory. But that is easier said than done. Dr. Bunk for his part made efficient cooperation really happen, and there are indeed few DLR programs which have not benefited in some way or other from the materials science institute's guidance and support. Management assistance was also given on a national level. For more than ten years Dr. Bunk was responsible for the conduct of the Federal Government's program on materials research, which brought together industry and universities.

Cooperation beyond the institutional boundaries of DLR is, of course, another must. Here, Wolfgang Bunk showed a firm hand in choosing partners both in industry and academia which in their exchanges with DLR took an active interest in the institution's scientific standing and economic well-being. The cooperative network developed by him from the very beginning also included partners abroad, especially in the United States of America and Japan. Dr. Bunk has in fact always been deeply convinced of the technological prowess of these two countries and invested many trips and a sabbatical leave in the development of closer links with renowned institutions. His personal reputation as President (1982 - 1984) and later Honorary Member of the German Society for Materials certainly helped to foster such contacts.

Prof. Bunk's contributions to the science of materials are reflected in the preceding pages of this impressive compilation. I should like to add a personal note by stating how sincerely I appreciate his unfailing readiness for dialogue and advice with sharp wit, humour and personal example.

Prof. Dr. W. Kröll
Chairman, Board of Directors
German Aerospace Research Establishment